SEPARATIONS CHEMISTRY

Revised and Expanded Edition

FEDOR MACÁŠEK
Professor Emeritus
Comenius University
Bratislava, Slovakia
and
Quality Manager
BIONT, AS
Bratislava, Slovakia

JAMES D. NAVRATIL
Professor Emeritus
Clemson University
Clemson, South Carolina, USA
and
Senior Technical Advisor
Hazen Research Inc.
Golden, Colorado, USA

Copyright © 2016 by Fedor Macášek; James D. Navratil.

Library of Congress Control Number:		2015917275
ISBN:	Hardcover	978-1-5144-1733-1
	Softcover	978-1-5144-1732-4
	eBook	978-1-5144-1730-0

All rights reserved. No part of this book may be reproduced or transmitted in any form or by any means, electronic or mechanical, including photocopying, recording, or by any information storage and retrieval system, without permission in writing from the copyright owner.

Any people depicted in stock imagery provided by Thinkstock are models, and such images are being used for illustrative purposes only. Certain stock imagery © Thinkstock.

Print information available on the last page.

Rev. date: 06/03/2016

To order additional copies of this book, contact:
Xlibris
1-888-795-4274
www.Xlibris.com
Orders@Xlibris.com

Acknowledgments

We owe special thanks to Nicole Navratil for her kind assistance on the second edition, to Sylvia Tascher Navratil for her skillfully enhancing of the original manuscript in all respects; to Katarina Moravkova for her dedicated help in the production of illustrations; to many colleagues for their sustaining encouragement; and last but not least to our families, Sylvia and Ljuba for their loving patience.

We are sincerely grateful to Nick Hazen, President of Hazen Research Inc. of Golden Colorado, for his kind and generous financial support.

Contents

1 PHYSICO-CHEMICAL CHARACTERISTICS OF SEPARATION 1

1.1 SEPARATION OF COMPOUNDS 1
1.1.1 Mechanical separation 8
1.1.2 Physicochemical separation 23

1.2 EQUILIBRIA IN SOLUTIONS 35
1.2.1 Metrics of solutions 36
1.2.2 Phenomenology of thermodynamical non-ideality of solutions 42
1.2.3 Chemical interactions. Complex equilibria 64
1.2.4 Electrostatic interactions. Electrolyte solutions 90
1.2.5 Weak intermolecular interactions. Non-electrolyte solutions 97
1.2.6 Specifics of microconcentrations 113

1.3 HETEROGENEOUS EQUILIBRIA 116
1.3.1 Phase equilibria 117
1.3.2 Chemical equilibria and distribution in biphasic systems 130

1.4 DYNAMICS OF SEPARATION 158
1.4.1 Non-equilibrium thermodynamics 158
1.4.2 Convective transport 171
1.4.3 Transport in external field. Electrophoresis 202
1.4.4 Interphase transfer. Physico-chemical hydrodynamics 214
1.4.5 Combined convective and interphase transport. Chromatography 236

1.5 MEMBRANE PROCESSES 255
1.5.1 Passive transport 255
1.5.2 Carrier-facilitated transport 268

1.6 PHOTOCHEMICAL SEPARATION 282

1.7 SPECIFIC NUCLEAR PHENOMENA IN SEPARATION 287
    1.7.2 Effects of nuclear composition (heterogeneous isotope exchange) 287
    1.7.3 Effects of nuclear transformations 292

1.8 GENERALIZATION AND CLASSIFICATION OF SEPARATION PROCESSES 302

2 OPERATION AND OPTIMIZATION OF SEPARATION 309

2.1 ELEMENTARY SEPARATION AND SEPARATION UNIT 311
    2.1.1 Equilibrium processes 314
    2.1.2 Rate and local equilibrium processes 332

2.2 CHEMICAL CONTROL OF SEPARATION 343
    2.2.1 Initial and equilibrium parameters 349
    2.2.2 Actual and measured parameters; cybernetic principles 358

2.3 MANY-FOLD SEPARATION 368
    2.3.1 Stage separation 370
    2.3.2 Column separation 382

2.4 COUNTERCURRENT SEPARATIONS 399
    2.4.1 Cascades of equilibrium units 403
    2.4.2 Continuous transfer 419

REFERENCES 431

LIST OF SYMBOLS 545

ABOUT THE AUTHORS 555

1 Physico-chemical characteristics of separation

1.1 SEPARATION OF COMPOUNDS

Cinderella could not go to the ball unless she carried out her stepmother's demand that she should separate an intimate mixture of lentils and ashes. Her friends, the turtle doves and tame pigeons came in response to her urgent request to pick out
    "the good lentils into the pot
    the rest into your crop"
The birds finished the job in one hour with an efficiency of 100%

Adapted from "Aschenputtel"
by the Brothers Grimm

Natural substances, complex reaction mixtures and biological materials in particular represent systems that are complex in their chemical, phase and morphological compositions. The task of separation science is to isolate the compounds or components of these systems either for preparative or analytical identification purposes. Thus **separation** is the process of spatial displacement and division of the components of mixtures.

A **component** may be present as a particle, i.e. a discrete piece of material, the lower limit being a molecule. There is no maximum limit to the size, and the component may form a **continuous phase** occupying a major part of the system volume. Although in stable mixtures the particles forming a **dispersed phase** are generally solids differing in size and shape, they may be solids dispersed in a liquid (suspension) or gas (smoke), drops of liquid dispersed in another liquid (emulsion) or in gas (fog) or small bubbles of gas dispersed in a liquid. When the components are of molecular size the continuous mixtures are called **solutions** (liquid, gas or solid solutions).

The major component of a mixture is called the **matrix or solvent**, minor components can be considered as **macro components, micro components (impurities) or solutes**. When the separation proceeds with removal of a matrix, the process is called the **absolute preconcentration**; the opposite process, removal of impurities from matrix, is **purification**. Typical examples of the latter processes are the removal of water or another solvent to leave a non-volatile residue of mutually unseparated solutes, and the distillation of water to obtain pure water. If one component is separated in preference to other macro- or micro components, the **relative preconcentration** of the component occurs. For example, selective precipitation of calcium from spring water occurs with the addition of alkali metal carbonates, leaving many other salts dissolved in water.

Separation is necessary for many practical purposes when a pure component is desired. Separation procedures are an important part of most chemical operations in the **chemical, pharmaceutical and clinical laboratories**, and in the **chemical and petrochemical industry** (*A.S. Michaels, 1968; C.J. King, 1980; J.M. Douglas, 1988; H.Z. Kister, 1992; S. Kulprathipanja, 2002; R. Vazquez-Duhalt, R. Quintero-Ramirez, 2004; P.C. Wankat, 2012; A.A. Gaile, 2012*), the separation systems are often the major factor in capital and operating costs, requiring large equipment and vast amounts of energy (*A.S. Michaels, 1968*). For instance, liquid-solid separation is the single most expensive unit operation in **ore mills**, accounting for about half of the costs in most circuits; by far the largest single variable cost in the corn **sweetener industry** is in evaporating waste steep waters (*A.S. Grandison, M.J. Lewis, 1996*). **Desalination** of sea water, which is the most feasible fresh water supply source in arid countries, consumes worldwide many GW of power (*E.D. Howe, 1974; I.S. Al-Mutaz, 1986; L.K. Wang, J.P. Chen, Y-T. Hung, 2011*). The choice of a proper separation process is frequently the factor that makes for success in the **analytical procedure** (*J.Stary, M. Kyrs, M. Marhol, 1975; Yu.A. Zolotov, 1978; T.J. Bruno, 1990; L. Moskvin, L. Caritsyna, 1991; N.M.Kuzmin, Yu.A. Zolotov, 1998; H.Y. Aboul-Enein, 2003; J. Rydberg, 2004; L.R. Snyder, J.J. Kirkland, J.L. Glajch, 2012; L.M.L. Nollet, F. Toldra, 2013*), **organic synthesis and biotechnology** (*A. Weissberger, 1950, 1951; R.W. Rousseau, 1987; L.E. Hood, L.M. Smith, 1988; T. Milton, M.T.W. Hearn, B. Anspach, 1990; C. Horvath, L.S. Ettre, 1993; N.S. Tavare, 1995; K. Scott, R. Hughes, 1996; K. Valko, 2000; M.R. Ladisch, 2001; H.J. Issaq, 2002; G. Subramanian, 2007; H-J. Huang, S. Ramaswamy, U.W. Tschirner, B.V. Ramarao, 2008; S. Rizvy, 2010; A. Van Nieuwenhuijzen, J. Van der Graaf, 2011; Z. Deyl, 2011*), **processing of raw materials and preparation of pure materials** (*K.J. Bachmann, 1995; W.L.F. Armarego, C.L.L. Chai, 2013*), **nuclear fuel cycle** (*J.L. Jenkins, 1984; G.R. Choppin, Kh.M. Khankhasaev, 1999; K.L. Nash, G.J. Lumetta, 2011*) and **waste treatment** (*T.J. Veasey, R.J. Wilson, D.M. Squires, 1993; R.B. Long, 1995; C. Comninellis, M. Doyle, J. Winnick, 2011; E. Worrell, M. Reuter, 2014*).

The discovery of many new chemical species and the development of a number of new technologies has depended on the development of the appropriate separation techniques, and in turn the search for new materials has raised challenges for new separation techniques (Table 1).

Table 1 —Chronology of separation techniques applied or developed for enrichment, purification, discovery or preparation of materials.

Material, problem	Separation technique	Reference
rose water, ethers	distillation	"Papyrus Ebers" 1500 B.C.
gold	wet amalgamation	I.A. von Born, 1786
morphine	extraction, precipitation	F.W. Serturner, 1806
cystine	precipitation	W.H. Wollaston, 1810
optical antipodes	mechanical	L. Pasteur, 1848
metal complexes	solvent extraction	W. Skay, 1867
metal ions	sulfide precipitation	F.W. Clarke, 1870
lanthanides	fraction crystallization	C. Auer von Welsbach, 1885
gold	leaching	J.S. MacArthur, W. Forrest, 1887
thiamine	precipitation	C. Eijkman, 1894
polonium, radium	coprecipitation, cocrystalization	M. Curie-Sklodowska, 1891
inert gases	distillation	W. Ramsay, 1898
adrenalin	precipitation	J. Takamine, 1903
chloroplast pigments	adsorption chromatography	M.S. Tswett, 1903
isotopes (various)	various (negative)	B.B. Boltwood, 1907; O. Hahn, 1907

Material, problem	Separation technique	Reference
lutetium	crystallization	G. Urbain, 1907
colloids	ultrafiltration	H. Bechholz, 1907
metal ions	chelates precipitation	O. Baudisch, 1909
histamine	precipitation	H.H. Dale, G. Barger, 1910
amylase	bio specific sorption	E. Starkenstein, 1910
aneurin (vitamin B$_1$)	precipitation	B.C.P. Jansen, W.F. Donath, 1926
neon isotopes	gas diffusion	J.G. Aston, 1920
hafnium	crystallization	D. Coster, G. Hevesy, 1923
urease	precipitation	J.B. Summer, 1926
vitamin C	precipitation	A. Szent-Györgyi, 1928
pepsin	crystallization	J.H. Northrop, 1930
deuterium	distillation	H. Urey, F.G. Brickwedde, G.M. Murphy, 1932
heavy water	electrolysis	G.N. Lewis, R.T. McDonald, 1933
progesterone	precipitation	A.F.J. Butenandt, N. Westphal, 1934
antibodies	precipitation	H. Heidelberger, F.E. Kendall, 1935
mercury isotopes	photochemical oxidation	K. Zuber, 1935

Material, problem	Separation technique	Reference
hydrocarbons	gas chromatography	A. Eucken, H. Knick, 1936
serum globin	electrophoresis	A. Tiselius, 1937
nitrogen isotopes	chemical exchange	H.C. Urey et al., 1937
penicillin	solvent extraction	H.W. Florey, E.B. Chain, 1938
enantiomeric camphor derivatives	column chromatography	G.M. Henderson, H.G. Rule, 1939
amino acids	partition chromatography	A.J.P. Martin, R.L.M. Synge, 1941
plutonium	micro precipitation	B.B. Cunningham, L.B. Werner, 1942
amino acids	paper chromatography	A.H. Gordon, A.J.P. Martin, R.L.M. Synge, 1943
sulfite waste	ion exchange	O. Samuelson, 1943
promethium	ion exchange	J.A. Marinsky, L.E. Glendenin, C.D. Coryell, 1945
blood *in vivo*	dialysis	W.J. Kolff, H.T.J. Berk, 1946
^{235}U	diffusion	J.F. Hogerton, 1945
antigens	gel precipitation	J. Oudin, 1946
nuclear uranium	solvent extraction	R.I. Moore, 1951
semiconducting germanium	zone melting	W.G. Pfann, 1952

Material, problem	Separation technique	Reference
antibodies	electrophoresis	P. Grabar, C.A. Williams, 1953
mendelevium	ion exchange	A. Ghiorso et al., 1955
[^{14}N]DNA, [^{15}N]DNA	gradient ultracentrifugation	M.S. Meselson, F.W. Stahl, J. Vinograd, 1957
^{235}U	gas centrifuges	W. Groth, 1957
metal ions	isotachophoresis	B.P. Konstantinov, O.V. Oshurkova, 1963
biological cells	fluidic	M.V. Fulwyler, 1965
enantiomers	gas-liquid chromatography	E. Gil-Av (Zimkin), B. Feibush, R. Charles-Sigler, 1966
104 element	thermo chromatography	I. Zvara et al., 1967
peptides, proteins	affinity chromatography	R. Axén, J. Porath, S. Ernback, 1967
rubidium isotopes	laser photo excitation	R.V. Ambartsumian, V.S. Letokhov, 1971
enantiomers	chiral chromatography	V.A. Davankov, S.P.V. Rogozhin, 1971
nucleotides sequetioning	gel electrophoresis	A. Maxam, U. Gilbert, 1977; F. Sanger et al., 1977
chromosomes	pulse-field electrophoresis	C.R. Cantor, C.L. Smith, 1986
magnetic cells	high gradient magnetic field	S. Miltenyi, W. Muller, W. Weichel, A. Radbruch, 1989

Material, problem	Separation technique	Reference
micro particles and cells	standing acoustic waves in microfluidics	Xiaoyun Ding et al., 1992
breast cancer cells	dielectrophoresis	F.F. Becker, Xiao-Bo Wang, Ying Huang, R. Pethig, J. Vykoukal, P.R.C. Gascyone, 1995
embryonic stem cells clones	cell sorting	K.S. Sidhu, B.E. Tuch, 2006

1.1.1 Mechanical separation

The compounds that are present in the form of macro particles (phases) in discontinuous mixtures with other bodies can be separated mechanically; such separation is also usually a final stage in chemical separation (*A.F. Orlicek, A.E. Hackl, P.E. Kindermann, 1964; A.G. Kasatkin, 1971; W.L. McCabe, J. Smith, 1976; A. Rushton, A.S. Ward, R.G. Holdich, 2000*).

It is necessary to define mechanical separation of components. The definition for the efficiency of mechanical separation of phases I and II from a feed F into two fractions (streams or portions), a product P and a waste W is (*H.W. Cremer, 1956; K. Rietema, 1957*):

$$E = |n_I^W/n_I^F - n_{II}^W/n_{II}^F| = |n_I^P/n_I^F - n_{II}^P/n_{II}^F| \tag{1.1}$$

in which: n_I^F is the amount of phase I in the feed

n_I^F is the amount of phase I in the feed
n_{II}^F is the amount of phase II in the feed
n_I^P is the amount of phase I in fraction P
n_{II}^P is the amount of phase II in fraction P

n_I^W is the amount of phase I in fraction W

n_{II}^W is the amount of phase II in fraction W

For instance (Figure 1), the efficiency of separation of solids (II) from suspension in a liquid (I) by filtration depends not only upon the fraction of solid recovery (n_{II}^P/n_{II}^F) but also upon the liquid retained in the filter cake (n_I^P/n_I^F) the efficiency being the highest when $n_{II}^P/n_{II}^F = 1$ and $n_I^P/n_I^F = 0$ (a full solids recovery in a dry cake).

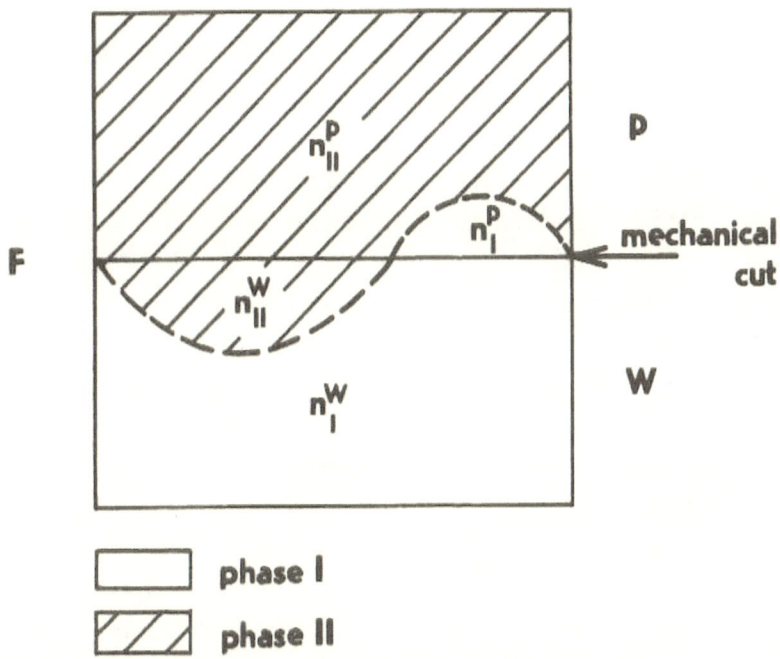

Figure 1. Scheme of efficiency of mechanical separation.

The mutual separation of solid particles is possible only if they differ in size, density, electric, optical or magnetic properties (A.M. Gaudin, 1939; H. Robel, P. Vogel, 1985).

The separation of particles according to their ability to pass through holes or slots is known variously as **sieving, screening** or **grading**. The choice of the screening medium depends on the size and shape of the particles (*K.W. Tromp, 1937*). Circular holes separate according to the medial dimension (the average of the breadth and the length) while long slots separate by the smallest dimension (the thickness). An "edge-effect" of tapered, funnel-shaped holes, which was supposed to circumvent the second law of thermodynamics (*D.H. Deutsch, 1981*) in the case of rod-shaped particles does not generate asymmetry of transport at a molecular level but may work through other effects with macroscopic particles (*F.A. Greco, 1983*). The screens should be chosen with a high proportion of screen surface area available as aperture, called the density of aperturing or screening area (substantial material should be left between the holes to withstand impacts during sieving). The earliest forms of screen were woven textile fabrics (woolen cloth and silk) and recently nylon has been used. The textile screens are of very regular aperture with standard aperture dimensions within the range 50 – 1500 µm and a screening area of 20-70%. Woven wire screens are produced with apertures ranging from about 50 µm up to 1 cm and a screening area of 30-80%. The size of apertures is often expressed by the number of meshes per linear inch, and approximately

$$\text{hole diameter (mm)} \approx \frac{16}{mesh} \tag{1.2}$$

so that, for instance, the "100 mesh" passes particles with a radius of 0.08 mm or less. Accurate wire screens of small apertures (100 mesh and smaller) are difficult to produce.

To induce particles to pass through the screen aperture and to prevent clogging of the apertures, a variety of grading and screening machines (reels, trammels, sifters and plansifters) with single or multi-decked arrangements of sieves are used.

Sieving is a statistical process and there is always the possibility for a particle of "near-mesh" size either to pass or not to pass through the sieve. The rate of sieving is controlled by the size of the material in relation to the aperture, the proportion of small and large particles in the mixture, the amount of material on the screen and the oscillatory or rotating motion of the screen induced by shaking or tapping. Higher rates occur with coarse rather than fine materials. Particles of size substantially smaller than the screen aperture and materials which contain rounded particles of limited size range give the greatest rates of screening. Detailed expressions are necessarily empirical and are limited in their predictive value.

Solid particles can be separated also by **sedimentation** in a column of fluid or **elutriation** in an upward current of fluid at high velocities. The two processes are similar, the relative velocity of fall of the spherical particles being determined by **Stokes' equation** (*G.G. Stokes, 1851*).

$$w = \frac{d^2}{18\eta}(\rho_1 - \rho_2)g \tag{1.3}$$

where w is the free falling velocity (m s^{-1}), d is the particle diameter (m) and ρ_1 is the density of particle (kg m^{-3}) in the medium of density ρ_2, η dynamic viscosity (Pa s), and g is the gravitational acceleration (9.81 m s^{-2}). This last factor is replaced by $\omega^2 r$ (angular rotor velocity in radians per second, i.e. 2π times the number of revolutions per second ω, and rotor radius r) in centrifugal fields —see section 1.4.3. The ratio $\omega/\pi^2 r$ (s) can be used for the characterization of a sedimenting particle and the value 10^{-13} s = 100 fs is used as "Svedberg unit" ("S") for biopolymers and colloidal partides subjected to **centrifugation** (*T. Svedberg, J.P. Nichols, 1923; T. Svedberg, K.O. Pedersen, 1940; T. Gerritson, 1969; P. Sheeler, 1982; V. Piljac, G. Piljac, 1986*). Large molecules of polymers and biopolymers resemble rotational ellipsoids rather than spheres. With a strong asymmetry of their axes the falling velocity decreases, for instance at an axis ratio 1:10 for 1.4 or 1.7 times for a flattened or a prolonged ellipsoid respectively. High-speed centrifuges ($\omega^2 r \approx 20{,}000 - 60{,}000$ g) and ultracentrifuges ($\omega^2 r$ up to 600,000 g) are used for sedimentation of cells, cell organelles and macromolecules. E.g., the nuclei may be removed by low-speed centrifugation (700 g, 5 min), the mitochondria by further centrifugation (10,000 g for 10 min) and the ribosomes separated from soluble RNA species by ultracentrifugation (100,000 g for 90 min) (*J. Maddox, 1990*). The resolution of centrifugation separation can be increased using a medium with a density gradient (*M.S. Meselson, F.W. Stahl, J.R. Vinograd, 1957, 1958; C.A. Price, 1982*).

Isopycnic centrifugation is based on centrifugation of bioorganic components in a gradient solution the composition of which is chosen to avoid unexpected aggregation and denaturation of the biotic macromolecules. Beside sucrose and cesium chloride solutions, cesium trifluoroacetate, colloidal silica (*H. Pertoft et al., 1978*) and iodinated organic compounds such as metrizamide and Nycodenz® (*D. Rickwood, 1976, 1982, 1983*) are used.

For large particles (above 1 mm in water) and the high velocities of the upward stream used in **elutriation**, **Newton's law** of eddying or turbulent movement is valid in place of Stokes' law:

$$w = \left(\frac{4}{3} d \frac{\rho_1 - \rho_2}{\rho_2} g\right)^{1/2} \tag{1.4}$$

(see also equation 1.474).

If the solids differ in size or density, a suitable liquid medium, which has a density close to one of the solid particles, is chosen for use in gravity **settling (*W.A. Deane, 1920; S.A.K. Jeelani, S. Hartland, 1985*)**. One of the most ancient techniques in this field is gold washing, i.e. the separation of gold and shale in a stream of water. The separation of solids into two or more fractions in the sub sieve region (usually below 50 μm, the so-called slimes) may be based on the settling of solid particles in their suspension in a fluid medium — **wet classification**. According to the laws of sedimentation, the coarsest particles settle at a comparatively rapid rate while the finest remain on top, with a gradation in size between the extremes. **Clarification** takes place without a clear line between the settling solids and the supernatant liquid. In some suspensions, however, at a critical point the transitional zone is minimal and a compact "compression zone" of settling appears.

The simplest type of non-mechanical **classifier** consists of a conical vessel placed in a stream of pulp, the feed flowing in on one side and overflowing on the other. The coarser material sediments quickly to the bottom and the fine fraction goes to the overflow. Centrifuges and hydrocyclones (see later) use centrifugal force rather than gravity. An upward current of fluid ("hydraulic water") is used in hydraulic classifiers. Using sedimentation and elutriation it is possible to classify the slimes (fine solids in the range 100 to 1 µm). When the dispersed particles are in the range 0.1–0.001 µm in mean diameter and there are no particular conditions for their aggregation, Brownian motion keeps them in a permanently suspended state (colloidal solutions) and they will settle only in centrifuges.

A combination of hindered settling and flowing stream selection is used in the **tabling** separation of solids. A pulsating stream of water is used in **jigging**. In both techniques the denser and larger particles move further across the flow stream (*A.M. Gaudin, 1931*).

A more precise and sophisticated small scale separation in a perpendicular flow-driving force system, **field-flow fractionation (FFF)** of particles in the range 0.01 – 30 µm has been developed (*J.C. Giddings, 1973; J.C.Giddings, F.J.F. Yang, M.N. Myers, 1974*). This range is of particular interest for biomedical analysis (*J.C. Giddings, H.N. Myers, K.D. Caldwell, 1980; K.D. Caldwell et al., 1979, 1981; K.D. Caldwell, 1986; S.K.R. Williams, K.D. Caldwell, 2011*) for the separation of proteins, viruses, cell organelles and whole cells. FFF is an alternative to the conventional field-induced fractionation process, in which the separation occurs along the axis perpendicular to the applied field (it combines the advantages of field-based non-elution methods with the convenience of column elution techniques). The components are compressed against one wall of a capillary or slot ($d = 0.05 - 0.5$ mm) channel by an external centrifugal field and carried down by a laminar carrier flow, the less compressed zones moving the fastest (Figure 2). The ratio of the zone velocity w and average carrier velocity $\langle w \rangle$ is (*E. Grushka et al., 1973*)

$$\frac{w}{\langle w \rangle} = 6\xi[\coth(1/2\xi) - 2\xi] \qquad (1.5)$$

where ξ is the dimensionless ratio of characteristic (average) thickness (h) of the particle distribution layer to the channel thickness, $\xi = h/d$. The square bracketed function in equation 1.5 rapidly approaches unity as w increases ξ and $w/\langle w \rangle \approx 6\xi$. Obviously ξ diminishes both with the diameter and density of the particles, because h is inversely proportional to the velocity of sedimentation given by equation 1.3. By careful balancing of the settling field and the channel flow it is possible to separate, for example, human and chicken erythrocytes (6 and 9.5 μm respectively) in a matter of minutes whereas centrifugation takes about one hour.

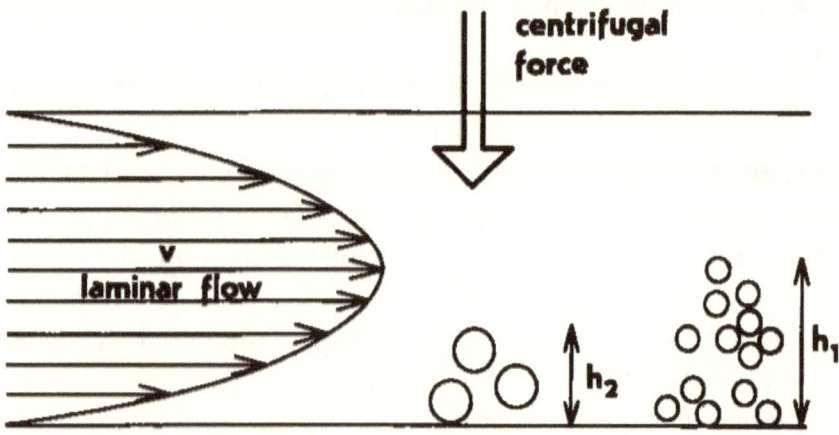

Figure 2. Field-flow fractionation channel.

Laminar flow is important in the continuous- and stopped-flow microanalytical devices (*J. Ruzicka, E.H. Hansen, 1988*), **lab-on-chip** separation technologies (*J.P. Kutter, Y.Fintschenko, 2005; K.E. Herold, A. Rasooly, 2009*) and especially in **cytometers** to count and sort biological cells without their clogging (*P.J. Crosland-Taylor, 1953; P.P.A. Suthanthiraraj, S.W. Graves, 2013*). The cell stream, hydrodynamically focused in capillary is possible to break by axial vibration of the nozzle into uniform-sized droplets that encapsulate single cells. The cells are excited by laser and from light scattering (adsorption and fluorescence) as many as 5000 cells per second can be individually identified. Then, a drop-charging signal for each cell is triggered by an electronic system that determines the appropriate plus or minus charge. In electrostatic field of current machines two or four populations of undamaged cells can be retrieved, and the originally uncharge cells form further fraction (*M.J. Fulwyler, 1965; T.G. Pretlow, T.P. Pretlow, A.M. Cheret, 1987; D. Recktenwald, 1997*).

Another sedimentation method consists of using a liquid medium with a density between the densities of the solids to be separated, so that the heavier fraction of solids settles and the group of lower density comes to the surface. This technique is often used for **densimetric ("sink-or-float") separation** of minerals in rocks, coal washing, etc. The dense media used are various aqueous solutions, suspensions and heavy organic liquids (e.g. calcium chloride, cesium chloride or saccharose solutions and barites or magnetite suspensions).

Separation of high density nonmagnetic particles can be performed in super paramagnetic ferrofluids, i.e. colloidal suspensions of ferromagnetic particles in organic liquids. On magnetization, these ferrofluids exhibit an apparent density which enables the densest materials to levitate (*R.E. Rosensweig, 1955; S.E. Khalafalla, 1973*). Non-uniform field gradients in such liquids can be used for magnetic grid filters (*H. Fay, J.M. Quets, 1980*).

A difference between the density of solid particles can be artificially achieved by using surface-active reagents which enhance the adhesion of bubbles of air to one variety of fine particles (e.g. metal oxides), and then lift them in the form of a froth to the surface of the agitated pulp. This separation technique, **froth flotation** (*I.N. Plaksin, S. V. Bessonov, 1948; F. Sebba, 1962, 1972, 1987; R. Lemlich, 1972; J. Leja, 1982; A. Clarke, D.J. Wilson, 1983; J.H. Harwell, J.F. Scamehorn, 1989; J.A. Finch, G.S. Dobby, 1990; R.S. Ramachandra, 2004*), is widely used for the enrichment of nonferrous medium-grade ores and for the production of super clean coal or glass particles from finely ground solid waste fractions. Such flotation is ineffective for particles larger than 0.20–1 mm (10 mm in the case of coal), nor is it applicable to colloidal particles. In flotation cells, a rapid removal of the froth is desirable and can be achieved by the use of froth paddles or wipers freely suspended from a slowly revolving frame which assists the overflow. Very fine particles are often not recovered because of what has been termed "low collision efficiency". To be more effective, **micro bubble flotation**, a further development of froth flotation, can be used (*F. Sebba, 1987*). A theoretical model for predicting the collection efficiency of nanoparticles was proved on colloidal silica particles with diameters in the range 40-160 nm (*A.V. Nguyen, P. George, G.J. Jameson, 2004*).

The optical properties of some solids (e.g. fluorescence of diamonds) are used by mechanical separation for selective observation and detection of individual pieces in combination with pneumatic matching, their removal in a **stream of air** (*T.J. Veasey, R.J. Wilson, D.M. Squires, 1993; A.F. Tirmyaev et al., 2007*).

High intensity **magnetic separation** is limited to components possessing ferromagnetic properties. Usually fields of about 1000 A m^{-1} are used for the beneficiation of iron, manganese, titanium and some other ores. Eddy magnetic currents produced by permanent magnets of alternating polarity and high frequency eddy current devices are used for separation of aluminum from solid wastes. Ferrites, the ferrimagnetic crystalline materials, has been used for separation of a wide variety of substances from aqueous waste by magnetic means (*S. Shimiza, 1977; T.E. Boyd, M.J. Cusick, J.D. Navratil, 1986*). Biological cells labeled by magnetic nanoparticles MACS® are parting in high gradient field (*S. Miltenyi, M.Muller, W.Weichel, A.Radbruch, 1990*). Recently the widest array of leukocyte subsets, stem cells, and connective tissue cells can be addressed by this technique (*C. Esser, 1998*).

Magnetizable beds, e.g. calcium alginate-magnetite are used in bioaffinity separation (*M.A. Burns, D.J. Groves, 1985; R.F. Masseyeff et al., 1992*).

Electrostatic separation of differentially charged solids is based on the phenomenon that the fresh surfaces of broken materials are electrically charged (for example, organic components of coal positively, and mineral matter negatively) and can be collected on charged rotating disc electrodes (*H. Feibus, 1986*).

Sieving is also applied to separate solids from gaseous media. In the form of more-or-less stable **solid-gas mixtures**, the aerosols come into consideration and due to the small size of solid matter particles in the gaseous dispersions (below $5 - 20$ μm, and for non-sedimenting dispersions less than 1 μm) denser filtration media should be applied, usually in bag or frame filters. High-efficiency particulate air (HEPA) filters from glass or glass/asbestos fibers reach 99.99% efficiency for particles larger than 0.1 μm (*J.A. Paulhaus, 1972*). The distance between the fibers is large compared with the size of the particles, which are deposited not by the screening action, but because on striking the fiber, they are statistically retained by the process of adhesion (*F. Loffler, 1980*).

Filters should be periodically cleaned by occasional flow reversal ("blow-back"), or changed. From this point of view, cyclones are more convenient separation units (*K. Rietema, C.G. Verver, 1961*). A **cyclone** consists of an upper cylindrical section (diameter 5-50 cm) into which gas is pumped tangentially.

Centrifugal forces in the resulting vortex thrust the solid particles onto the walls of the lower, conical section where they collect and can be removed through the apex opening. Cyclones are much preferred for the removal of the major part of coarser solids (above 10 μm) because they give a low pressure drop and do not require cleaning.

For analytical purposes (aerosol fractionation) the separation of solid particles from the gaseous stream directed from a jet against a wall in **cascade impactors (centripeters)** is used. The efficiency of separation of accelerated particles by impact with the wall depends upon the flow and distance of the wall (*N.A. Fuchs, 1984; L. Thedore, A.J. Buonicore, 1978*).

Water scrubbing and barbotage columns are also an efficient means of removing solids from a gas, but they result in dilute suspensions of solids in water that are difficult to treat.

Microscopic solid particles often carry a net electric charge, due to an ionized atmosphere, and can be separated by **electrostatic filters**. Charging of the particles can be increased by electrodes producing a corona discharge. Electro precipitators work at electric fields of $2 - 3$ kV cm^{-1} and reasonable migration velocities are obtained for the particles of $1 - 100$ μm diameter, the efficiency increasing with diminishing particle size. High collection efficiency (95-99%) is achieved by the use of high-duty design of the receiving electrodes ("chute-type" electrodes) from which the dust falls undisturbed by the gas stream into a hopper underneath the electrodes. Electro deposition is widely used for cleaning the gases of coal power plants and air conditioning (*M. Cranford, 1976; C. Comninellis, M. Doyle, J. Winnick, 2011*).

Solid-liquid mixtures are often encountered as a result of chemical precipitation and crystallization. Settling in a gravitational and centrifugal field is based on the same principles discussed previously for the classification of solid particles. Using gravitational settling, the sediment often contains a significant fraction of wetting liquid and has a friable consistency (*W.A.*

Deane, 1920; J.V.N. Dorr, P.L. Franklin, 1945; A. Rushton, A.S. Ward, R.G. Holdich, 2000).

Liquid-solid separation in pilot and industrial practice is achieved by cyclones and thickeners based on sedimentation (D.A. Dahlstrom, C.A. Cornell, 1971; M.Ungarish, 1993). A sedimentation process is called **clarification** when used for obtaining a clear liquid from a dilute pulp (1-5% of solids), e.g. carbonated sugar solutions. When the main objective is to remove as much liquid as possible from highly concentrated pulps (15-30%) to obtain thickened solids (for example, the dewatering of cement slurry) the term **thickening** is applied. The invention of **thickeners** allows sedimentation to be operated on a continuous basis which has been improved by **continuous counter-current decantation (CCD)** systems. Industrial thickeners are shallow cylindrical settling tanks which range in diameter from 5 – 30 m and have a centrally located mechanism with radial arms equipped with plough-type blades. Each revolution of the blades slowly sweeps the area of the bottom to produce a positive mechanical means (a gradual consolidation of settled particles) for discharging thickened sludge to a centrally located outlet. In washing type thickeners, the wash water enters at the bottom and flows countercurrent to the solids that are progressively devoid of soluble components. Settling characteristics may vary greatly and thickener unit areas required for various pulps range from 0.3 to 30 m^2 per ton of solids per day (J. V.N. Dorr, 1906; P.L. Franklin, 1946).

Some pulps require special treatment to destabilize the suspension and aggregate the settled particles. **Flocculation** is promoted mechanically by gentle agitation and also by a rise in temperature; and chemically by the *in situ* formation of a gel-like substance (e.g. aluminum hydroxide after chemical dosing with aluminum sulphate). This last procedure is preferred only in the case of clarification.

Liquid cyclones (hydroclones) with a diameter from 2 to 50 cm are used to separate solids in the range 2 to 200 μm, finer particles leaving the hydroclone with most of the water overflow. Coarser solids are removed via the apex valve opening at the bottom (*D. Bradley, 1965*).

In some processes, expensive liquid-liquid and solid-liquid separation steps are reduced when the above process is combined with another recovery step; that is the solids enter the chemical process directly in the suspension. Typical examples are **solvent-in-pulp** or **resin-in-pulp** recovery of metals by solvent extraction or ion exchange respectively, or **electrolysis from leach slurries** (*G.M. Ritcey, 1986*).

For a more complete separation, **filtration** is utilized where the liquid must be removed in the presence of a filter. In the laboratory, cellulose, glass, asbestos and stainless-steel filters are most commonly encountered; in industry, textile and metal filters are employed (*F.A. Gooch, 1878; M. Dittrich, 1904; L. Moser, W. Maxymovicz, 1924; A.F. Orlicek, A.E. Hackl, P.E. Kindermun, 1954; N.P. Cheremisinoff, D.S. Azbel, 1983; A.Rushton, A.S. Ward, R.G. Holdich, 1990*).

Vacuum, pressure, leaf, drum and disc **filters** have been employed in industry for many years. More efficient belt vacuum filters have been in recent use because of their increased washing efficiency and lower soluble losses.

For **microfiltration** of colloidal particles, **membrane filters** of 0.005 to 3 µm pore sizes are available (*H. Bechhold, 1907; C.J. Van Oss, 1970*). The filters are necessary to filter viruses (0.03 − 1 µm), and bacteria (0.5− 20 µm) which flourish in dilute biotechnological solutions (beers). High uniformity of the pore size is achieved in plastic filters, "nucleoporous" filters, prepared by bombardment with high-energy heavy ions in cyclotrons. Since the open area is often low (0.1-1% of surface), the housing geometries for membranes are designed to ensure high total area and low feed pressures; hollow fibers, tubular cartridges, thin channels and more recently spiral (double) wound and pleated filters are in current use (*G.B. Tanny, D. Hauk, 1980; C.J.D. Fell, 1980*). The hollow fibers modules are designed like shell-and-tube heat exchangers in which the feed flows down the filters' interior and filtrate is collected from the outside. In the spiral wound the suspension being separated is filtered through the loosely rolled envelope. These devices plug more easily, however, than conventional filters though the microfiltration proceeds without formation of an outer layer of filtration cake.

For large amounts of solids the rate of filtration depends considerably on the layer of solid deposited on the filter - the **"filtration cake"**. The filtrate flow rate, dV/dt, at any instant can be expressed as

$$\frac{dV}{dt} = -\frac{A\Delta p}{\eta(R_0+R_i)} \tag{1.6}$$

where V is the filtrate volume (m³) at time t(s), A is the filter area (m²). Δp is the pressure drop across filter (Pa = kg m^{-1}s^{-2}), η is the filtrate dynamic viscosity (Pa s = kg m^{-2}s^{-1}), and R_o and R_i are the constant and instantaneous resistance (m^{-1}) of the filter and filtration cake, respectively. The resistance of the incompressible cake can be expressed as

$$R_i = \alpha\frac{m}{A} = \alpha\frac{wV}{A} \tag{1.7}$$

where m is the mass (kg) of cake (proportional to the volume of suspension and weight concentration of solids w) and α is the specific cake resistance (m kg^{-1}) which can be evaluated for instance by the relationship derived for porous masses (*P. C. Cannan, 1937, 1938*).

$$\alpha = \frac{5(1-\varepsilon)a}{\varepsilon^2\rho} \tag{1.8}$$

where ε is the cake porosity (dimensionless), a is the specific surface area, i.e. the area of particle per unit volume of solids (m^{-1}), and ρ is the density of solids in the cake (kg m^{-3}) — see equations 1.463 and 1.464. For instance, the biomass from bacterial broth has a specific cake resistance of about 2×10^9 m kg^{-1}.

Almost all cakes formed of biological materials, however, are compressible and the cake resistance is a function of the pressure drop, α being proportional to $(\Delta p)^s$, where s ranges from 0.1 − 0.8 (*D.R. Sperry, 1928; H.P. Grace, 1953; M. Better et al., 1988*).

The pressure drop Δp necessary for filtration, increasing with the mass of filtration cake, is created by a vacuum at the output (hence the maximum Δp is 0.1 MPa), or by pressure at input (about 1 MPa in industrial units). According to equation 1.8, the finer the solid, the higher is the resistance of the filtration cake.

The residues of liquid from the filtration cake are removed by **drying** at elevated temperature, centrifugation, in a gas stream or in a vacuum. In spray dryers a feed slurry is directly sprayed into a hot dry gas (*K. Kroll, 1959; R.E. Treybal, 1968; M. V. Lykov, 1970; R.B. Keey, 1972*).

Gas-liquid separation is rarely important. As long as fine dispersions —fogs — do not arise, the phases form a continuity spontaneously and the interface is hence easily distinguished. Separation is not difficult owing to the low shearing stresses between the phases. Cyclones are the most widely used devices for the disengagement of phases. **De-gasification** of liquids is performed efficiently by passing the liquid through sintered glass or metal filters into an evacuated chamber.

Immiscible **liquid-liquid** phases are also easy to separate providing care is taken to avoid high shearing stresses leading to the break-up of droplets (e.g. in cyclones). The bulk behavior can be derived from a fundamental consideration of interdrop and drop/interface coalescence (*T.K. Sherwood, R.L. Pigford, 1952; C. Hanson, 1968; S.A. Jeelani, S. Hartland, 1985*). Immiscible liquids in general can be easily separated by passing them through a porous metal filter wet with the liquid to be filtered or through hydrophobized cellulose filters that retain the aqueous phase. Following stable emulsion formation, surface-active additives, a contact with specially designed grids or columns filled with coalescers, or high alternating electrical fields are applied to break the emulsions. Gravity settling in mixer-settler units or solvent extraction columns is the simplest performance of separation but a partial coalescence and formation of small secondary drops can be responsible for settling difficulties. The interfacial tension and the viscosity of the disperse phase are the primary factors in determining the size of the secondary drops. The improved method of "absolute" two-liquid separation involves centrifugation up to 15000 g in discontinuous or continuous mode (*H. Reinhardt, J. Rydberg 1969; J. Rydberg, G. Skarnemark, 1986*), where it takes about 1 second to separate > 1 µm droplets from a liter volume of emulsion. Liquid-liquid separation is also sensitive to contamination of the interface with third phase solids, a phenomenon which often occurs in real separation systems (*E.K. Dursma, C. Hoede, 1967; S. Winitzer, 1973, G.M. Ritcey, 1980; G.A. Yagodin, V. V. Tarasov, S. Yu. Ivakhno, 1982*).

1.1.2 Physicochemical separation

A homogeneous mixture (mostly liquid and gaseous) of a molecular dispersion cannot be separated mechanically because any mechanical force applied leads to an equal displacement of every component of its volume elements. If a solid surface is present, some gradients of solute concentration may appear in fluid streams (see section 1.4.1.) but usually such an effect is insufficient to produce a significant separation.

A generalized force, regardless of its origin, that exerts an influence on a certain amount of substance (force per mole) and causes its displacement can be considered as a gradient of the universal force field, chemical potential.

The **chemical potential** is a thermodynamic characteristic of a solute in a given matrix and external field under isothermic-isobaric conditions and is known as the partial **molar Gibbs free energy** (*J. W. Gibbs, 1875*)

$$\left(\frac{\partial G}{\partial n_i}\right)_{T,p,n_j} = \bar{G}_i = \mu_i \tag{1.9}$$

This represents the Gibbs energy released ($\mu < 0$) or consumed ($\mu > 0$) after the addition of an infinitesimal amount of the i-th component ∂n_i to the mixture of other components (in molar amounts n_j) at a given temperature, pressure and other external parameters. If the component is separated (∂n_i is negative), the chemical potential has the opposite sign as on mixing. The chemical potential (J mol^{-1}) is an intensive property (such as temperature or pressure) of a substance whereas the Gibbs energy (J) is an extensive property (such as volume).

The Gibbs energy at the thermodynamic temperature T, pressure p and volume V is related to the enthalpy H, entropy S and internal energy U as shown in the following relationship

$$G = H - TS = U + pV - TS \tag{1.10}$$

At constant temperature, the change in Gibbs energy with change of enthalpy and entropy becomes

$$\Delta G = \Delta H - T\Delta S \tag{1.11}$$

Two opposite processes, mixing and separation of substances, are described by the same thermodynamic function:

$$\Delta G_{sep} = -\Delta G_{mix} \tag{1.12}$$

In contrast to many other chemical reactions where $|\Delta H| \gg |T\Delta S|$, both the enthalpy and entropy terms of the Gibbs energy are important in separation processes; in the case of ideal mixtures (see later) the magnitude of ΔG_{sep} is determined solely by the $T\Delta S$ term.

Separation is spontaneous, as is any other process, only if it is accompanied by a decrease in the Gibbs energy. Because we are usually concerned with thermodynamically stable mixtures, some work must be introduced during a separation process to cover the increase of Gibbs energy which corresponds to the separated components. The minimal energy for separation and pressure-volume work (pV) in a reversible process is equal to ΔG_{sep}:

$$E_{sep}^{min} = -\Delta G_{mix} = -\Delta H_{mix} + T\Delta S_{mix} \quad (1.13)$$

Any real separation process is far from reversibility, especially when many operations involving heating and material transport (boiling, cooling, condensing, pumping, compressing, etc.) are involved. For the irreversible separation process there is always

$$E_{sep} > -\Delta G_{mix} \quad (1.14)$$

The **thermodynamic efficiency of separation**

$$\eta = \frac{E_{sep}^{min}}{E_{sep}} = -\frac{\Delta G_{mix}}{E_{sep}} \quad (1.15)$$

rarely reaches several percentage points (see section 2.6).

Let us briefly consider the general relationship between those thermodynamic quantities that are important for separation chemistry. If a system consists of a single component, the chemical potential is simply the Gibbs energy per mole of pure substance at a given pressure and temperature. For a closed two-component system, where the numbers of moles of components A and B are n_A and n_B, the equation valid for every intensive thermodynamic property yields

$$dG = \mu_A dn_A + \mu_B dn_B \quad (1.16)$$

or, for a dosed multicomponent system

$$dG = \sum \mu_i \, dn_i \tag{1.17}$$

If the chemical potential (partial molar derivatives of G) remains constant, then on simultaneous addition of every component in the same ratio as in the mixture, i.e. by enlarging the system, equation 1.17 may be integrated, yielding **a rule of chemical potential additivity**:

$$G = \sum n_i \, \mu_i \tag{1.18}$$

The full differential of equation 1.18 gives

$$dG = \sum (\mu_i dn_i + n_i d\mu_i) \tag{1.19}$$

Because this equation was obtained without any additional conditions, subtracting equation 1.17 from equation 1.19 gives the **Gibbs-Duhem** equation (*P. Duhem, 1886*)

$$\sum n_i \, d\mu_i = 0 \tag{1.20}$$

This equation is useful for the experimental investigation of multicomponent systems, because it gives the dependence of the chemical potential of one component upon another. For instance, in two-component systems

$$n_A d\mu_A + n_B d\mu_B = 0; \tag{1.21}$$

therefore μ_A can be calculated from the equation

$$\mu_A = \int -\frac{n_B}{n_A} d\mu_B \tag{1.22}$$

and vice versa (see section 1.2.1.). In this way, it is for instance possible to calculate the chemical potential of a solute A from the total change of chemical potential of solvent B found from partial pressure or osmotic pressure measurements.

The rule of additivity, expressed by equation 1.18, was obtained under the condition of μ_i = constant. This occurs when the separation (or ad-mixing) of a component does not affect the physic-chemical composition of other components of the mixture, and it takes place when the components are either immiscible and thus remain in the system in separate phases or they are miscible but without any resultant energetic effects ($\Delta H_{mix} = 0$). A typical plot of the components of Gibbs energy for binary mixtures (systems) is shown in Figure 3. In the first case the separation is spontaneous, but in the second case ("ideal mixtures") the minimal energy for separation is $E_{sep} = T\Delta S_{mix}$.

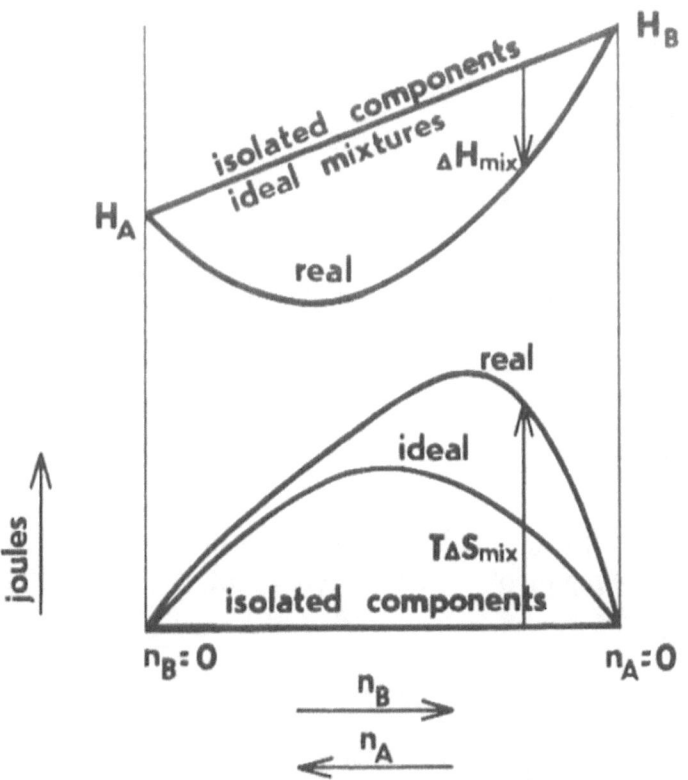

Figure 3. Enthalpy and entropy of two-component system versus number of added or subtracted moles of components A and (n_A, n_B).

The **entropy of mixing** or separation ($\Delta S_{mix} = -\Delta S_{sep}$) can be rather precisely derived from the basic postulates of statistical mechanics. For N molecules of i sorts (independent components), each of which consists of N_i molecules, the number of indistinguishable arrangements (thermodynamic probability) is $W = N!/\Pi N_i!$. According to the **Boltzmann relation** *(L. Boltzmann, 1896; M. Planck, 1906)* the entropy change on mixing the N molecules is given by

$$\Delta S_{mix} = k \ln W \tag{1.23}$$

$$\Delta S_{mix} = k \ln \left(\frac{N!}{\Pi N_i!}\right) = k(\ln N! - \sum \ln N_i!) \tag{1.24}$$

(k is the Boltzmann constant, $k = 1.38 \times 10^{-23}$ J K^{-1}). Stirling's formula, $\ln N! = N(\ln N - 1)$, is valid for the large numbers N, and because

$$\frac{N_i}{\sum N_i} = \frac{N_i}{N} = x_i \tag{1.25}$$

is the mole fraction of the i-th component, equation 1.24 becomes

$$\Delta S_{mix} = k[(\ln N - 1) - \sum N_i (\ln N_i - 1)] = k[\ln N - \sum N_i \ln N_i - 1] =$$

$$= k \sum N_i \ln(N/N_i) = -kN \sum x_i \ln x_i \tag{1.26}$$

Considering that $k = R/N_A$ (R is gas constant, $R = 8.31$ J K^{-1}mol^{-1} and N_A is Avogadro's constant $N_A = 6.02 \times 10^{23}$ mol^{-1}), and $N/N_A = n$,

$$\Delta S_{mix} = -R \sum n_i \ln x_i = -R \sum n_i \ln(n_i/n), \tag{1.27}$$

or

$$\Delta S_{mix} = \sum \Delta S_i \tag{1.28a}$$

where

$$\Delta S_i = -Rn_i \ln x_i. \tag{1.28b}$$

The molar entropy of mixing (per moles of all components of the mixture) is derived from equation 1.26

$$\frac{\Delta S_{mix}}{n} = \Delta \overline{S}_{mix} = -R \sum x_i \ln x_i \tag{1.29}$$

For instance, the entropy of mixing for air (78% N_2, 21% O_2 and 1% Ar) is $\Delta \overline{S}$ = -R (0.78 ln 0.78 + 0.21 ln 0.21 + 0.01 ln 0.01) = 4.72 J K^{-1} mol^{-1}

For two-component systems ($x_A + x_B = 1$, or $x_A = x$ and $x_B = 1 - x$), according to equation 1.13, the minimal energy required to separate one mole of the mixture at temperature T is

$$\frac{E_{sep}^{min}}{n} = T\Delta \overline{S}_{mix} = -RT[x \ln x + (1-x) \ln (1-x)] \tag{1.30}$$

This function has a maximum at equimolar mixtures (then $x = 0.5$ and $\overline{S} = +5.76$ J K^{-1}mol^{-1}); for a mixture of macro- and micro components, when $x \ll 1$, it approaches zero.

Equation 1.30 represents the **minimal energy required** to separate one mole of a binary mixture. If the energy necessary to isolate a certain amount (n_A) of one of the components, present in a mole fraction $x = n_A/n$, is to be calculated, then

$$\frac{E_{sep}^{min}}{n_A} = \frac{E_{sep}^{min}}{n} \frac{1}{x} \tag{1.31}$$

and from equation 1.30

$$\frac{E_{sep}^{min}}{n_A} = -RT[\ln x + \frac{1-x}{x} \ln(1-x)] \tag{1.32}$$

According to this equation, the energy for the total separation of the same amount of component A depends strongly on its original concentration mixture x (Figure 4). For instance, at

300 K the separation from a rich mixture where $x = 0.1$ needs $E_{sep} = -5.73(\log 0.1 + 9 \log 0.9) = -5.73(-1 - 0.413) = 8.1$ kJ mol^{-1}, but at $x = 10^{-5}$ it becomes $10^5 \log (1 - 10^{-5})] = -5.73(-5 - 0.434) = 31$ kJ mol^{-1}) (i.e. 3.8 times more) per mole of the isolated component A. Equation 1.30 gives 0.81 and 3.1×10^{-4} kilojoules per mole of the mixture (effectively per mole of purified matrix) and 8.1 or 31 kJ of total energy since 10 or 10^5 moles of mixture must be separated to obtain the same amount of A.

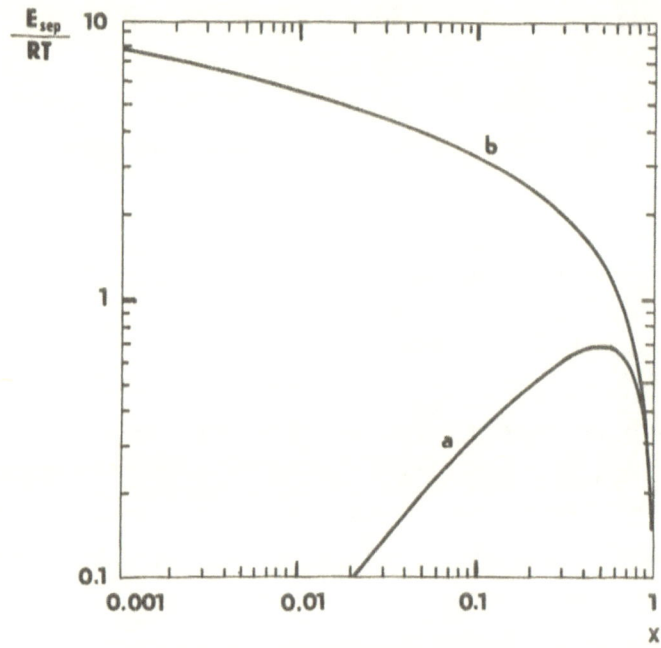

Figure 4. Separation work (in RT units) as a function of mole fraction x: a) per mole of mixture — equation 1.30, b) per mole of component A — equation 1.32.

Particular effort needs to be applied to separate trace impurities and purify a matrix (*V.A. Kireev, 1942; M. Benedict, 1947*) which is also connected with the performance of the process in real equipment (see sections 2.4 and 2.6): an infinite size of separation device and an infinite process time would be necessary for an absolutely complete separation (section 2.4.1.). This should be realized when referring to chemical purity and the efficiency of separation operations — part 2. Moreover, at extremely low concentrations (where x is below 10^{-17}), fluctuations of thermodynamic values can occur (see section 1.2.5).

Once again, it should be emphasized that the separation work calculated according to equations 1.30 or 1.32 is the minimal energy. For example, on mixing 0.5 mole of benzene and 0.5 mole of toluene at 25°C the Gibbs free energy increases by approximately 1.7 kJ due to the positive entropy effect ($+5.8$ J K^{-1}). In the real process of separation of the substances, however, the energy consumption is about 20 times higher. For instance on rectification, simple evaporation of the amount of the mixture requires about 32 kJ and the heat of condensation needs to be carefully exploited to make the process more economical (S.A. Bagaturov, 1961; F.E. Becker, A.I. Zakak, 1985).

For the purposes of separation chemistry it is extremely useful to have the chemical potentials expressed explicitly in terms of the concentrations of the components in the mixture. Differentiating equation 1.11 and considering equation 1.9, the chemical potential of the i-th component is

$$\mu_i = \frac{\partial \Delta G_i}{\partial n_i} = \frac{\partial \Delta H_i}{\partial n_i} - T \frac{\partial \Delta S_i}{\partial n_i} \tag{1.33}$$

If the enthalpy of the component in the mixture is the same as for the isolated component,

$$\frac{\partial \Delta H_i^{id}}{\partial n_i} = \bar{H}_i = \frac{\Delta H_i}{n_i}, \tag{1.34}$$

and if

$$\frac{\partial \Delta S_i^{id}}{\partial n_i} = \frac{\Delta S_i}{n_i}, \tag{1.35a}$$

then the mixtures are called ideal solutions. Using equation 1.28a, it follows that

$$-T\frac{\partial \Delta s_i^{id}}{\partial n_i} = RT \ln x_i \qquad (1.35b)$$

The ideal mixtures obey the **ideal gas equation**

$$pV = nRT \qquad (1.36)$$

i.e. the volume V of ideal mixture at given pressure p and temperature T depends only on the total number of moles (n). Then, equation 1.33 becomes the explicit expression for the **chemical potential of the i-th component** as a function of its concentration in ideal solution:

$$\mu_i^{id} = \underbrace{\mu_i^*}_{\text{enthalpic term}} + \underbrace{RT \ln x_i}_{\text{entropic term}} \qquad (1.37)$$

where μ_i^* is the molar enthalpy ($\overline{H_i}$) of the pure i-th component ($\mu_i = \mu_i^*$ if its mole fraction is unity) at a given temperature and pressure. Equation 1.37 is a principal equation used to calculate separation processes which clearly expresses the enthalpic and entropic terms of the chemical potential in terms of the amount of the component in the mixture.

There is no separation (transport) of the i-th component between two parts I and II of a system, where

$$\mu_i^I = \mu_i^{II} \qquad (1.38)$$

i.e. when the same Gibbs energy is released (or consumed) by addition of the same amounts of component to system I and system II. This is a general condition of **thermodynamic equilibrium**. On the other hand, a **moving force of separation** of the i-th component from part I to part II can be expressed as a quantity proportional to the difference

$$\vartheta_i \propto \mu_i^I - \mu_i^{II} \tag{1.39}$$

This is the spatial gradient of the chemical potential and it represents the force (in newtons, N) having an effect on one mole of a substance, for example along the z-axis of Cartesian coordinates

$$\left[\frac{\partial \mu}{\partial z}\right] = \frac{\text{J mol}^{-1}}{\text{m}} = \frac{\text{N m mol}^{-1}}{\text{m}} = \text{N mol}^{-1} \tag{1.40}$$

We thus arrive at the dynamic aspects of separation.

Just as in gravitational fields, a body may move spontaneously from a point of higher potential to one of lower value, so also the chemical separation of a substance proceeds in the direction of decreasing chemical potential. The factors that influence the chemical potential (Figure 5) will be considered in detail later (Chapter 1.2). Each of these factors represents a particular contribution to the process and, if separation and preconcentration as a non-spontaneous process is to be achieved, the overall interactions should ensure even an "up-hill" concentration movement of a substance due to internal and external forces (e.g. an outside agency). An efficient separation requires the creation of a steady and favorable concentration gradient of the substance even if the gradient of chemical potential disappears, i.e. at chemical equilibrium, and it should be the transfer of the separated substance that is responsible for chemical potential equalization. To separate the substance in such a way, the necessary initial gradient of chemical potential should be created by a suitable choice of interacting species or external fields and in particular by chemical composition or physicochemical effects. This should be seen as the ultimate objective of **separation chemistry**.

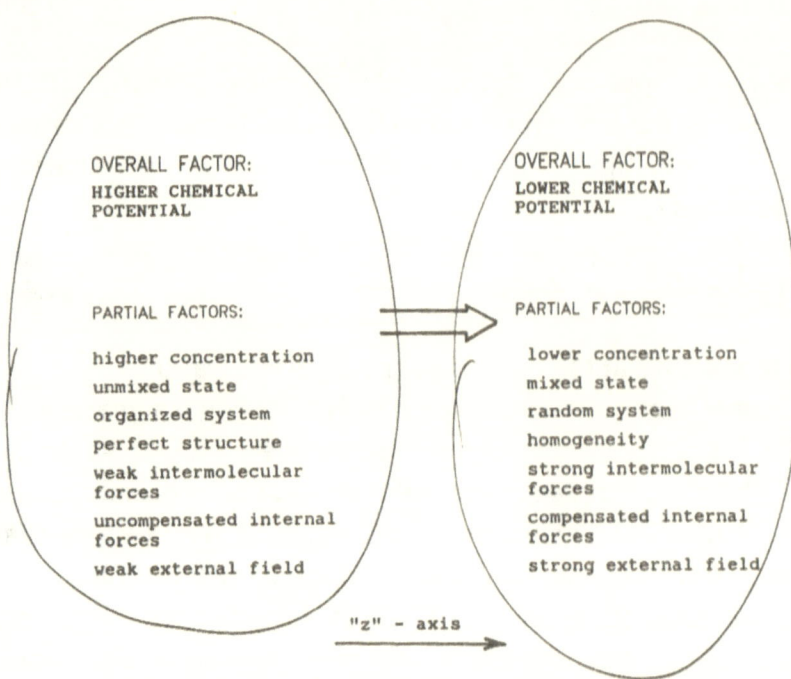

Figure 5. Direction of spontaneous movement of a substance

Movement of the substance in the direction of the negative gradient of chemical potential represents the true spatial displacement of a substance with a velocity $u(ms^{-1})$ generally given by

$$u = -\frac{1}{f}\frac{\partial \mu}{\partial z} \tag{1.41}$$

where f stands for the **molar frictional coefficient** (N s m^{-1} mol^{-1} = kg s^{-1} mol^{-1}). For spherical molecules with radius r, the frictional coefficient is determined by **Stokes' equation** (*G.G. Stokes, 1851*)

$$f = N_A 6\pi\eta r \tag{1.42}$$

(see equation 1.3).

In water at 25°C, where $\eta = 10^{-3}$ Pa s $= 10^{-3}$ kg m^{-1}s^{-1}, the frictional coefficient for spherical molecules of diameter 1 nm has the value 1.13×10^{13} kg s^{-1}mol^{-1} = 1.13×10^{13} J m^{-2} s mol^{-1}. This means, for example, that to move one mole of a substance at a velocity 10^{-3} m s^{-1}, a chemical potential gradient of about 10^{10} J mol^{-1} m^{-1} needs to be created.

The question, of course, is how to create a sufficient and selective chemical potential gradient so that the substance will separate. In principle, molecular forces and external fields should be established so that

(i) the components will form individual phases (e.g. crystals — solution) or
(ii) the components will distribute between two phases (e.g. liquid and steam, liquid — sorbent, liquid — liquid) or
(iii) the components will become spatially distributed when an external field is applied (e.g. in a centrifugal or electrical field).

1.2 EQUILIBRIA IN SOLUTIONS

Kinetic factors influencing the behavior of solutes in mixtures are not negligible and can play an important role in the dynamics of the separation process, as will be shown in section 1.4. However, for most of the mixtures that enter the separation process, their components are in thermodynamic equilibrium. Therefore, in this section only the thermodynamic differences between isolated components and their homogeneous mixtures will be discussed (heterogeneous systems are included in section 1.3).

The previous section dealt with ideal mixtures, where no interactions between molecules of components occurred. However, this concept covers only ideal-gas mixtures, and it is clear that even simple formation of a condensed phase is impossible without intermolecular interactions. The main task of solution theory is to derive the properties of solutions (mixtures) from the properties of individual components. Such an aim is important even for the simplest, two-components systems which can exist in an infinite number of combinations of miscible chemical individuals. Thus the property of any particular mixture should often be derived from a limited number of experimental data (e.g. molar enthalpy, viscosity, density, electric susceptibility, light scattering and refraction, etc.) that are available for individual components or their chemical analogues.

1.2.1 Metrics of solutions

In practice, the composition of solutions is expressed in various ways.

Previously we have used the mole fraction x_i (dimensionless) as the number of moles of the i-th substance divided by total number of moles of all other components (x) of a mixture:

$$x_i = \frac{n_i}{\sum n_i} \tag{1.45}$$

where the subscripts i refer to solvent S and other components A, B, C .... It is evident that the sum of the mole fractions for all the components is unity:

$$\sum x_i = 1 \tag{1.46}$$

The mass fraction w_i (dimensionless) is the ratio of mass of a substance $i(G_i)$ to the mass of mixture (G)

$$w_i = \frac{G_i}{G} \tag{1.47}$$

and usually is expressed by **weight percent**, wt% $= 100\, w_i$, but also as weight per thousand part, wt‰ $= 1000\, w_i$, **weight-to-weight** ratio (w/w) or in **parts**: parts per million, 1 ppm $= 10^6\, w_i$, parts per billion, 1 ppb $= 10^9\, w_i$ or part per trillion, 1 ppt $= 10^{12}\, w_i$. Obviously,

$$1 \text{ ppm} = 10^{-4} \text{ wt \%} \qquad (1.48\text{a})$$

$$1 \text{ ppb} = 10^{-7} \text{ wt \%} \qquad (1.48\text{b})$$

$$1 \text{ ppt} = 10^{-10} \text{ wt \%} \qquad (1.48\text{c})$$

An example of an unusual, and sometimes misunderstood form of concentration unit is "mg %" used in clinical medicine: it means mg of solute per 0.1 dm³ of solution, i.e. it equals 10^{-3} wt % or 10 ppm.

Volume fraction v_i (dimensionless) is expressed as the fraction

$$v_i = \frac{V_i}{V} \qquad (1.49)$$

where V_i is the volume of i-th component measured in the pure state and V is the total volume of solution. This is also often given in **volume percents**, vol% = $100 v_i$ or as a **volume-to-volume ratio** (v/v).

The commonest unit is the **molar concentration, molarity**, c_i (mol dm^{-3}, mol L^{-1}),

$$c_i = \frac{n_i}{1000 V} \qquad (1.50)$$

where V is the volume of solution (m³). Between molarity and mole fraction there is the relation

$$c_i = x_i c_t \qquad (1.51\text{a})$$

where

$$c_t = \frac{\sum n_i}{1000 V} = \sum c_i \qquad (1.51\text{b})$$

is the **total molarity** or **molar density** of the solution (total number of moles of each component in the volume V). Further, molarity and volume fraction are related by

$$c_i = \frac{v_i}{1000\overline{V}_i} \tag{1.51c}$$

where $\overline{V}_i$ is the molar volume (m³ mol⁻¹) of component i.

The molarity of solvent ($i = 1$ or $i \equiv S$) is

$$c_S = \frac{0.001\rho - \sum_{i=2}^{k} c_i M_i}{M_s} \tag{1.52}$$

where ρ is the density of solution (kg m⁻³) and M's stand for molar masses of mixture components and M_s is the molar mass of solvent (kg mol⁻¹). For diluted solutions $c_t \approx 1/\overline{V}_S$, i.e. it is equal to reciprocal molar volume of solvent (molarity of solvent).

The accepted definition of mole (the numerical value of which has been preserved the same as that of "molecular weight") instead of kmole, and preservation of the traditional definition of molarity as the number of moles of solute in one liter, mol L⁻¹ or mol dm⁻³, left certain inconveniences in the use of the molar concentration scale (otherwise there would be a simple transfer, mol dm⁻³ = kmol m⁻³), particularly in calculations where both volume and density should be given in basic *SI* units (m³ and kg m⁻³). The coefficient 10³ or 10⁻³ often appears in this connection in many relations such as the equations 1.50–1.52.

Because the highest experimental precision is achieved by the weighing of solution components, the concentration of a component is also expressed by the **molal concentration (molality)** m_i (mol kg⁻¹) defined as

$$m_i = \frac{n_i}{G_S} \tag{1.53a}$$

where G_s is the mass of solvent (matrix). Unlike the molarity, molality does not depend on temperature. For the solvent

$$m_S = n_S/G_S = 1/M_S \tag{1.53b}$$

To make use of the chemical concentration units more concisely in description, the following **abbreviations** are used for y-molal or y-molal (y is a numerical value) solutions respectively:

"y M A" for the solution of y-moles of A in 1 dm³ of solution, $c_A = y$ mol dm^{-3}, i.e. simply "0.8 M NaCl" instead of $c_{NaCl} = 0.8$ mol dm^{-3}, or

"y m (Italic type) A" for the solution of y moles of A in 1 kg of solvent, $m_A = y$ mol dm^{-3}, e.g. "0.8 m NaCl" instead of $m_{NaCl} = 0.8$ mol kg^{-1}

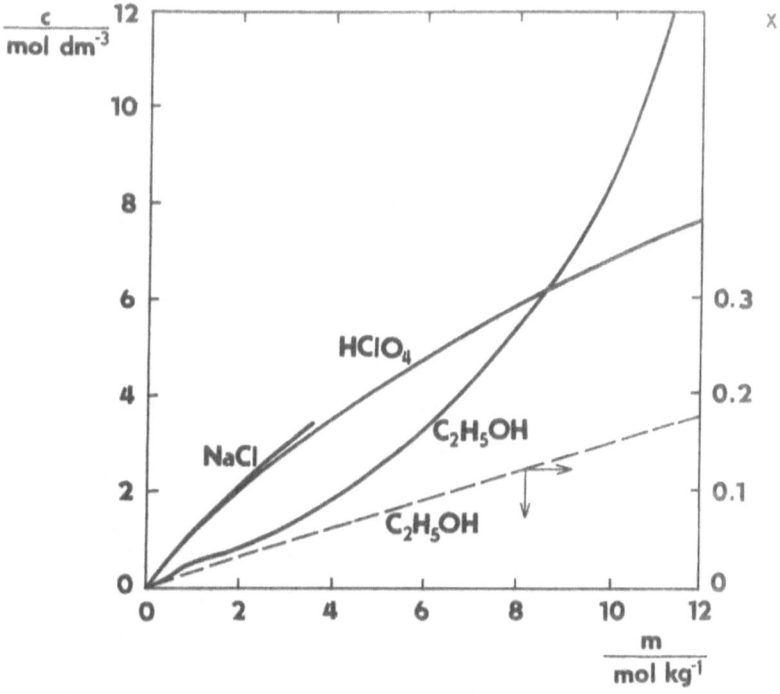

Figure 6. Relationships between various (x, c, m) concentration scales for typical aqueous solutions (HClO$_4$ — water, NaCl — water, ethanol — water).

For two-component (solute A and solvent S) solutions, or generaly k-component solutions (S ≡ 1, A ≡ 2, B ≡ 3 ...) useful relationships between x, c and m can be derived. Because the density of the solution (kg m^{-3}) is

$$\rho = \frac{G_A + G_S}{V} \tag{1.54a}$$

or

$$\rho = \frac{\sum_{i=1}^{k} G_i}{V} = 1000 \sum_{i=1}^{k} c_i M_i \tag{1.54b}$$

the mole fraction is related to the molality and molarity by

$$x_A = \frac{c_A}{(0.001\rho - M_A c_A)/M_S + c_A} = \frac{m_A}{1/M_S + m_A} \tag{1.55a}$$

or

$$x_i = c_i \frac{c_S + \sum_{i=2}^{k} c_i M_i}{0.001\rho c_i} = \frac{m_i}{1/M_S + \sum_{i=2}^{k} m_i} \tag{1.55b}$$

where M_A (M_i) and M_S are molar masses (kg mol^{-1}) of solute A (solutes i) and solvent respectively. Further,

$$c_A = \frac{0.001\rho - M_A}{M_S/M_A - M_A m_A} = \frac{0.001\rho - x_A}{M_A x_A + (1 - x_A) M_S} \tag{1.56}$$

$$m_A = \frac{M_S c_A}{M_A 0.001(\rho - M_A c_A)} = \frac{x_A}{(1 - x_A) M_S} \tag{1.57}$$

These relations are often necessary to recalculate data from various literature sources (Fig. 6). The density of real solutions should be known to treat molarity concentration data because the volume of a mixture differs from the sum of the volumes of its individual components, except for ideal solutions (compare with equation 1.35a). Volume contraction or expansion on mixture formation is a typical phenomenon of thermodynamical non-ideality of the solutions, though not the most important one from the viewpoint of separation procedures.

In nuclear chemistry and biochemistry some special units are used to express the compositions of solution of radioactive or bioactive species.

Osmolality (osM) is the total ionic molar concentration of sodium chloride which has the same osmotic pressure (section 1.5.) as the solution in question. Solutions of equal osmolality are **isotonic solutions**. For example, 0.25 M glucose has the same osmotic pressure as 0.15 M NaCl for which 0.15 M NaCl ≡ 0.30 osM NaCl. This means that 0.25 M glucose ≡ 0.30 osM glucose.

The concentration of radioactive substances is based on the proportionality between the molar amount and the activity (radioactivity) A expressed in the number of decays of the substance per second (1 becquerel = 1 Bq = 1 s^{-1} or in older units Curie = 1 Ci = 3.7 × 10^{10} s^{-1} = 3.7 × 10^{10} Bq). **Specific (volume) activity**

$$a_V = \frac{A}{V} \text{ Bq m}^{-3} \text{ (Ci dm}^{-3}) \tag{1.58}$$

has the meaning of concentration, and a linear relation exists between them:

$$a_V = a_m c_A \tag{1.59}$$

where the coefficient of proportionality a_m (Bq kmol^{-1}, Ci mol^{-1}) is the molar activity of the particular substance (a function of the radioactive decay constant λ, s^{-1}).

The concentration of enzymes is expressed either through the enzyme unit (IUPAC, 1961)

$$1\, U = 1\, \mu\text{mol min}^{-1} \tag{1.60}$$

which is the amount of enzyme which catalyzes a transformation of 1 μmol of specific substrate per minute, or through the newer analogical unit **katal**, abbreviation **kat** (IUPAC, 1972),

$$1 \text{ kat} = 1 \text{ mol s}^{-1} \tag{1.61}$$

This means that the concentration of enzyme is expressed in the units as

$$1 \text{ kat L}^{-1} = 1 \text{ mol dm}^{-3}\text{s}^{-1} = 6 \times 10^7 \text{ U dm}^{-3} \qquad (1.62)$$

It should be stressed that the product mol dm^{-3} in relation (1.62) concerns the concentration of substrate, not that of enzyme. Because there is no simple relation between the enzyme units and the molar amount of the enzyme (its catalytic activity depends upon the reaction conditions and the nature of the substrate), there is no simple relationship between the molar concentration of an enzyme and its concentration expressed by enzyme units. Since 2001 the unit katal has been adopted by IUPAC as SI unit for <u>any catalyst</u> quantification. "Unusual" multiples of kat are Ekat (attokat, 10^{18} kat), Zkat (zettakat, 10^{21} kat) and Ykat (yottakat, 10^{24} kat).

1.2.2 Phenomenology of thermodynamical non-ideality of solutions

Thermodynamic non-ideality, which is a measure of the mutual influence of the components of mixtures, is rather important in separation science.

In practice, there are two different approaches to express the non-ideality of solutions:

(i) establishing the deviation from the additivity of extensive thermodynamic parameters (volume, entropy, enthalpy or Gibbs energy) of components in real mixtures, from the rule of additivity — equation 1.18.

(ii) introducing correction factors for the deviation of the ideal chemical potential dependence on mixture composition given by equation 1.37. The first approach is preferred in general thermodynamics of mixtures. The **molar excess functions** are defined as

$$\bar{V}^E = \Delta\bar{V}_{mix} - \Delta\bar{V}_{mix}^{id} \tag{1.63}$$

$$\bar{H}^E = \Delta\bar{H}_{mix} - \Delta\bar{H}_{mix}^{id} \tag{1.64}$$

$$\bar{S}^E = \Delta\bar{S}_{mix} - \Delta\bar{S}_{mix}^{id} \tag{1.65}$$

$$\bar{G}^E = \Delta\bar{G}_{mix} - \Delta\bar{G}_{mix}^{id} \tag{1.66}$$

From the definition of ideal solutions — equations 1.34 and 1.35a — it follows that $\overline{\Delta V^{id}} = 0$ and $\overline{\Delta H^{id}} = 0$ and the excess functions become

$$\bar{V}^E = \Delta\bar{V}_{mix} \tag{1.67}$$

$$\bar{H}^E = \Delta\bar{H}_{mix} \tag{1.68}$$

For "normal", non-associated solutions, there is the relation (*J.H. Hildebrand, R.L. Scoit, 1950*)

$$\Delta\bar{V}_{mix} = \frac{\alpha}{1+\beta T}\Delta\bar{H}_{mix} \tag{1.69}$$

where α and β are temperature and pressure expansion coefficients, respectively:

$$\alpha = -\frac{1}{\bar{V}}\left(\frac{\partial \bar{V}}{\partial T}\right)_p \tag{1.70a}$$

$$\beta = -\frac{1}{\bar{V}}\left(\frac{\partial \bar{V}}{\partial p}\right)_T \tag{1.70b}$$

The expansion coefficients of mixtures can be approximately calculated from the single components coefficients

$$\alpha = \alpha_A v_A + \alpha_B v_B \tag{1.71a}$$

$$\beta = \beta_A v_A + \beta_A v_B \tag{1.71b}$$

where v's are volume fractions of the components. If $\bar{V}^E$ is positive (consumption of heat on mixing) the deviation from ideality is conventionally considered to be positive and vice versa. For ethanol-water mixtures, a positive deviation from ideality indicates a stronger interaction between $C_2H_5OH - H_2O$ molecules than between either $C_2H_2OH - C_2H_5OH$ or $H_2O - H_2O$ pairs. Extreme volume changes are observed in concentrated electrolyte solutions. For instance, the apparent molar volume of $[Zn(H_2O)_6]^{2+}$ ions increases from $V^\infty = 0.060$ to $V^0 = 0.083$ dm^{3}mol^{-1} respectively on going from infinitely dilute solutions to the "anhydrous" state (*V. Jedinakova, J. Celeda, 1975*) obeying to a first approximation the relation

$$V = V^\infty + (V^0 - V)\sqrt{c/c^0} \tag{1.72}$$

which is a modification of the empirical **Masson rule** (*O. Masson, 1929*) and where c^0 is the concentration of the electrolyte corresponding to the hypothetical "water-free" solution.

The aim of theory is the prediction of the excess enthalpy of mixing which will be considered in more detail in sections 1.2.3–1.2.5.

Excess entropy of mixing for the components of different size can be calculated more easily, as was done for solutions of macromolecules (*M.L. Huggins, 1942, 1970; P.J. Flory, 1942*). The **Flory-Huggins equation** for mixtures of molecules A and B, the volume of which is given by the ratio

$$r = V_B/V_A \tag{1.73}$$

is isomorphous with equation 1.29, namely

$$\Delta \bar{S}_{mix} = -R(x_A \ln v_A + x_B \ln v_B) \tag{1.74}$$

but as regards the logarithmic functions there are volume fractions of the components (v_A, v_B) instead of mole fractions. Because of the relation $x_A = rv_A/(rv_A + v_B)$, the excess entropy is obtained as

$$\Delta \bar{S}^E = R\{\ln[1 + (r-1)x_B] - x_B \ln r\} \tag{1.75}$$

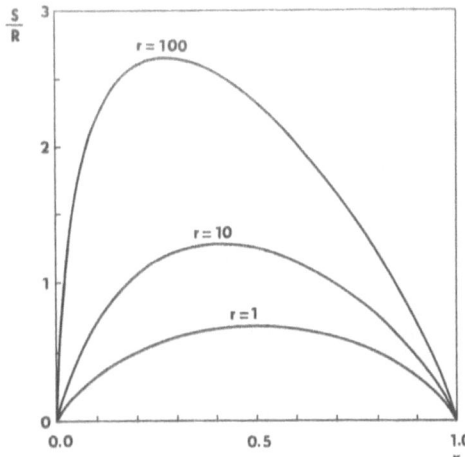

Figure 7. Excess entropy — equation 1.75 — of mixtures of molecules of unequal volume.

The value is positive if $r > 1$, however its value is significant only for large molecules (macromolecules), when $r \gg 1$ (see Figure 7). For instance, if $r = 2$ and $x_B = 0.4$, then $S = R(0.336 - 0.227) = 0.45$ J K^{-1} mol^{-1}, i.e. it is only 0.15 kJ mol^{-1} at 300 K.

It can be said that the expression 1.29 for the ideal entropy of mixing for mixtures of molecules of equal volume ("isomegetic solution") was derived from the postulate of the ideal solution

$$\left(\frac{\partial \Delta S_{mix}^{id}}{\partial n_i}\right)_{p,T,n_j} = \frac{\Delta S_{mix}}{n_i} \tag{1.76}$$

However, from 1.28b it can be seen that, rigorously, (because $\partial n_i = \partial n$ if $n_j =$ const) the entropic term is

$$\frac{\partial \Delta S_{mix}}{\partial n_i} = R(\ln x_i + 1 - x_i) = \frac{\partial S_{mix}^{id}}{\partial n_i} + R(1 - x_i) \qquad (1.77)$$

and therefore even the expression 1.29a is valid only if $x_i \approx 1$, i.e. for the majority component (matrix) which dilutes other components. This is the solvent and, for such dilute solutions, **Raoults law** is valid for solvent behavior on evaporation (*F.M. Raoult, 1887*):

$$p_S = x_S p_S^0 \qquad (1.78)$$

i.e. the partial vapor pressure of the solvent (p_S) above the solution is proportional to the mole fraction of the solvent (x_S) and the proportionality constant is identical to the vapor pressure of pure solvent (p_S^0). Even the behavior of non-ideal solutions might obey Raoult's law if the solvent mole fraction approaches unity and if the pure solvent (matrix) corresponds to its standard state (the vapor pressure of the solvent obeys Raoult's law in dilute solutions, where the vapor pressure obeys Henry's law - see below). This gives another reason for the choice of pure solvent as the standard state in many thermodynamic calculations.

For solutions where a component is present as the solute at minor concentrations, another empirical rule was established; it describes the partial pressure of the component above dilute solutions and is known as Henry's law (*W. Henry, 1803*):

$$p_A = x_A k_A \qquad (1.79)$$

where k_A is referred to as the **Henry's law constant**. For instance, for oxygen solutions in water at 25°C we have $k_{O_2} = 917$ MPa, i.e. at the normal partial pressure of oxygen in air, which is $p_{O_2} = 2.1 \times 10^4$ Pa, we have $x_{O_2} = 2.29 \times 10^{-5}$ (which corresponds to a 1.27×10^{-3} M solution of oxygen in water). For ideal solutions

$$k_A = p_A^0 \tag{1.80}$$

and Raoult's and Henry's law become identical. Obviously, Henry's law should hold for the solute (micro component) in the same range in which Raoult's law is valid for the solvent (macro component).

The second approach of describing real solutions, preferred in separation science, ensues from the expression for chemical potential. As proposed by G.N. Lewis (*G.N.Lewis 1907; G.N. Lewis, M. Randall, 1923*), it is essential to operate with a relation isomorphous with equation 1.37 giving a simple logarithmic dependence between concentration or another optional quantity of the mixture composition and the chemical potential which is of crucial importance in thermodynamic calculations. Once again, there are two convenient ways how to achieve a description of real mixtures which do not follow the ideal chemical potential function:

(a) replacing the mole fraction of the i-th component in equation 1.37 by an arbitrary quantity related to concentration and defined by G.N. Lewis as the **thermodynamic activity** a (dimensionless) of the i-th component,

$$\mu_i = \mu_i^o + RT \ln a_i \tag{1.81}$$

where μ_i^o is the **standard chemical potential** of the component i.e. the chemical potential when the thermodynamic activity is defined as unity ($a_i = 1$) in any specific concentration scale and at arbitrary composition which can be chosen at will (see below). For the activity of gaseous compounds the term **fugacity** was originally and often used as an equivalent of activity *(G.N.Lewis, 1901)*.

(b) introducing a correction parameter (g_i) as a multiplier of the entropic term of chemical potential (*N. Bjerrum, 1918*)

$$\mu_i = \mu_i^o + g_i RT \ln x_i \tag{1.82}$$

This correction is suitable to express the deviations from ideality at $x_i \approx 1$, i.e. to describe the behavior of the matrix (solvent). The parameter g_i is called the **osmotic coefficient**.

Now we shall consider the phenomenological relations of thermodynamic activity (*or* simple activity) and the osmotic coefficient to other parameters, namely to the concentration of the solution component expressed in units that have been chosen (molarity, molality, weight fractions, etc.) in metrics of solutions.

According to the concept of thermodynamic activities, the choice of **standard state** is crucial. The chemical potential depends only on the real composition of the solution and is independent of the concentration scale and standard state at a given temperature and pressure. A temperature of 25°C (298.13 K) is usually chosen as the standard temperature and the standard pressure of 1 bar $= 10^5$ Pa is recomended (IUPAC, 1982). However, as shall be shown, the values of the standard chemical potential and activity (or osmotic coefficients), i.e. "enthalpic" and "entropic" terms of chemical potential, will be, in general, affected by the choice both of standard state and concentration scale.

In principle, as a standard state, any solution (mixture) or single component state can be chosen and then the following equality is postulated

$$a_i^{st} = x_i^{st} \tag{1.83}$$

to keep the standard chemical potential (enthalpic term) equal both in the activity and concentration scale:

$$\mu_i^o = \mu_i^{id} \tag{1.84}$$

Between activity and concentration of real solutions a simple linear proportionality is used

$$a_i = \gamma_i x_i \tag{1.85}$$

but the coefficient of proportionality, the **activity coefficient** γ, depends on the actual composition of the mixtures. For the standard state

$$\gamma_i = 1 \tag{1.86}$$

Because x is dimensionless, the activity coefficient γ is also dimensionless. In other concentration scales (molar, molal) γ has the inversional dimension of concentration ($dm^3\ mol^{-1}$, $kg\ mol^{-1}$) because the thermodynamic activity was defined as a dimensionless parameter (if the activity had the same dimension as the concentration, then γ would also be dimensionless —all of which is a question of convention).

Because equations 1.81 or 1.82 are formulated for an identical value of the chemical potential, from these equations and equation 1.85, a relation between activity coefficient and osmotic coefficient ensues as follows

$$g_i = \frac{\ln a_i}{\ln x_i} = 1 + \frac{\ln \gamma_i}{\ln x_i} \tag{1.87}$$

In practice, to define a standard state, the following conventions are the most typical:

Convention I:
The standard state is the state of the pure i-th component ($x_i^{st} = 1$) which means that the enthalpic terms in equations 1.37, 1.81 and 1.82 are identical with the molar enthalpy ($J\ mol^{-1}$) of the i-th component,

$$\mu_i^* \equiv \mu_i^0 \equiv \overline{H}_i \tag{1.88}$$

This means that absolute activity coefficient is given by the difference expressing the excess chemical potential

$$RT \ln \gamma_i = \mu_i - \mu_i^{id} = \mu_i^E \qquad (1.89a)$$

where

$$\mu_i^E = \frac{\partial \bar{G}^E}{\partial n_i}, \qquad (1.89b)$$

in correspondence with equation 1.9.

For example, at 300 K even the slight difference $\mu_i^E = 4$ kJ mol^{-1} due to weak interactions between components of the solution causes a twofold change in γ_i (Figure 8).

This convention is used usually for mixtures of organic compounds (nonelectrolyte solutions). However, it is unsuitable for electrolyte solutions because the pure state (crystals, gas) strongly differs from that in solution (solvated ions and molecules).

Convention II:
Generally, instead of a pure component, a state in any solution of defined composition can be chosen for the standard state. According to the commonest alternative convention (Convention II), the hypothetical standard state is chosen as the solution of unit concentration (molar, molal) in which the i-th solute exhibits the same properties as in an infinitely dilute solution, i.e. $\gamma_i = 1$ at $x = 0$,

$$\gamma_i(c_i = 1) = \gamma_i(c_i = 0) = 1 \qquad (1.90)$$

and

$$a_i^{st} = c_i^{st} = 1 \qquad (1.91)$$

or

$$a_i^{st} = m_i^{st} = 1 \qquad (1.92)$$

It should be stressed that the infinitely dilute solution itself cannot be used as a standard state, because at $x \to 0$, $\mu \to -\infty$ (Figure 8).

To distinguish thermodynamic activities and the activity coefficients defined by this convention — **standard or concentration coefficients** — we will denote them, at least in this chapter, by a point in superscript, $\dot{a}$ and $\dot{\gamma}$. The standard chemical potential in this convention will be $\dot{\mu}$,

$$\mu_i = \dot{\mu}_i + RT \ln \dot{a}_i \tag{1.93}$$

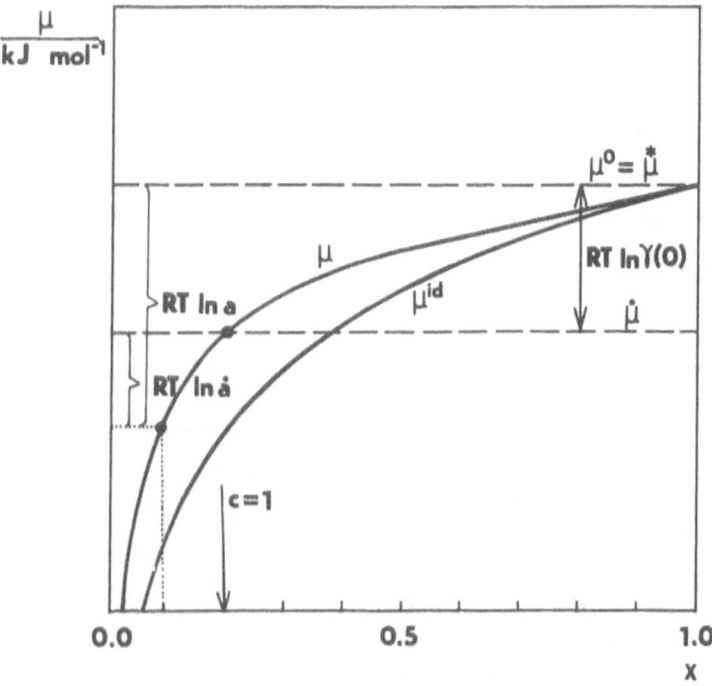

Figure 8. Ideal and real chemical potentials (μ^{id} and μ), and their enthalpic (μ^0 and $\dot{\mu}$) and entropic ($RT \ln a$, $RT \ln \dot{a}$) terms according to Convention I and Convention II.

This chemical potential should have the same value as expressed by equation 1.81 and therefore

$$RT \ln \gamma_i = \mu_i - \dot{\mu}_i - RT \ln c_i \qquad (1.94)$$

As a result, the relation between absolute and standard activity coefficients becomes

$$RT \ln \frac{a}{\dot{a}} = \dot{\mu} - \mu^* \qquad (1.95)$$

or

$$RT \ln \frac{\gamma}{\dot{\gamma}} = \dot{\mu} - \mu^* \qquad (1.96)$$

The difference in the right-hand side of the equations represents a constant: the energy to transfer a solute from the pure state to a standard solution. Hence, the ratio $\gamma/\dot{\gamma}$ depends upon the choice of standard state and does not depend on the composition (concentration) of mixtures. For instance, when comparing the activity coefficients at infinite concentrations, $\gamma(0)$ and $\dot{\gamma}(0)$ and at any other concentrations, from equation it follows that

$$\frac{\gamma}{\dot{\gamma}} = \frac{\gamma(0)}{\dot{\gamma}(0)} = \gamma(0) = \exp\left(\frac{\dot{\mu}-\mu^*}{RT}\right) \qquad (1.97)$$

because the standard coefficient at zero concentration, $\dot{\gamma}(0) \equiv 1$, according to the definition used in Convention II (Figure 9). It should be stated that the deviation of the absolute coefficient from unity represents a **positive** ($\gamma > 1$) or **negative** ($\gamma < 1$) **deviation from ideality**. The $\ln \gamma$ and $\ln \dot{\gamma}$ have opposite signs in any particular case. However, due to another choice of standard state, the non-unity concentration activity coefficients ($\dot{\gamma} \neq 1$) indicate in fact only a deviation from infinite dilute solution behaviour, i.e. from Henry's law (*A.M. Rozen, 1967*). The equality $\dot{\gamma} = 1$ should not mean that the solute is in an ideal solution and this is more often a rare coincidence.

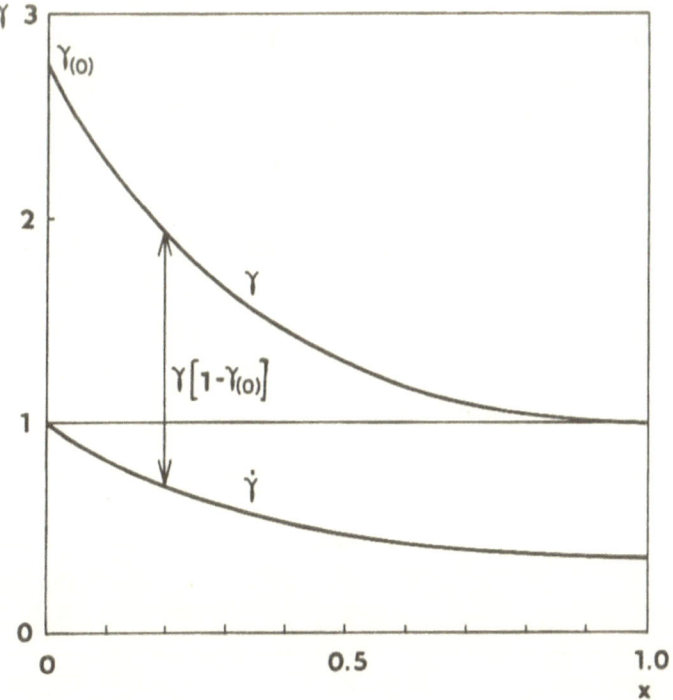

Figure 9. Absolute (γ) and standard ($\dot{\gamma}$) activity coefficients.

The ideality by Convention I corresponds to Raoult's law 1.78 and the unit standard activity coefficient corresponds to Henry's law 1.79.

In two-phase systems a further convention is necessary to consider the overall deviation from ideal behavior on the distribution of solute between the two phases (F. Macasek, 1979) unless Convention I or other unifying definitions are used — see section 1.3.2.

Negative deviations from ideality indicate that the solute interacts more strongly with solvent molecules than in other intermolecular interactions ($\Delta H < 0$) and vice versa. For solutions of macromolecules, dilution with strongly interacting solvents also indicates that intramolecular interactions (between various segments of the polymer chain) are weakened. Negative non-ideality is difficult to distinguish from chemical interactions and can often be described by chemical (pseudo chemical) equilibria — see below. However, chemical compound formation is evident only when there is a strong decrease of activity coefficients: chemical interaction is characterized by an enthalpy change $200 - 500$ kJ mol^{-1} and coordination bond formation gives about 25 kJ mol^{-1} for mono-functional ligands. In the latter case this means that the activity coefficients of interacting molecules decrease 22,000 times at 300 $K (RT = 2.5$ kJ mol$^{-1})$!

According to the type of non-ideality of solution, *J.H. Hildebrand* (1929) proposed the following **thermodynamic classification of solutions**:

Type of solution	$\Delta \overline{G}_{mix}$	
	enthalpic term $\Delta \overline{H}_{mix}$	entropic term $\Delta \overline{S}_{mix}$
perfect	0	$\Delta \overline{S}^{id}_{mix}$ (Eq. 1.35)
athermic	0	$\Delta \overline{S}_{mix}$ (Eq. 1.74)
regular	$\Delta \overline{H}^{E}$	$\Delta \overline{S}^{id}_{mix}$ (Eq. 1.35)
real	$\Delta \overline{H}^{E}$	$> \Delta \overline{S}^{id}$

The athermic, and moreover the regular solutions represent useful abstractions because they serve as a first approximation for a real solution model: simple relations were found to calculate activity coefficients in these solutions — see below.

The relation between activity coefficients in different concentration scales is also of great practical importance. According to the conventions discussed, several activity data can be met in practice:

$$a_x = \gamma x \qquad (1.98a)$$

$$a_c = \gamma c \qquad (1.98b)$$

$$a_m = \gamma m \tag{1.98c}$$

every of which also depends on the choice of standard state. To distinguish activity coefficients in various concentration scales, the activity coefficient corresponding to molar fractions, equation 1.98a is called the **rational activity coefficient** (dimensionless); sometimes the symbols f or γ_x are used to distinguish it. In the molar concentration scale, the **molar activity coefficient** (dm³mol⁻¹), y or γ_c is used; for molality the **molal (practical) activity coefficient** (kg mol⁻¹), γ or γ_m, is used.

According to equations 1.55–1.57, the following relations can be established between the absolute activity coefficients of solute A in solvent S for different scales:

$$\gamma_x = \gamma_c \frac{c_A}{x_A} \frac{1}{0.001\rho_S m_S} = \gamma_m \frac{m_A}{x_A} \frac{1}{m_S} \tag{1.99}$$

$$\dot{\gamma}_x = \dot{\gamma}_c \frac{0.001\rho + c_A(M_S - M_A)}{0.001\rho_S} = \gamma_m(1 + m_A M_S) \tag{1.100}$$

where the parameters have the same meaning as in equations $1.55 - 1.57$ (ρ is in kg m⁻³) and the activity coefficients refer to component A. However, if the composition is expressed in certain concentration units, the activity coefficients γ which correspond to the concentration scale should be used. This is why the activity coefficients can be given without more precise specification, simply indicating them as γ's.

Between the rational, practical and molal activities, faint by equations 1.98, is the relation

$$a_x = a_c \frac{M_S}{0.001\rho_S} \frac{1}{1 \text{ dm}^3 \text{ mol}^{-1}} = a_m M_S \frac{1}{1 \text{ kg mol}^{-1}} \tag{1.101}$$

Thus for aqueous solutions, $M_S = 0.018$ kg mol⁻¹ and $M_S/0.001\rho_S = 0.018$ dm³ mol⁻¹.

Though the existence of those various systems does not promote clarity in the thermodynamics of solutions, all of the systems described are widely used and appropriate care should be taken, especially when comparing and applying various literature and handbook data.

Specific problems arise when the activities of **electrically-charged species** (ions, electrons) are defined.

First of all, when an electrolyte consisting of a cation $M^{z_M^+}$ and anion $L^{z_L^-}$ is dissolved in a polar solvent, the following electrolytic dissociation occurs

$$M_{\nu_+}L_{\nu_-} \rightleftarrows \nu_+ M^{z_M^+} + \nu_- L^{z_L^-} \qquad (1.102a)$$

When salt $M_{\nu_+}L_{\nu_-}$ is a strong electrolyte (dissociation in solution is complete), the concentration of cations and anions is ν_+ and ν_- times higher than the concentration of dissolved compound (m, when expressed in molality). When the dissociation proceeds to a **degree of dissociation** α, which is the fraction of dissociated molecules $M_{\nu_+}L_{\nu_-}$, the molality of the resulting species is

$$m(M^{z_M^+}) = \alpha \nu_+ m \qquad (1.102b)$$

$$m(L^{z_L^-}) = \alpha \nu_- m \qquad (1.102c)$$

$$m(M_{\nu_+}L_{\nu_-}) = (1-\alpha)m \qquad (1.102d)$$

(the same is true for molality c and molar amounts n of the ions). This means that every molar of the dissolved electrolyte produces $\alpha(\nu_+ + \nu_-)$ moles of ions and $(1-\alpha)$ moles of undissociated molecules. The question now is, how can the chemical potential μ_{ML} of the electrolyte in solution be expressed. Because μ is an intensive property of the solution, the additivity of the chemical potentials of the ionic and neutral species should be preserved. The **chemical potential of an electrolyte** is equal to the sum

$$\mu_{ML} = \alpha \nu_+ \mu_+ + \alpha \nu_- \mu_- + (1-\alpha)\mu_0 \qquad (1.103)$$

where μ_+, μ_- and μ_0 denote the chemical potentials of the cations, anions and neutral (undissociated) molecules respectively. Because this equality is valid also for the standard state of the species (Convention II), all $\overset{\circ}{\mu}$ values can be eliminated, which results in the equality

$$RT \ln \dot{a}_{ML} = \alpha v_+ RT \ln \dot{a}_+ + \alpha v_- RT \ln \dot{a}_- + (1-\alpha)RT \ln \dot{a}_0 \quad (1.104)$$

If the dissociation is complete ($\alpha = 1$)

$$\ln \dot{a}_{ML} = v_+ RT \ln \dot{a}_+ + v_- RT \ln \dot{a}_- \quad (1.105)$$

then the **activity of a strong electrolyte** is obtained as the product

$$\dot{a}_{ML} = \dot{a}_+^{v_+} \dot{a}_-^{v_-} \quad (1.106)$$

According to equations 1.102a,

$$\dot{a}_+ = v_+ \gamma_+ m \quad (1.107a)$$

$$\dot{a}_- = v_- \gamma_- m \quad (1.107b)$$

represent the activities of individual ions and therefore

$$\dot{a}_{ML} = (v_+^{v_+} v_-^{v_-})(\gamma^{v_+}\gamma^{v_-})m^{(v_+ + v_-)} \quad (1.108)$$

For instance, the activity of K_2SO_4 ($v_+ = 2, v_- = 1$) will be defined as the product $\dot{a}_{K_2SO_4} = (2^2 1^1)(\gamma_+^2 \gamma_-^1)m^{(2+1)} = 4\gamma_+^2 \gamma_- m^3$. However, the estimation of **individual activity coefficients of ions** (γ_+, γ_-) is connected both with theoretical and experimental problems.

No solutions containing only cations or anions can be obtained to measure their properties: real solutions of electrolytes are electrically neutral. Even if this were not so, the chemical potentials of the ions would depend on the size and shape of the bulk of the solution representing a charged body.

The influence of internal and external fields (gravitational, centrifugal, etc.) on the chemical potential is hidden in the enthalpic term μ_1^0. However, the electrostatic component of the enthalpy of an ion with z_i (dimensionless number) elementary electric charges is conventionally excluded from the enthalpic term and, instead, the **electrochemical potential** $\tilde{\mu}$ of the ion is defined as

$$\tilde{\mu}_i = \mu_i + z_i F\varphi \qquad (1.109a)$$

where

$$z_i = \nu_+ z_M = -\nu_- z_L \qquad (1.109b)$$

Now, μ_i is the "chemical part" of the potential, given by equation 1.93, i.e. the chemical potential in the absence of electric charges. By analogy, the **electrochemical activity coefficient** is considered as the product

$$\tilde{\gamma}_i = \dot{\gamma}_i \, \gamma_i^{\text{elstat}} \qquad (1.110)$$

where $\dot{\gamma}$ is the chemical and γ_i^{elstat} the electrostatic activity coefficient (for uncharged species $\gamma_i^{\text{elstat}} = 1$). The second term on the right-hand side of equation 1.109a is the product of the number of elementary charges z_i, Faraday's constant F (9.65×10^4 C mol^{-1}) and the **inner (Galvani) electric potential** φ (V). This represents the work (Gibbs energy increase) necessary to transfer one mole of the ions from infinity to a solution volume element at a given potential,

$$z_i F\varphi = -\Delta G_i^{\text{elstat}} \qquad (1.111)$$

For instance, the increase of chemical (electrochemical) potential of an ionic compound that is $\Delta\mu_i = 1$ kJ mol^{-1} corresponds to the transfer of 1.04×10^{-2} moles of singly-charged ions ($z_i = 1$) through a potential difference $\varphi = 1\,V$; the transfer of one mole of the ions is equivalent to a Gibbs energy increase of $9.65 \times 10^4\,C\,V = 96.5\,kJ$.

The Galvani potential is not directly measurable and therefore only the difference between the potentials in a given and standard state can be obtained. As the standard state, the potential of the **normal (standard) hydrogen electrode** (φ_{NHE}) working at unit activity of hydrogen ions ($a_{H+} = 1$) and unit fugacity of gaseous hydrogen has been chosen. The electrochemical potentials for the two scales are equal:

$$\dot{\mu}_i + RT \ln (m_i \dot{\gamma}_i \gamma_i^{elstat}) + z_i F \varphi_{NHE} = \dot{\mu}_i + RT \ln (m_i \dot{\gamma}_i) + z_i F \varphi \quad (1.112)$$

The **electrostatic part of the activity coefficient**, expressing the energy for transfer of an ion from a normal hydrogen electrode potential to a real one (compare with equation 1.89),

$$RT \ln \gamma_i^{elstat} = z_i F E \quad (1.113)$$

where

$$E = \varphi - \varphi_{NHE} \quad (1.114)$$

is the measurable electrode potential towards a normal hydrogen electrode, NHE (for electrons it is the normal reduction potential). Since the NHE is rather difficult to prepare it is usually replaced by a saturated calomel electrode, SCE ($E^0_{SCE} = 0.2415\,V$ at $25°C$).

The potential difference 1.113 still cannot be measured for individual ions, for the reasons discussed earlier, but only for a given electrolyte as a whole. On the dissociation of 1 mole of electrolyte, according to 1.102, the change of Gibbs energy is due to an increase of chemical potential by ion formation and a decrease of the chemical potential of undissociated molecules — see equation 1.17:

$$-\Delta \bar{G} = \alpha v_+\mu_+ + \alpha v_-\mu_- - (1-\alpha)\mu_0 \qquad (1.115)$$

Because $\Delta G = -zFE$ (equation 1.111),

$$zFE = \alpha v_+(\mu_+ + RT \ln \dot{a}_+) + \alpha v_-(\mu_- + RT \ln \dot{a}_-) - (\alpha - 1)(\mu_0 + RT \ln \dot{a}_0) \qquad (1.116)$$

and, if $\dot{\gamma} = 1$, i.e. $\dot{a} = (\alpha - 1)m$ but $\alpha \neq 1$, considering relations 1.102 and 1.107, we have

$$zFE = zFE_0 + \alpha RT \ln \left[(v_+^{v_+} v_-^{v_-}) \frac{\alpha^{(v_+ + v_-)}}{\alpha - 1} (\dot{\gamma}_+^{v_+} \dot{\gamma}_-^{v_-}) m^{(v_* + v_- - 1)} \right] \qquad (1.117a)$$

or, for full dissociation ($\alpha = 1$),

$$zFE = zFE_0 + RT \ln \left[(v_+^{v_+} v_-^{v_-})(\dot{\gamma}_+^{v_+} \dot{\gamma}_-^{v_-}) m^{(v_* + v_-)} \right] \qquad (1.117b)$$

where E_0 is the electromotive force (EMF) for the solution in which the following product of activities is unity:

$$\frac{(a_+^0)^{\alpha v_+}(a_-^0)^{\alpha v_-}}{(a_+^0)^{(\alpha - 1)}} = 1 \qquad (1.118)$$

(superscript 0 on the activity symbols denotes the standard state). Measurement of the potential

$$E = E_0 + \frac{RT}{zF} \ln \left[(v_+^{v_+} v_-^{v_-})(\dot{\gamma}_+^{v_+} \dot{\gamma}_-^{v_-}) m^{(v_+ + v_-)} \right] \qquad (1.119)$$

which is the potential of the electrolyte solution versus the standard electrode, can give information only on the product of the activity coefficients, $\dot{\gamma}_+^{v_+} \dot{\gamma}_-^{v_-}$. Moreover, when incomplete dissociation occurs, the coefficients calculated from 1.119 also include a degree of dissociation, i.e. they would not be "pure" electrostatic activity coefficients. Hence, instead of individual coefficients γ_+ and γ_-, their geometric mean is used as the **mean ionic activity coefficient** $\dot{\gamma}$,

$$\dot{\gamma}_\pm = \sqrt[v]{\dot{\gamma}_+^{v_+} \dot{\gamma}_-^{v_-}} \qquad (1.120)$$

where

$$\nu = \nu_+ + \nu_- \tag{1.120a}$$

Now, the expression for **mean ionic molal activity** of a strong electrolyte follows from equations 1.106 and 1.107

$$\dot{a} = (\nu_+^{\nu_+}\nu^{\nu_-})\dot{\gamma}_\pm^\nu \; m^\nu \tag{1.121}$$

and 1.119 can be written as the **Nernst equation (*W.Nernst, 1889*)**

$$E = E_0 + \frac{RT}{zF} \ln a_\pm \tag{1.122}$$

It can be shown that for ionic equilibria a product of the type 1.121 always appears in equilibrium constants and therefore the mean ionic activity $a_\pm$ is a convenient parameter to be used in calculations concerning the separation of ionic compounds. In the case of undissociated species present in solutions, suitable relations are derived from equation 1.115 (*A.K. Covington et al., 1955; H.E. Wirth, 1971*).

Obviously in practice, only a few combinations appear for electrolyte activity expressions:

type of electrolyte ($z_M:z_L$)	examples	ν	$\nu_+^{\nu_+}\nu_-^{\nu_-}$
1:1 2:2	$HClO_4$ UO_2SO_4	2	1
2:1 1:2 4:2	$CaCl_2$, H_2SO_4 $Th(SO_4)_2$	3	4
3:1 1:3	$LaCl_3$ $K_3Fe(CN)_6$	4	27
3:2 4:1	$La_2(SO_4)_3$ $Th(NO_3)_4$	5	108 256

for instance, the mean ionic activity of an m-molal solution of $CaCl_2$ is $a_\pm = 4\dot\gamma_\pm^3 \; m^3$. The expressions for the molarity concentration scale will be essentially the same, but the difference in definition of standard state should not be neglected in calculations.

The transfer from mean ionic activity coefficients to individual ones can be done only theoretically or by postulating a proportion between the coefficients of at least one pair of ions. The equality of activity coefficients for K^+ and Cl^- (*D.A. McInnes, 1919*), Cs^+ and Cl^- (*R.G. Bates, B.R. Staples, R.A. Robinson, 1970*) or tetraphenylarsonium Ph_4As^+ and tetraphenylborate Ph_4B^- ions (*A.J. Parker, 1976*) has been proposed, for example.

Generally it is also accepted that the measured potential of the cell NHE-SCE is connected with the activity of hydrogen ions, pH $= - \log a_{H+}$, regardless of the anions present.

Empirical equations for the phenomenological description of individual ions can be derived from the "specific interaction theory" (G. Scatchard, 1936; E.A. Guggenheim, 1966), e.g.

$$\log \gamma_M = \log \gamma_\pm^{elstat} + \varepsilon(M^+, L^-)m \tag{1.123}$$

where $\varepsilon(M, L)$ is an interaction coefficient of the ions, e.g. $\varepsilon = 0.12$ for HCl (I. Aparicio, 1985). Otherwise, the mean ionic activity coefficients of the electrolyte can be used to calculate the thermodynamic activity of individual ions according to equations 1.107. For instance, in K_2SO_4 solutions we have $\dot{a}_{K^+} = 2\dot{\gamma}_\pm m$ and $\dot{a}_{SO_4^{2-}} = \dot{\gamma}_\pm m$.

For ions in mixtures of m_1- and m_2- molal electrolytes, a weighed average may be used to express the mean activity coefficients in the mixture:

$$\gamma_\pm = \sqrt{\frac{m_1}{m}(\gamma_\pm)_1^2 + \frac{m_2}{m}(\gamma_\pm)_2^2} \tag{1.124}$$

where $(\gamma_\pm)_1$ and $(\gamma_\pm)_2$ are the mean activity coefficients in electrolytes 1 and 2 at molalities m_1 and m_2 respectively (V.I. Levin et al., 1972).

The expression of ionic activities on the mole fraction scale (absolute activities, Convention I) is used only in special cases (unified scales in two-phase systems). Recalling equation 1.97, the relation between them will correspond to

$$a_\pm = (v_+^{v_+} v_-^{v_-}) \gamma_\pm^v \gamma(0)^v m^v \tag{1.125}$$

i.e. it differs from $\dot{a}_\pm$ (equation 1.121) by a constant coefficient $\gamma(0)^v$. The absolute activity coefficient is of the order $\gamma(0) = 10^{-3} \div 10^{-2}$ (strong negative non-ideality) and characterizes the association of electrolytes (A.M. Rozen, M. V. Ionin, 1971; Yu.G. Frolov, 1981).

The theoretical consideration of parameters influencing activity coefficients both of non-electrolytes and electrolytes will be given in sections 1.2.3.–1.2.5.

1.2.3 Chemical interactions. Complex equilibria

In separation chemistry, the following types of reversible reaction (interactions) are often used for process performance:
(a) chemical interactions (200 – 500 kJ mol^{-1})
  (i) complex equilibria
  (ii) oxidation-reduction reactions
  (iii) precipitation
(b) weak interactions (below 50 kJ mol^{-1})
  (i) crystallization and dissolution
  (ii) adsorption
  (iii) clathrate equilibria
  (iv) flotation
  (v) biospecific (serological, immunological) reactions.

Obviously, other types of interaction including ionic interactions may occur in all the processes simultaneously. Most of the reactions used in separation science can be described by the formation of an easily-separated product (A_3) from two reactants (A_1, A_2) as a principal equilibrium reaction

$$\nu_1 A_1 + \nu_2 A_2 \rightleftarrows A_3 \qquad (1.126)$$

with a **thermodynamic equilibrium constant (reaction quotient)**

$$K^0 = \frac{a_3}{a^{\nu_1} a^{\nu_2}} \qquad (1.127)$$

according to **Guldberg-Waage law of mass action** (*G.M. Guldberg, P. Waage, 1867*).

Even rather complicated systems can be described when removing from consideration complications introduced by side reactions or using various simplifying considerations and corrections for activity coefficients, such as "effective concentrations" and "effective stoichiometric coefficients", instead of activities $a_1 - a_3$ and coefficients $v_1 - v_3$, as illustrated later.

Examples are as follows:

	A_1	A_2	A_3
complex equilibrium	central atom (M)	ligand (L)	complex (M_mL_n)
clathrate equilibrium	guest molecule	host phase	clathrate
oxidation-reduction	oxidized form (Ox)	electron (e)	reduced form (Red)
precipitation	ion	counter ion	precipitate
flotation	surface-active (SAC)	counter ion	micelle
serologic reaction	antigen (Ag)	antibody (Ab)	associate
enzymologic reaction	substrate, inhibitor	enzyme	associate

Even though biospecific reactions of complicated macro biomolecules are described in this way (*G. Scatchard, 1949; E. Wasserman, P. Lalegerie, M. Bailly, 1981; L. Levine et al. 1971, 1981; B. Perlmutter-Hayman, 1986*), the best mathematical models were developed for complex equilibria (*F. Rossotti, H. Rossotti, 1961; L.G. Sillen, 1961; N. Ingri, L.G. Sillen, 1964; M.T. Beck, 1970; L. Sucha, S. Kotrly, 1972*).

Two introductory remarks are necessary concerning the applicability of the theoretical treatment for practical separation processes, starting from the basic process 1.126:

(i) The thermodynamic equilibrium constant does not give a picture of the abundance of separated species in various forms (e.g. A_1 and A_3) in a medium different from the standard state, i.e. without knowledge of activity coefficients. Moreover, the availability of the thermodynamic parameters $(K^0, \gamma's)$ is usualy scarce.

(ii) To get the expected, unknown composition of separated mixtures at equilibrium (in spite of this being just a limited state of a real non-equilibrium process) a predictive calculation based on the initial state, characterized by certain known compositions of the mixtures (e.g. by the molar concentrations of reactants, c_1 and c_2) must be possible. The second problem will be discussed in practical detail in section 2.2.1. At this point a short discussion is given to illustrate the general approach to the treatment of reaction 1.126.

To avoid the first problem, instead of the thermodynamic constant 1.127 the **apparent constant (equilibrium quotient, concentration product)**

$$K_c = \frac{[A_3]}{[A_1]^{\nu_1}[A_2]^{\nu_2}} \tag{1.128}$$

is used. This product is related to the thermodynamic equilibrium constant thus

$$K^0 = \frac{\gamma_3}{\gamma_1^{\nu_1}\gamma_2^{\nu_2}} K_c \tag{1.129}$$

This means that when the ratio $\gamma_3/\gamma_1^{\nu_1}\gamma_2^{\nu_2}$ is kept constant (the composition of the mixture does not change too seriously to change the ratio), the concentration product is really a practical constant that can be used in calculations. However, using literature data, the last may differ by several powers of ten and also the actual conditions may differ substantially from the standard ones to which the constant was related — see further.

To handle the second task, the relation between the initial and equilibrium concentrations needs to be found from material (mass) balances:

$$c_1 = [A_1] + \nu_1[A_3] \tag{1.130a}$$

$$c_2 = [A_2] + \nu_2[A_3] \tag{1.130b}$$

this can be solved together with equation 1.128 (see section 2.2.1.). Non-unity stoichiometric coefficients ν_1 and ν_2 will cause problems when the mathematical analytical solution of the non-linear equation system is to be found (the numerical computation, however, is without serious problems). In the simplest case, when the direct reaction 1.126 is bimolecular, $\nu_1 = \nu_2 = 1$, from equations 1.128 and 1.130 we get for the equilibrium concentration of the product A_3 the relation

$$[A_3] = K_c[A_1|[A_2] = K_c(c_1 - [A_3])(c_2 - [A_3]) \tag{1.131}$$

which gives a quadratic equation

$$[A_3]^2 - (c_1 + c_2 + 1/K_c)[A_3] + c_1 c_2 = 0 \tag{1.132}$$

with a solution

$$[A_3] = \tfrac{1}{2}\left(c_1 + c_2 + 1/K_c - \sqrt{[c_1 + c_2 + 1/K_c]^2 - 4c_1 c_2}\right) \tag{1.133}$$

After this, the concentrations $[A_1]$ and $[A_2]$ are easily found from equations 1.130 as well. Because the concentrations c_1 and c_2 are usually precisely established, the success of such calculations depends on the reliability of the constant K_c used and adequacy of its use at a given composition of mixtures (i.e. in the absence of side reactions and weak interactions which may seriously change the thermodynamic activity of both reactants, product and therefore the apparent K_c value) should always be checked.

In many practical situations, limited partial solutions of equation 1.133 can be useful:
1) at a great excess of one reactant, e.g. A_2:
$$c_1 \gg c_2 \text{ and } c_2 \gg [A_3]$$
from 1.131 we get

$$[A_3] = \frac{c_1 c_2 K_c}{1 + c_2 K_c} \qquad (1.134)$$

or, when $c_2 K_c \gg 1$, simply

$$[A_3] = c_1 \qquad (1.135)$$

i.e. the amount of product is given by the amount of reactant A_1 present in a minority, i.e., "sub equivalent" or "sub stoichiometric" amounts.

2) at the equivalence of reactants (their presence in the ratio corresponding to the stoichiometry of the product), generally when

$$v_2 c_1 = v_1 c_2 = c \qquad (1.136)$$

then from 1.133 we have

$$[A_3] = c + 1/(2K_c) - \sqrt{[c + 1/(2K_c)]^2 - c^2} \qquad (1.137a)$$

It can be shown that at high values of $K_c c$, at this equivalency point, there is a break in the concentration of interacting species A_1 and A_2 and the use of this region for separation purposes is inadvisable because of its high sensitivity to the mixture composition and it is difficult to control (in turn, this sensitivity is used for analytical purposes) – Fig.10.

3) <u>at arbitrary reactant ratios</u> the parameters of separation reaction can be found by various numerical and algebraic methods of linearization. E.g. for $v_1 = v_2 = 1$ the eqns. (1.128) and (1.130) give

$$\frac{[A_3]}{c_2-[A_3]} = c_1 K_c - K_3[A_3] \qquad (1.137b)$$

From which the K_c can be evaluated e.g. when $[A_3]$ or $R_2=[A_3]/c_2$ is measured:

$$\frac{R_2}{1-R_2} = \frac{c_1}{c_2} K_c - K_c R_2 \qquad (1.137c)$$

(K.A. Hasselbach, 1917; F.J.Rossotti, H.Rossotti, 1961).

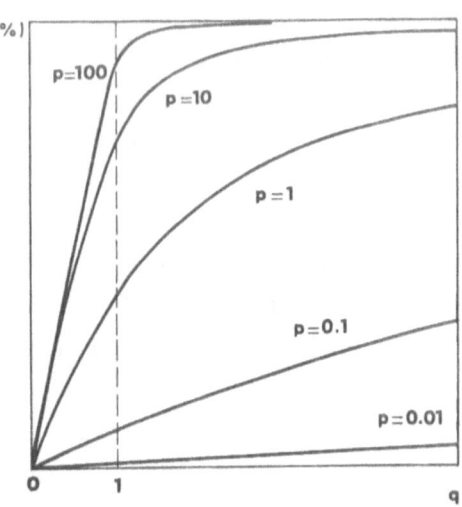

Figure 10. Yield (parameter $R = 100[A_3]/c_1$) of product A_3 of bimolecular interaction $A_1 + A_2 \rightleftarrows A_3$ for various equilibrium constants and initial concentration parameters —equation 1.133.

For the stoichiometry $v_1 \geq 1$, the logarithmic plot is convenient. Issuing from (1.128)-(1.130)

$$\frac{[A_3]}{[A_2]} = K[A_1]^{v_1} \qquad (1.137d)$$

$$[A_2] = c_2 - [A_3] \qquad (1.137e)$$

and then

$$\log \frac{R_2}{1-R_2} = \log K + v_1 \log[A_1] \qquad (1.137f)$$

Stoichiometric coefficient v_1 can be established if data for a set of equilibrium concentrations $[A_1]$ are achieved (*K.A. Hasselbach, 1917*); see also discussion to eq. (1.144).

The relative concentration (e.g. the yield of one reactant found in the product) is a more important parameter for separation chemistry than the absolute concentrations of all species at equilibrium. R_2 represents a yield of bonding of reactant A_2 in product A_3, and in two-phase system $\frac{R_2}{1-R_2} = D_m(A_2)$ is its mass distribution ratio.

The number of independent parameters in practical calculations according to equation 1.133 can be easily reduced from three (K_c, c_1 and c_2) to two, for instance $p = K_c c_1$ and $q = c_1/c_2$. From Figure 10 it can be seen that at the same equilibrium constant and reagent ratio q, more concentrated solutions are more favorable for separation processes with chemical reaction: both p and R increases simultaneously with increasing concentration of reactants. Conversely, there is no need to apply too great an excess of reagent when the yield is sufficiently high. Many chemical separation processes lack optimization in this respect.

The chemical interaction of a solute means the decrease of its concentration from a total (initial) value to the equilibrium one (e.g. from c_i to $[A_i]$) and therefore the decrease of its thermodynamic activity with respect to other reactions. Phenomenologicaly the ratio can be defined through the activity coefficient

$$[A_i] = \gamma_i c_i \qquad (1.138)$$

and γ_i can be calculated from equilibrium constants and material balances as shown above or through empirical relations (*W. Davis, Jr. et al., 1970*).

For instance, in the case of reaction 1.126 and an excess of A_2, the activity coefficient γ_1 can be derived (compare with equation 1.134) as inversely proportional to concentration of the second reactant, c_2:

$$\gamma_1 = [A_1]/c_1 = 1/(1 + \nu_1 c_2 K_c) \tag{1.139}$$

(K_c includes the weak interactions activity coefficients according to equation 1.129). However, chemical interactions are usually expressed explicitly through the equilibrium constants and mass balances.

In a more complicated mixture there is a larger set of equilibria and the calculations become more troublesome. To make them easier to treat, they can be presented in matrix form (*Y. Zeldovitch, 1938; S. Ya. Pshezhetskii, R.N. Rubinstein, 1947; S. R. Brinkley, 1947; S. H. Storey, F. Zeggeren, 1970; A.A. Buguevskii, B.A. Dunai, 1971; C.L. Cooney, H.Y. Wang, D.I.C. Wang, 1977*).

General equilibria, by analogy with 1.126, should include all reactants $A_1 \ldots A_j$ and all products $A_{j+1} \ldots A_q$,

$$\nu_1 A_1 + \nu_2 A_2 + \cdots + \nu_i A_j \rightleftarrows \nu_{j+1} A_{j+1} + \cdots + \nu_q A_q \tag{1.140}$$

and the general mass balance consists of the set of *j* equations of the type

$$c_i = \sigma_{i,j}[A_j] + \sigma_{i,j+1}[A_{j+1}] + \cdots + \sigma_{i,q}[A_q] \tag{1.141}$$

where $\sigma_{i,j}$ is the abundance coefficient giving the number of formula units of the *i*-th component (molecule or its fragment) in the j-th compound (A_j). Particularly, $\sigma_{i,i} = 1$.

The **apparent equilibrium constant** of reaction 1.140 can be written as

$$K = \prod_{j=1}^{q} [A_j]^{\nu_j} \tag{1.142a}$$

(for reactants v_j has a negative sign) and the balance 1.141 is for each of the k components

$$c_i = \sum_{j=1}^{q} \sigma_{i,j} [A_j] \tag{1.142b}$$

It can be easily verified that the constant 1.142a of the overall reaction 1.140 can be presented as the product of the constants of the partial equilibria K_r,

$$K = \prod K_r \tag{1.143}$$

To calculate k unknown equilibrium concentrations $[A_1]$ ... $[A_q]$, besides the set of k mass balances 1.141, there should be $(q - k)$ independent equations known. This usually means that instead of one gross constant 1.143 there should be a set of $(q - k)$ partial equilibrium constants K_r.

Again, the mathematical treatment of the equations is easier when presented in matrix form. The latter is obtained after taking logarithms of the general expression 1.142a:

$$\ln K_r = \sum v_{r,j} \ln [A_j] \tag{1.144}$$

where the summation refers to the species (indexed by j's) involved in particular equilibrium reaction (indexed by r's). This means that, in matrix symbols, that

$$\ln \mathbf{K} = \mathbf{N} \cdot \ln \mathbf{A} \tag{1.145a}$$

where $\mathbf{N}$ is a $r \times q$ matrix of stoichiometric coefficients of the j-th species in the r-th equilibrium reaction,

$$\mathbf{N} = \begin{bmatrix} v_{11} & \cdots & v_{1q} \\ \vdots & \ddots & \vdots \\ v_{r1} & \cdots & v_{rq} \end{bmatrix} \tag{1.145b}$$

($v_{r,j} < 0$ if the j-th species enters the r-th equilibrium from the left), K is the vector (column) of equilibrium constants K_f and A is the vector of equilibrium concentrations $[A_j]$ to be found. Equation 1.144 is used together with the mass balances 1.142b that in matrix form become

$$\mathbf{C} = \mathbf{S}\,\mathbf{A} \tag{1.146a}$$

where S is the $k \times q$ matrix of the abundance coefficients $\sigma_{i,j}$,

$$\mathbf{S} = \begin{bmatrix} \sigma_{11} & \cdots & \sigma_{1q} \\ \vdots & \ddots & \vdots \\ \sigma_{k1} & \cdots & \sigma_{kq} \end{bmatrix} \tag{1.146b}$$

C is the vector (column) with k elements of total concentrations c_1 and **A** is the vector with q elements of equilibrium concentration $[A_j]$. The methods for the mutual solution of equations 1.145 and 1.146 are well described (*A. Rolston, 1965*).

We shall illustrate this approach using the protolytic equilibria in a solution of a multifunctional acid, e.g. citric acid, H_3Cit with a strong base, e.g. sodium hydroxide, NaOH. The concentration of Na^+ being considered a constant, the equilibrium of all species present can be written as follows:

$$3H_3Cit + OH^- \rightleftarrows H_2Cit^- + HCit^{2-} + Cit^{3-} + 5H^+ + H_2O \tag{1.147}$$

Such an equation of equilibrium is equivalent to equation 1.140, for it contains all species involved in the equilibrium. The mass balance of the citrate moieties is

$$c_{cit} = [H_3Cit] + [H_2Cit^-] + [HCit^{2-}] + [Cit^{3-}] \tag{1.148a}$$

The mass balance of protons is calculated by considering OH^- as equivalent to the complex having a deficit of one proton (*D. Dyrssen et al., 1968*),

$$c_{H+} = 3c_{cit} - c_{NaOH} = 3[H_3Cit] + 2[H_2Cit^-] + [HCit^{2-}] + [H^+] - [OH^-] \quad (1.148b)$$

Comparing equations 1.148 with the universal one 1.141, the symbolism corresponds to $c_1 \equiv c_{Cit}$ and $c_2 \equiv c_{H+} (k = 2)$. The concentration of water during reaction is supposed to be constant and no mass balance of H_2O is established. Now the unknown equilibrium concentrations of species $A_1 \equiv [H_3Cit]$, $A_2 \equiv [H_2Cit^-]$, $A_3 \equiv [HCit^{2-}]$, $A_4 \equiv [Cit^{3-}]$, $A_5 \equiv H^+$, $A_6 \equiv OH^-$ ($q = 6$) are to be calculated. Therefore, $q - k = 6 - 2 = 4$ independent partial equilibrium constants (K_i) should be known. Let them be

$$K_1 = \frac{[H_2Cit^-][H^+]}{[H_3Cit]} \quad (1.149a)$$

$$K_2 = \frac{[HCit^{2-}][H^+]^2}{[H_3Cit]} \quad (1.149b)$$

$$K_3 = \frac{[Cit^{3-}][H^+]^3}{[H_3Cit]} \quad (1.149c)$$

These constants are usually well-documented in the literature as dissociation constants. The last necessary constant,

$$K_4 = [H^+][OH^-] \quad (1.149d)$$

is the **ionization constant (autoprotolysis constant) for water** which is $K_4 = K_w = 10^{-14}$ mol^2 dm^{-6} at 25°C.

According to equilibria 1.149, the matrix of stoichiometric reaction coefficients is

$$N = \begin{vmatrix} -1 & 1 & 0 & 0 & 1 & 0 \\ -1 & 0 & 1 & 0 & 2 & 0 \\ -1 & 0 & 0 & 1 & 3 & 0 \\ 0 & 0 & 0 & 0 & 1 & 1 \end{vmatrix} \quad (1.150)$$

and from the mass balances 1.148, the matrix of the abundance coefficients is

$$S = \begin{vmatrix} 1 & 3 \\ 1 & 2 \\ 1 & 1 \\ 1 & 0 \\ 0 & 1 \\ 0 & -1 \end{vmatrix} \tag{1.151}$$

The universality of such an approach is of great advantage for computer treatment. The choice of proper and reliable equilibrium constants is very significant, however, as we have discussed in relation to equation 1.133. Several critical compilations in this field are available. (*K.B. Yatsimirskiy, V.P. Vasilyev, 1959; L.G. Sillen, A.F. Martell, 1964, 1971; A.A. Bugaevskii, 1967; V.A. Navarenkov. V.P. Antonovich, Ye.M. Nevskaya, 1979; F. Hogfeldt, 1980; D.J. Legget, 1985; M. Lederer, 1992*).

Going back to the thermodynamic formulation of chemical equilibrium, the former is not defined through the apparent equilibrium constant 1.141 but as the state when the Gibbs free energy of the reaction mixture is at a minimum (equation 1.17), i.e.

$$dG = \sum \mu_i dn_i = 0 \tag{1.152}$$

More comprehensive calculation methods of chemical equilibria using minimization of free energy approach are necessary for inclusion of temperature factors (*F. Van Zeggeren, S.H. Storey, 1970*) which are necessary e.g. for combustion systems and therefor were developed at the NASA over more than 40 years (*S. Gordon, B.J. McBride, 1994*).

A universal **progress variable** for reaction 1.140, i.e. the **extent of reaction**, is the proportional change of the number of moles of any reactant (product) in the reaction

$$d\xi = \frac{dn_i}{\nu_i} \tag{1.153}$$

(v_i is negative if it refers to the reactant, i.e. when $dn_i < 0$). Then, substituting for dn_i in equation 1.152 gives **Van't Hoff's isotherm** *(J.H. Van't Hoff, 1884)*:

$$dG = \sum v_i \, \mu_i \, d\xi \qquad (1.154)$$

and a **general condition for chemical equilibrium** is obtained as

$$\sum v_i \, \mu_i = 0 \qquad (1.155)$$

or, according to equation 1.81,

$$\sum [\, v_i \mu_i^0 + RT \ln(a_i)^{v_i}\,] = 0 \qquad (1.156)$$

Since the chemical potentials of the substances are equal to their molar Gibbs free energy values, the free energy change $\overline{\Delta G}$ (kJ mol^{-1}) for the reaction 1.140 is

$$\sum v_i \, \mu_i^0 = -RT \ln \prod (a_i)^{v_i} \qquad (1.157)$$

or

$$\overline{\Delta G}^0 = -RT \ln K^0 \qquad (1.158a)$$

where K^0 is the **thermodynamic equilibrium constant**

$$K^0 = \prod_{i=1}^{q} (a_i)^{v_i} \qquad (1.158b)$$

The constant depends upon the choice of standard state of the interacting species. The standard value $\overline{\Delta G}^0$ is the change in Gibbs energy when a certain number of moles of reactants are converted into a certain number of moles of products in their standard state, i.e. when the extent of reaction given by equation 1.153 is $d\xi = 1$ mol.

Vice versa, any constant can be considered as a thermodynamic one when the very equilibrium composition at which it was established, is declared to be the standard state for each component included in the reaction (V.V. Fomin, 1967).

The relation between the concentration product K_x (apparent equilibrium constant) and the thermodynamic constant on the mole fraction scale K_x^0 — compared with equation 1.129 — is derived from the definitions

$$K_x = \prod_{i=1}^{q}(x_i)^{\nu_i} \tag{1.159}$$

and

$$K_x^0 = \prod_{i=1}^{q}(a_i)^{\nu_i} = \prod_{i=1}^{q}(\gamma_i x_i)^{\nu_j} \tag{1.160}$$

so that

$$K_x^0 = Y_x K_x \tag{1.161}$$

where

$$Y_x = \prod_{i=1}^{q}(\gamma_i)^{\nu_i} \tag{1.162}$$

is the product of the rational activity coefficients. If the standard states of the components are chosen to be close to ideal or quasi-ideal behavior (i.e. Convention I is used for the matrix and major components and Convention II for the minor solutes) the product Y will be close to unity. However, its value at real composition of quasi-ideal mixtures in various standard states is shifted by a constant difference — see discussion to equation 1.96. The scale of mole fractions is usual for macro components while the molar or molal scale is more convenient for micro components. The relation between the constants in the various concentration scales follows from

$$K_c = \prod(c_i)^{\nu_i} \tag{1.163}$$

and therefore, considering equation 1.51,

$$K_x = K_c(c_t)^{\Sigma \nu_i} \tag{1.164}$$

where c_t is the total molar concentration 1.51b and the exponent is the sum of the stoichiometric coefficients of reaction 1.140. For instance, if on the formation of associate (3) arising from hormone (1) and receptor (2) in dilute aqueous solution, the equilibrium constant at $T = 300K$ was found to be $K_c = \frac{[A_3]}{[A_1][A_2]} = 10^{10}$ mol^{-1}dm^3, the constant in the mole fraction scale, when c_i is practically determined by molarity of water (55.5 mol dm^{-3}), is $K_x \approx K_c(55.5)^{(-2+1)} = 1.8 \times 10^8$ and

$$\overline{\Delta G} = -8.31 \times 300 \times \ln(1.8 \times 10^8) = 47.4 \text{ kJ mol}^{-1}$$

The equilibrium constants in various concentration scales will not differ when reaction proceeds at a constant number of moles, e.g. $[A_1] + [A_2] = [A_3] + [A_4]$, the solvent as a component not being an exception. The constants of such equilibria have the same numerical values and are dimensionless. In various applications there are a few practical devices used to eliminate the thermodynamic non-ideality of solutions and to make the mathematical description of equilibria more consistent with real behavior:

(i) the calculation of effective concentrations or stoichiometric coefficients for a principal equilibrium equation (i.e. for the controlling reaction of the separation process),

(ii) the use of experimentally available equilibrium activities of the species present (e.g. the activity of hydrogen ions),

(iii) the use of direct measurement of equilibrium concentrations of compounds present (e.g. by spectrometric methods).

In the first approach, the activity coefficients are calculated from the mass balances of side reactions — equation 1.139. In this way calculated activity coefficients often carry special names, as the **Fronaeus function** *(S. Fronaeus, 1950, 1951)* or **Ringbom coefficients** *(A. Ringbom, 1958)* -see below. Their meaning consists of determining the fraction of principal reactant that is available for the controlling reaction.

Thus, for a reaction of type 1.126, instead of the exact thermodynamic constant 1.127 on the one hand and too rough an approximate concentration constant 1.128 on the other, an **effective (conditional, mixed)** constant

$$\overline{K} = \frac{[A_3]}{(\overline{[A_1]})^{v_1}(\overline{[A_2]})^{v_2}} \qquad (1.165)$$

is used, where $\overline{[A_1]}$ is postulated to consist of free A_1 species and all other bound species A_1 except those in product A_3:

$$\overline{[A_1]} \equiv c_1 - v_1[A_3] \qquad (1.166)$$

This balance is formally the same, as in 1.130. This means, that if

$$\alpha_1 = \frac{[A_1]}{\overline{[A_1]}} = \frac{[\text{free } A_1]}{[\text{free } A_1] + [A_1{}^{\text{in side}}_{\text{products}}]} \qquad (1.167)$$

is the ratio ($\alpha \leq 1$) of free (from the point of the principal reaction equilibrium) reactant A_i to all its forms except those transformed to product A_3, the mass balance 1.130 becomes universally formulated as simply as:

$$[A_i] = \alpha_i(c_i - v_i[A_3]) \qquad (1.168)$$

Now, the relation between the conditional constant $\overline{K}$ and the concentration product 1.128 can be established as

$$\overline{K} = K_c \alpha_1^{v_1} \alpha_2^{v_2} \qquad (1.169)$$

This is formally the same result as obtained by analogy directly from equation 1.161, when $\gamma_1 = \alpha_1$, $\gamma_2 = \alpha_2$ and $\gamma_3 = 1$, but neither K_c nor $\overline{K}$ is a true constant and depend on the mixture composition. Nevertheless, such effective "constants" are widely used in calculations of separation processes (*M. Tanaka, 1963; M. Kyrs, 1965; A.M. Rozen, 1967 etc.*). The values of the constants can be predicted from side reaction constants and minor chemical interactions.

For instance, if antigen (Ag) reacts in a **biospecific reaction** with antibody (Ab) by formation of the complex AgAb,

$$Ag + Ab \rightleftarrows AgAb \tag{1.170}$$

it may be not the only reaction it enters. In the presence of other reactive antibodies (Ab$_j$) there are a number of equilibria

$$Ag + Ab_i \rightleftarrows AgAb_i \tag{1.171}$$

and the activity of the antibody in the principal immunologic reaction 1.170 will be diminished by the coefficient

$$\alpha_{Ab} = \frac{[Ab]}{[Ab] + \sum [AgAb_i]} \tag{1.172}$$

This value depends upon the concentration of competing antibodies and partial equilibrium constants of their reactions 1.171, and more generally on the binding capacities of proteins, which may vary even from sample to sample (*D. Geiseler, M. Ritter, 1982*).

Another use of the effective concentration approach is to segregate the activity factors in the thermodynamic constant in such a way that the ratio of one reactant (A_1) and product (A_3) is excluded,

$$K^0 = \frac{[A_3]}{[A_1]^{v_1}} \frac{\gamma_3}{(\gamma_1)^{v_1}(a_2)^{v_2}} \tag{1.173}$$

The product

$$a_2^* \equiv a_2(\gamma_1)^{v_1/v_2}/\gamma_3^{1/v_2} \qquad (1.174)$$

is called the **effective activity** of reactant A$_2$. This means that exactly at a given activity a_2^*, a certain ratio between product and one reactant is achieved in the reaction characterized by the thermodynamic constant K^0:

$$\frac{[A_3]}{[A_1]^{v_1}} = K^0(a_2^*)^{v_2} \qquad (1.175)$$

When the activity a_2 itself is well established then the effective constant which includes the activity coefficients γ_1 and $\gamma_2 iB$ considered for the same reason:

$$\frac{[A_3]}{[A_1]^{v_1}} = K^*(a_2)^{v_2} \qquad (1.176)$$

where

$$K^* = K^0(\gamma_1)^{v_1}/\gamma_3 \qquad (1.177)$$

Such an approach was used for the formulation of the effective activity of hydrogen ions in protolytic equilibria, the **Hammett acidity function** (*L.P. Hammett, A.J. Deyrup, 1932*) or effective activity of complexing ligands (*Y. Marcus, C.D. Coryell, 1959*).

The use of non-integral (empirical, semi empirical or theoretical) stoichiometric coefficients v_i in effective equilibrium constants is less common. They are useful especially for stepwise reactions, typical examples being the complexation of metals (M) by ligands (L). It can be demonstrated with **complex equilibria**, particularly for an excess of ligand, that the description of fairly complicated reaction systems through one or two equilibrium concentration parameters is useful both for theoretical mathematical treatments and experimental control of separations.

Traditionally, the equilibrium constants of complex equilibria are called stability constants. The overall **equilibrium** constant of the complexation reaction

$$M + nL \rightleftarrows ML_n \tag{1.178}$$

(a special case of the reaction 1.126 when

$$A_1 \equiv M, A_2 \equiv L, A_3 \equiv ML_n, \nu_1 = 1, \nu_2 = n, K^0 \equiv \beta_n^0;$$

the charges of reacting species being omitted) is the **thermodynamic stability constant**

$$\beta_n^0 = \frac{a_{ML}}{a_M a_L^n} \tag{1.179}$$

The activities in equilibrium 1.178 can be calculated, e.g. for ionic equilibria (*V.V. Fomin, E.P. Maiorova, 1956; V.V. Fomin, 1967; M.C. Cognet, H. Renon, 1976; Yu. G. Frolov, D.A. Denisov, 1978; N.P. Komar, 1983*).

The complexation proceeds step-wise (*N. Bjerrum, 1921*)

$$ML_{i-1} + L \rightleftarrows ML_i \tag{1.180}$$

and the partial, **consecutive stability constants** are defined as

$$K_i^0 = \frac{a_{ML_i}}{a_{ML_{j-1}} a_L} \tag{1.181}$$

and

$$\beta_n^0 = K_1^0 K_2^0 \ldots K_n^0 = \prod_{i=1}^n K_i^0 \tag{1.182}$$

(compare with equation 1.143). The attachment of any further ligand L to the central atom M is less probable and from combinatorical principles (*J. Bjerrum, 1941; D. Dyrssen, L.G. Sillen, 1953*) their statistical ratio should be

$$\frac{K_i}{K_{i+1}} = \frac{(n-i+1)(i+1)}{i(n-i)} \tag{1.183}$$

This general tendency ($K_1 > K_2 > \cdots > K_n$) is exaggerated by molecular energetic factors and very often the difference between consecutive constants becomes so large that the complex formation proceeds as a pseudo-one-step process.

Mass balances 1.141 applied to metal and ligand mixtures in equilibrium 1.180 are correspondingly

$$c_M = [M] + [ML] + [ML_2] + \cdots + [ML_n] = \sum_{i=0}^{n}[ML_i] \tag{1.184a}$$

$$c_L = [L] + [ML] + 2[ML_2] + \cdots + n[ML_n] = \sum_{i=0}^{n} i\,[ML_i] \tag{1.184b}$$

(each complex ML_i contains i ligands L). Applying concentration analogues of thermodynamic constants, β_n, results in

$$[ML_i] = \beta_i[M][L]^i \tag{1.185}$$

and equations 1.184 are transformed to concise forms of mass balances:

$$c_M = [M] \sum_{i=0}^{n} \beta_i\,[L]^i \tag{1.186}$$

$$c_L = [L] + [M] \sum_{i=1}^{n} i\,\beta_i[L]^i \tag{1.187}$$

The function

$$X = \sum_{i=0}^{n} \beta_i\,[L]^i \tag{1.188}$$

is called the **complexity (Fronaeus function)**.

As an example, the complexation of metal ion M^{3+} with ligand L^- will be discussed. For instance, at a concentration of ligand $c_L = 10^{-2}$ mol dm^{-3} and a metal M concentration much smaller, $c_M \ll c_L$, the equilibrium concentration $[L] \approx c_L$. The equilibrium concentration of M^{3+} ions was found to be 3×10^4 times smaller, than its initial concentration c_M, i.e. $[M^{3+}] = c_M/3 \times 10^4$. According to equations 1.186 and definition 1.188, the complexity of the metal in the solution is $= 3 \times 10^4$. If the complexation is caused entirely by complex ML_3 formation, according to equation 1.188 $X \approx \beta_3[L]^3$, the first approximation of β_3 can be $\beta_3 \approx (3 \times 10^4)/(10^{-2})^3 = 3 \times 10^{10}$ mol^{-3}dm^9. It can be expected from this value, for a decrease of ligand concentration to $[L] = 10^{-3}$ mol dm^{-3}, that the complexation of metal drops dramatically down to $\approx (3 \times 10^{10})(10^{-3})^3 = 30$, because there is a strong dependence on the ligand concentration to the third power. The actual value obtained was somewhat higher, $X = 54$, evidently due to the presence of lower complexes. Approaching the complexity by the more complete equation, $X \approx \beta_2[L]^2 + \beta_3[L]^3$, from two equations

$$\beta_2(10^{-2})^2 + \beta_3(10^{-2})^3 = 3 \times 10^4 \text{ (for } [L] = 10^{-2}M\text{)},$$

and

$$\beta_2(10^{-3})^2 + \beta_3(10^{-3})^3 = 54 \text{ (for } [L] = 10^{-3}M\text{)}$$

the values $\beta_2 = 2.7 \times 10^7$ mol^{-2}dm^6 and $\beta_3 = 2.7 \times 10^{10}$ mol^{-3}dm^9 are obtained. Following the statistical probability of ligand bonding — equation 1.183 — there would be ratios

$$K_2/K_3 = (3 - 2 + 1)(2 + 1)/2(3 - 2) = 3 \text{ and}$$

$$K_1/K_2 = (3 - 1 + 1)(1 + 1)/1(3 - 1) = 3$$

Recalling equation 1.182 ($\beta_2 = K_1K_2$), the expected K_1 value would be $K_1 = \beta_2/K_2 = \beta_2/(K_1/3)$, i.e. $K_1 = \sqrt{3\beta_2} = 9 \times 10^3$ mol^{-1} dm^3. The latter value is identical with β_1. In this way we have seen that from two empirical data, two unknown stability constants could have been calculated and the third one was approximated statistically. The expected $K_1 = \beta_1$ value indicate that the presence of ML^{2+} complex is negligible when compared with concentration of other complexes: at a ligand concentration $[L] = 10^{-3}$ mol dm^{-3} the calculated percentage abundances of complexes ML_3, ML_2 and ML (from the equilibrium equations 1.185 and mass balance 1.184) are 49.1, 49.1 and 1.63% correspondingly (the rest, 0.17% relates to the free metal ion M^{3+}).

This demonstrates that usually only a few complexes coexist in solutions at a given ligand concentration and complexation can be successfully approached by one or two prevailing species. Vice versa, special care should be undertaken to obtain reliable data by their treatment, the minor species introducing large errors in calculated data. A wider variety of experimental methods and suitable conditioning of solutions both for determination and evaluation should always be ensured (*L. G. Sillen, 1961*).

A variety of other functions are in use to describe complicated equilibrium compositions of solutions. The ratio

$$\langle n \rangle = \frac{\sum_{i=0}^{n} i[ML_i]}{c_M} = \frac{\sum_{i=0}^{n} i\beta_i[L]^i}{[M]X} = \frac{c_M - [L]}{c_M} \tag{1.189}$$

defines the **average ligand number**, i.e. the average ratio L: M in a mixture of coexisting complexes (*J. Bjerrum, 1941, 1944*).

Another average property (*P*) of complex mixtures such as optical density or electrophoretic mobility can also be expressed through analogous expressions $\langle P \rangle = \sum P_i [ML_i]/\sum [ML_i]$, where P_i is the corresponding property of the complex — see equation 1.509. The average electrophoretic mobility, for instance, is an important controlling characteristic of the separation of complexed ions (*V. Jokl, 1964; J. Majer, V.A. Trinh, I. Valaskova, 1980*).

Systems with equal mean ligand numbers $\langle n \rangle$ are called **corresponding systems** (*J. Bjerrum, 1944*). Using the functions 1.188 and 1.189, the mass balances 1.186 and 1.187 in a system of complexes adopts a rather elegant and instructive form:

$$c_M = [M]X \tag{1.190}$$

$$c_L = [L] + \langle n \rangle c_M \tag{1.191}$$

and the equilibrium can be solved with the stability constant as follows

$$\frac{[ML_n]}{c_M} = \beta_n \frac{(c_L - \langle n \rangle c_M)^n}{X} \tag{1.192}$$

providing the equilibrium concentration [L] is known, e.g. at a great excess of ligand.

Without special effort, a balance including complexation with other ligands, e.g. hydroxo anions or masking agents (B), can be included in the complexity function X:

$$X = \sum_{i=0}^{n} \beta_i [L]^i + \sum_{j=1}^{m} \beta_j [B]^j + \cdots, \tag{1.193a}$$

or, because mixed complexes ML_iB_j are also favored statistically, from the equilibrium $[ML_iB_j] = \beta_{ij}[M][L]_n^i[B]^j$ we have

$$X = \sum_{i=0}^{n} \sum_{j=1}^{m} \beta_{ij} [L]^i [B]^j \tag{1.193b}$$

($\beta_{00} \equiv 1$). The same can be accomplished with respect to the polyprotonated ligand, including the protonation equilibria, characterized by the association constant (the reciprocal of the dissociation constant K_a)

$$K_i = \frac{[H_iL]}{[L][H]^i} \tag{1.194}$$

so that the balance 1.191 is extended to produce

$$c_L = \Pi[L] + \langle n \rangle c_M \tag{1.195}$$

where the function

$$\Pi = \sum_{i=0} K_i [H]^i \tag{1.196}$$

($K_0 \equiv 1$) is the ratio of concentrations of all forms of H$_i$L to the simplest one, L.

The equations 1.191 and 1.192 can be transformed to more general forms (compare e.g. *I.M. Kolthoff, E.B. Sandell, 1941, and H. Irving, T.B. Pierce, 1959*).

In aqueous solutions, the hydroxo and polynuclear complexes usually appear due to the reaction scheme

$$mM + nL + pOH \rightleftarrows M_mL_n(OH)_p \tag{1.197}$$

with the constant

$$\beta_{mnp} = \frac{[M_mL_n(OH)_p]}{[M]^m[L]^n[OH]^p} \tag{1.198}$$

and we get mass balances for the mixtures as

$$c_M = [M](1 + \langle m \rangle) X \tag{1.199}$$

$$c_L = \Pi [L] + \frac{\langle n \rangle}{1+\langle m \rangle} c_M \tag{1.200}$$

where

$$X = \sum_{m=1} \sum_{i=0} \sum_{j=0} \beta_{mij}[M]^{m-1}[L]^i[OH]^j \tag{1.201}$$

$$\langle m \rangle = \sum_{m=1} \sum_{i=0} \sum_{j=0} (m-1)\beta_{mij}[M]^{m-1}[L]^i[OH]^j \tag{1.202}$$

$$\langle n \rangle = \sum_{m=1} \sum_{i=0} \sum_{j=0} i\beta_{mij}[M]^{m-1}[L]^i[OH]^i \tag{1.203}$$

$$\langle p \rangle = \sum_{m=1} \sum_{i=0} \sum_{j=0} j\beta_{mij}[M]^{m-1}[L]^i[OH]^j \tag{1.204}$$

($\beta_{100} \equiv 1$). Though an "exact" solution of the equations in the matrix form 1.145–1.146 is possible, their use for calculations of complex equilibria is very convenient, in spite of the complicated appearance of these formulae, and appears useful especially using various simplifications (*F. Macasek, 1974; F. Macasek, D. Vanco, 1981*). In turn, the neglect of some important equilibria may lead to incorrect conclusions.

With unknown chemical details and deviations from ideality, the complexity 1.188 or coefficient 1.167 may be replaced by an arbitrary polynomial of the type

$$\alpha = \sum_{i=0}^{k} \overline{K_i} [L]^{\overline{v_i}} \qquad (1.205)$$

where $\overline{K_i}$ and $\overline{v_i}$ are non-integral coefficients fitted to experimental data. Usually it is sufficient to adjust 2-6 such constants ($k \leq 3$) to get a good approximation to describe complicated equilibria through one variable, the equilibrium concentration of L (*I. Svantesson, et al., 1979*).

In bioinorganic chemistry, the equilibria of ions with **macromolecular ligands** play an important role. Biological ligands (proteins, polysaccharides, etc.) are polyfunctional, e.g. proteins contain many binding sites, exchangeable carboxyl, amine and inlidazole groups (*G. Scatchard, 1949; I.M. Klatz, 1953*). According to G. Scatchard the interaction between multivalent ligand with M not as an equilibrium (1.178) but as that of k identical and independent "subligands" $(L)_i$

$$M + (L)_i \rightleftarrows (ML)_i$$

The equilibrium constant of univalent metal ions is K_1

$$(K_1)_i = \frac{[(LM)_i]}{[(L)_i][M]} \qquad (1.206)$$

If there are k types of binding site ($k \geq 3$ in the case of proteins) and the mole fraction capacity of any of these is $m_i (\sum m_i = 1)$ the degree of occupancy of the i-th type of group (R_i) is given by the molar amount n_i per mole of occupied sites, i.e. in the equilibrium

$$R_i = \frac{n_i}{m_i} = \frac{[(LM)_i]}{[(L)_i]+[(LM)_i]} \qquad (1.207)$$

where $[(LM)_i]$ is the equilibrium concentration of bonds between the i-th group and M, and $[(L)_i]$ is the concentration of vacant sites ($n_i \leq m_i$). The following amount of $L-M$ bonds of the i-th sort results:

$$n_i = \frac{m_i(K_1)_i[M]}{1+(K_1)_i[M]} \qquad (1.208)$$

The total degree of occupancy of exchangeable groups with metal M is given by the ratio of the total amounts of occupied and exchangeable groups (m), **non-linear Scatchard plot** of metal-protein and antibody-antigen interactions:

$$R = \frac{\sum n_i}{m} = \frac{1}{m}\sum_{i=1}^{k} \frac{m_i(K_1)_i[M]}{1+(K_1)_i[M]} \qquad (1.209a)$$

Because m_i and $(K_1)_i$ are constants, R has the character of a hyperbolic function i.e. a **saturation curve**.

These equilibria are further examples of isomorphous relations to eqns. 1.180–1.192 (when replacing a central atom of higher coordination number for the polyfunctional macromolecular ligand and a ligand for the central atom with unit bonding property). The **linear Scatchard plot** is most famous by its application for immunological reactions

$$\frac{B}{F} = k\,c_L K - KB \qquad (1.209b)$$

where $B\,(\equiv [ML]_k)$ is bound amount and $F\,(\equiv [M])$ free amount of antibody respectively.

The last equilibria are important also for heterogeneous systems, sorption equilibria and equation 1.209a can also be treated formally like the Langmuir absorption isotherm (section 1.3.2.).

1.2.4 Electrostatic interactions. Electrolyte solutions

The electrostatic component of the activity coefficient γ_i was shown to be given by the energy $kT \ln \gamma_i$ which is necessary to transform a neutral molecule to an ion with charge z_1 (compare with equations 1.109 and 1.113). If there is a single ion of charge z_2 in its vicinity, the **energy of the electrostatic interaction** u(J) between them is given by Coulomb's law

$$u_c = -\frac{z_1 z_2 e^2}{4\pi\varepsilon_0 \varepsilon_r} \frac{1}{r} \qquad (1.210)$$

where r is the distance between the charges, $\varepsilon_0 = 8.85 \times 10^{-12}$ $C^2 J^{-2} m^{-1}$ is the dielectric permittivity (dielectric constant) of a vacuum and ε_r is the relative permittivity of the medium (solvent).

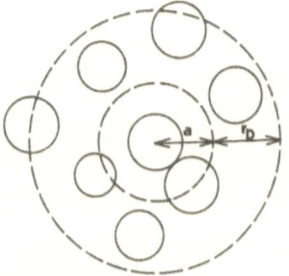

Figure 11. Ionic atmosphere model: a — average effective diameter of ions, r_D — Debye radius.

The ideal solution of an electrolyte is a completely random distribution of ions of different charge which is the opposite extreme of the ionic crystal lattice. The overall energy of electrostatic interactions in the ideal solution is zero (i.e. the full compensation of attractive and repulsive forces). However, the intensity of the electrostatic field in the vicinity of ions is great, about $10^9 \div 10^{10}$ V m⁻¹ and every ion is surrounded with an "atmosphere" of oppositely charged ions which diminishes the electrostatic field of the central ion (Figure 11). The charge density in the ionic atmosphere was derived by *P. Debye* and *E.*

Hückel (1923) using Boltzmann's distribution law. The average **increase in interaction energy due to the ionic atmosphere** with average radius r_D (Debye radius) was shown to be

$$\langle u^{DH} \rangle = -\frac{z_1 z_2 e^2}{4\pi\varepsilon_0\varepsilon_r} \frac{1}{r_D + a} \tag{1.211}$$

where e is the unit electrical charge ($1\,e = 1.6 \times 10^{-19}$ C), a is the smallest distance between the ions, i.e. the effective medium diameter of the ions and

$$\frac{1}{r_D} \equiv b = N_A \varepsilon \left(\frac{2I}{\varepsilon_0\varepsilon_r RT}\right)^{1/2} \tag{1.212}$$

N_A is Avogadro's constant (6.023×10^{23} mol^{-1}). Parameter I is an important characteristic value of electrolyte solutions, the **ionic strength**, defined by

$$I = \frac{1}{2}\sum c_i z_i^2 \tag{1.213}$$

The summation in this equation is performed over all types of ionic species present in solution. For instance, in 1 M K_2SO_4 we have $I = 1/2(2 \times 1^2 + 1 \times 2^2)$ mol dm^{-3} = 3 M.

In water at $25°C$, $\varepsilon_r = 78.54$ and consequently $r_D = 0.34/\sqrt{I}$ nm. For dilute solutions ($I < 0.01$)r_D is much greater than the individual diameters of ions (a for simple ions is about 0.2–0.5 nm) (Y. Marcus, 1977, 1983).

The electrostatic part of the chemical potential (J mol^{-1}) of the i-th ion is

$$\mu_i^{elst} = \langle u^{DH} \rangle N_A \tag{1.214}$$

and therefore

$$\ln \gamma_i^{elst} = \frac{\langle u^{DH} \rangle N_A}{RT} = -\frac{z_1 z_2 e^2}{4\pi\varepsilon_0\varepsilon_r kT} \frac{b}{1+ab} \tag{1.215}$$

Combining equations 1.211 and 1.215 and considering the expression 1.120 for the mean ionic coefficient, the **Debye-Hűckel formula** is derived as follows.

$$\log \gamma_\pm = -\frac{A|z_1 z_2|\sqrt{I}}{1+Ba\sqrt{I}} \qquad (1.216a)$$

where

$$A = \frac{e^3 N_A^{1/2}}{4\pi\sqrt{2}(\varepsilon_0 \varepsilon_r kT)^{3/2}} \qquad (1.216b)$$

and

$$B = \left(\frac{2e^2 N_A}{\varepsilon_0 \varepsilon_r kT}\right)^{1/2} \qquad (1.216c)$$

For aqueous solutions at 25°C, $A = 0.510$ dm$^{3/2}$mol$^{-1/2}$ and $B = 3.29 \times 10^7$ dm$^{1/2}$ mol$^{-1/2}$. Because $a = 2 \div 5 \times 10^{-10}$ m, the product $B \cdot a \approx 1$ dm$^{3/2}$ mol$^{-1/2}$. At low ionic strength ($I < 0.01$), $Ba\sqrt{I} \ll 1$ and the **limiting Debye-Hűckel law** is valid,

$$\log \gamma_\pm = -A|z_1 z_2|\sqrt{I} \qquad (1.217)$$

Because $a \propto T^{-3/2}$, the electrostatic activity coefficient increases slightly with temperature and this can be used e.g. for **hydrothermal precipitation** of hydroxides (*C.M. Criss, J.W. Cobble, 1964; J.W. Cobble, 1966*).

At higher concentrations ($I > 0.01 - 0.03$) mol dm^{-3} the ionic atmosphere of ions diminishes to such an extent that the fluctuation of charges due to thermal movement becomes significant and the Debye-Hűckel theory is not precise. A further disadvantage of it is that concentrated solutions are better organized and their microscopic dielectric permittivity may be much lower than the ε_f value of the solvent and the ionic interaction increases (*R.M. Noyes, 1962*). For instance, in a 0.1

M solution of a uni-valent electrolyte, the average distance between ions is about 1 nm; this is about 3.6 diameters of a water molecule and all molecules should "feel" the presence of ions. Further, the parameter a of the mean diameter of ions is rather uncertain and it can be only approximated by the sum of the crystallographic radii of ions. Therefore, equation 1.216 is valid at concentrations below $0.03 - 0.1$ mol dm^{-3} and parameter a should be found empirically.

At higher concentration, strong deviations from the Debye-Hückel law appear and activity coefficients generally increase (Figure 12). Several semi empirical formulae have been proposed to describe behavior in more concentrated solutions of electrolytes, in which the coefficients grow due to solvent activity decrease — the **salting-out effect.** A simple relation suitable in the region of $0.2 - 1$ M solutions has been proposed (*P. Van Rysselberghe, S. Eisenberg, 1939; C.W. Davies, J. Hoyle, 1951*)

$$\log \gamma_\pm = \log \gamma_\pm^{elst} + \alpha I \qquad (1.218)$$

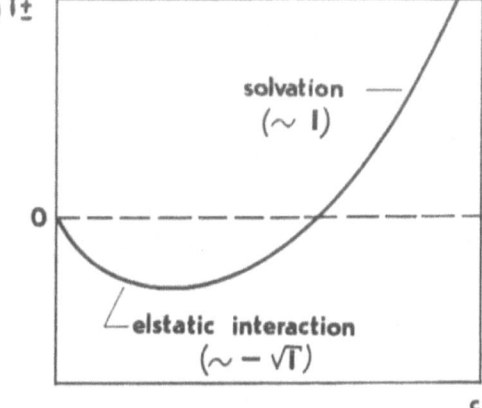

Figure 12. General course of concentration dependence of ionic activity coefficient.

where $\gamma_\pm^{elst}$ comes from equation 1.216a and coefficient $\alpha \approx 0.10 - 0.15|z_1 z_2|$. For uncharged species obviously $\gamma_\pm^{elst} = 1$ ($\log \gamma = \alpha I$) and therefore the coefficient α refers to the **salting-out coefficient** ($\alpha > 0$) or **salting-in coefficient** ($\alpha < 0$) — compare with equation 1.297.

At concentrations up to about 6 mol dm^{-3} there is a more general **Stokes-Robinson equation** for the activity coefficient at a molality of the electrolyte, m:

$$\log \gamma_{\pm} = \log \gamma_{\pm}^{elst} - \frac{h}{v} \log a_S - \log[1 - M_r(v-h)m] \quad (1.219)$$

(*R.H. Stokes, R.A. Robinson, 1948*). $\gamma_{\pm}^{elst}$ is the electrostatic activity coefficient, a_S is the activity of the solvent, M_r is the molar weight of solvent (for water $M_r = 0.018$ kg mol^{-1}), h is the solvation number (number of moles of solvent bound by 1 mole of electrolyte) and v is the stoichiometric coefficient according to equation 1.120a. Though the parameter h should be fitted to experimental data, as the parameter a in equation 1.216, the equation is of great importance for concentrated solutions where the solvation of ions is serious.

The solvation of anions is usually considered to be negligible ($h = 0$) and for cations it is usualy taken to be $h = 4$ for M$^+$ ions (coordination tetrahedral) and $h = 6$ for M^{2+} ions (coordination octahedral). However, the fitting of experimental results leads to much higher values for polyvalent electrolytes ($h = 14 \div 18$). Theoretical calculations (*H. Kistenmacher, 1974*) of the hydration of Li$^+$, Na$^+$, K$^+$, F$^-$ and Cl$^-$ in the gas phase lead to values 4, 5-6, 5-7, 4-6 and 6-7 respectively, the energetic difference between coordination spheres being rather small (several kJ mol^{-1}, i.e. water molecules enter and leave the coordination sphere easily). The calculations for the liquid phase, simulated by 50-200 molecules in ionic surroundings, gave coordination numbers of 4-6 for Li$^+$ and 4 for F$^-$.

The parameter h is related to the radius of the hydrated ion, (*Y. Marcus, 1983*) or, better, to the density of water molecules in the hydration spheres, ρ_i(m^{-2}) for the coordination number of the central ion N (*O.Ya. Samoilov, 1956; A.M. Rozen, 1967*).

$$h \approx \rho_i = \frac{N}{4\pi(r_i+r_w)^2} \tag{1.220}$$

where r_i is the crystallographic radius of the ion and $r_w = 0.138$ nm is the radius of H_2O molecules, since the Columbic energy of the ion-dipole interaction in the **Bernal-Fowler approximation** (*J.D. Bernal, R.H. Fowler, 1933*) is:

$$u_i^{BF} = 4\pi e \mu_w z_i \rho_i \tag{1.221}$$

where $\mu_w = 6.17 \times 10^{-30}$ C m is the dipole moment of water.

The multiply-charged small-sized ions have the largest density of electrostatic energy and, correspondingly, the largest hydration numbers (i.e. they are also the best salting-out agents).

Ion-dipole interactions are very important in the mechanism of complexation by solvation reagents used in the separation of ions, such as the crown ethers (*C.J. Pedersen, 1967; Y. Marcus, L.E. Asher, 1978*).

There are several experimental methods for determining the solvation numbers but they use various aprioristic suppositions and do not distinguish clearly between solvation and ion association phenomena (*J.F. Hinton, E.S. Amis, 1971*).

Concentrated solutions resemble crystal structures and can be modelled by the hydration of ionic lattices (*J. Celeda, D.G. Tuck, 1974*).

In ionic crystals, together with the Coulombic attraction of ions, the weak repulsive interaction of ions becomes important (see section 1.2.5). It is included in the Born equation for the energy of ions in a crystal lattice (*M. Born, 1920*).

$$u^B = -M\frac{z_1 z_1 e^2 1}{4\pi\varepsilon_0 \varepsilon_r r} + \frac{be^2}{r^n} \tag{1.222a}$$

or (*M. Born, J. Mayer, 1932*)

$$u^B = -M\frac{z_1 z_1 e^2 . 1}{4\pi\varepsilon_0 \varepsilon_{,r}} + b\exp(-r/a) \tag{1.222b}$$

where a and b are constants, and M is the Madelung constant which depends on the geometry (structure) of the crystal lattice (E. Madelung, 1918). Usually $M = 1.7 - 5$ (in the literature this parameter is tabulated for the commonest values of z_1 and z_2). From the condition of equilibrium interionic distance, $du/dr = 0$, b can be eliminated and we have

$$u^B = -M \frac{z_1 z_1 \varepsilon^2}{4\pi\varepsilon_0\varepsilon_r}\left(1 - \frac{a}{r}\right) \qquad (1.223)$$

The ratio a/r is usually in the range $0.08 - 0.13$ and therefore the weak interactions represent about 10% of the electrostatic interactions of the ions. In lattices of identical structure and equally-charged ions the energy of interaction will increase with decreasing ion size, e.g. in the series RbCl < KCl < NaCl < LiCl.

In separation science, all solutions are multicomponent and the above-mentioned relations can be used in practice only with an excess of one electrolyte. The problem of the ionic activities of individual components in solutions consisting of many electrolyte components has still not been solved satisfactorily.

Equations 1.216a and 1.217 exhibit the **rule of constant ionic strength**, i.e. the activity coefficients of ions should be the same in different solutions of equal ionic strength. For instance, if a reaction which depends on the concentration of hydrogen ions is to be investigated, solutions of various acid-salt compositions of mixtures with the same ionic strength are used, e.g. 0.1 MHCl with 0.1 M NaCl, i.e. 0.1 M(H, Na)Cl. At higher concentrations, however, significant deviation from the rule is observed (Fig.12). The first approximation of the **Harned rule**, can be expressed by the linear change of $\log \gamma_1$ on the ionic strength I_2 of the second component,

$$\log \gamma_1 = \log (\gamma_1)_{I_1=I} - \alpha_{12} I_2 \qquad (1.224)$$

where the first term on the right is the value of $\log \gamma_i$ in a pure solution of the first component at an Ionic strength $I = I_1$, and

α_{12} is the so-called **Harned coefficient** (*H.S. Harned, W.J. Hamer, 1935*). The latter was successfully applied in separation equilibria calculations (*I.L. Jenkins, H.A. McKay, 1954; B.A. Soldano, et al. 1955; F. Macasek, L. Matel, M. Kyrs, 1978*). Its value is connected with the hydration parameters of ions, equation 1.220, and the rule is reasonably symmetrical, i.e. $\alpha_{12} \approx -\alpha_{21}$. The Harned coefficient can be derived from single electrolyte data using the linear part of equation 1.218 ($\log \gamma_i \sim \alpha_i I$),

$$\alpha_{12} = \frac{1}{2}(\alpha_1 - \alpha_2) \tag{1.225}$$

(*A.M. Rozen, 1967*). If $\alpha_1 = \alpha_2$, the ionic strength rule is observed.

Further progress in the theory of electrolytes depends on the success of the theory of concentrated binary and multicomponent electrolyte solutions (*H.A.C. McKay, 1953; L.A. Broomley, 1973*). Hence, the effort to predict the activity coefficient of an individual ion in concentrated solution is understandable (*J. Kielland, 1937*). The concept of hydration sphere density (see equation 1.220) can be fruitful (*A.S. Solovkin, 1969, 1970; V.S. Shmidt, et.al., 1975; M. Ya. Zelvenskii, A.S. Solovkin, 1980; A.S. Solovkin, Yu. N. Zakharov, 1980*). Other attempts follow from semi empirical relations and adjustable parameters, as in the approximation (*A. Ferse, 1978*):

$$\gamma = k_1 \exp\left(-\frac{A\sqrt{m}}{k_1}\right) + (1 - k_1) \exp(k_2 m) \tag{1.226}$$

where A is the Debye-Hückel parameter and k_1 and k_2 are constants found by fitting experimental data on mixtures of uni-valent electrolytes. Some other works were mentioned at equation 1.123.

1.2.5 Weak intermolecular interactions. Non-electrolyte solutions

As was demonstrated in the previous section, weak intermolecular interactions play a minor role in electrolyte solutions, the ion-dipole interactions and solvation bringing the most significant contribution to the behavior of ions. In non-electrolyte solutions, obviously, the weak intermolecular interactions are the major type of interaction and the dominant source of non-ideality in the thermodynamic behavior of the components.

Conventionally, those intermolecular interactions whose energy does not exceed 40 kJ mol^{-1} (0.4 eV per molecule) and where the molecules do not approach nearer than 0.2 nm are called **weak** or **Van der Waals interactions**. By weak interactions, as distinct from chemical bond formation, there is no substantial change of geometry and electronic state of the molecule. The exception is in solutions of macromolecules, where the weak interactions, due to their great number, significantly change the form of the macromolecule and determine, for example the tertiary structure of proteins.

Below the energy limit 40 kJ mol^{-1} are located, for example, also the coordination bond (about 25 kJ mol^{-1} for monofunctional ligands) and hydrogen bond formation (15 ÷ 40 kJ mol^{-1}). The dinucleotide aromatic base pair stacking energies varies from 15 to 50 kJ mol^{-1}. Within a broad definition, electron donor-acceptor (charge-transfer) complexes are also regarded as Van der Waals (denoted *VdW*) systems, but because they are sufficiently stable at normal temperatures and can often be isolated as individual compounds, their formation is usually viewed as a chemical interaction and described by chemical (pseudo-chemical) equilibria producing **VdW molecules** with a shallow energy minimum (0.01 ÷ 40 kJ mol^{-1}):

$$A + B \rightleftarrows A \cdots B \qquad (1.227a)$$

and also

$$A + A \rightleftarrows A \cdots A \qquad (1.227b)$$

$$B + B \rightleftarrows B \cdots B \tag{1.227c}$$

For example, the gas phase dimerization energy of water molecules is about 22 kJ mol^{-1}, but that of argon atoms is only about 1 kJ mol^{-1} (*P. Hobza, R. Zahradnik, 1980*).

For these reasons we recognize as weak interactions those whose energy is below $10 \div 15$ kJ mol^{-1}. This energy is close to the entropic (athermic) effect on formation of small molecules (about $5 \div 15$ kJ mol^{-1} at $25°C$) but sufficiently higher than the energy of the thermal motion of molecules (about $RT = 2.5$ kJ mol^{-1} or 0.025 eV per molecule at $25°C$).

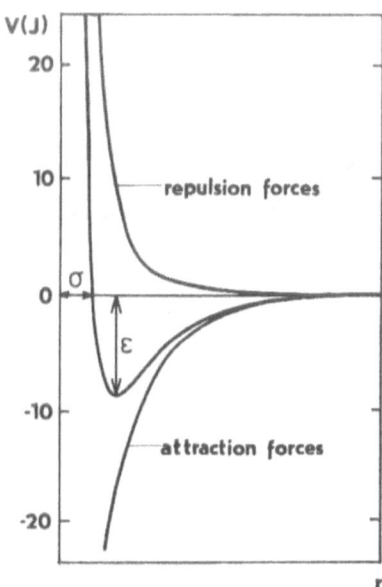

Figure 13. Potential energy versus intermolecular distance.

Weak interactions are sufficiently powerful to enable a subtle mechanism of compound separation (*L.R. Snyder, 1974, 1983; C.J. Van Oss, D.R. Absolom, A. W. Neumann, 1979, 1981*). Interactions of 15 kJ mol^{-1} at $= 300$ K, according to equation 1.89, cause a change of activity coefficient $\ln \gamma = -\Delta\mu/RT = -15000/2480 \approx -6$, i.e. $\gamma = 0.0024$. Hence, γ is diminished 400-times compared with the standard state (ideal solution). The role of theory is to predict accurately Van der Waals forces and to calculate activity coefficients of non-electrolytes in solutions.

The potential energy of a pair of molecules as a function of intermolecular distance (Figure 13) exhibits a region of **attraction** at longer distances and a region of **repulsion** at shorter approaches. Though the repulsion is caused by electrostatic forces between electron shells, it cannot be described by a simple Coulombic function. The repulsion energy is usually approximated by an exponential function (*M. Born, J.G. Kirkwood, 1931*)

$$u^{rep} = b \exp(-r/a) \tag{1.228}$$

where r is intermolecular distance and a and b are constants (a is 0.03–0.04).

At larger distances the interaction can be described by three types of force. The interaction of permanent and induced dipoles is electrostatic in character, but dispersion forces (their name comes from the fact that they are described by the same parameters as light dispersion) are of a quantum-mechanical nature.

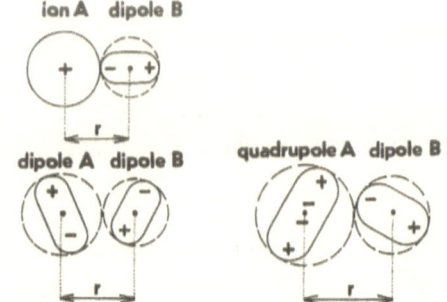

Figure 14. Ion-dipole, dipole-dipole and quadrupole-dipole interaction arrangement.

The field intensity produced by a dipole is $E \sim \mu/r^3$ and the **dipolar interaction** of molecules A and B with permanent dipole moments μ_A and μ_B (Figure 14) depends upon their mutual arrangement and can be averaged by the relationship (J.E. Lennard-Jones, 1931)

$$\langle u^{\text{dip}} \rangle = - \frac{2N_A}{3RT(4\pi\varepsilon_0)^2} \frac{\mu_A^2 \mu_B^2}{r^6} \qquad (1.229)$$

where $\langle u^{\text{dip}} \rangle$ is the mean interaction energy (J mol^{-1}) of randomly oriented dipoles and r is the mean distance between the centers of the dipoles. The constant, $2N_A/3RT(4\pi\varepsilon_0)^2$ at $T = 300$ K equals 5.64×10^{39} J C^{-4} m^2. Dipolar interactions are strongest among multipole interactions. For instance on the interaction of dipoles $\mu_A = \mu_B = 33.35 \times 10^{-30}$ C m (the 3.335×10^{-30} C m is the "Debye unit"), which is typical for quaternary ammonium salts, at a distance 1 nm and antiparallel orientation the maximal interaction energy is about 50 kJ mol^{-1} and the mean energy, according to equation 1.229, is 9.7 kJ mol^{-1}.

Averaged **dipole-quadrupole interactions** are proportional to $\mu^2 Q^2/r^8$ and those of quadrupole-quadrupole to $Q^2 Q^2/r^{10}$ (Q is the quadrupole moment, C m^2), i.e. they are very short-range. Interactions of octopoles and higher multipoles are not taken into account. The relation between u^{dip} and activity coefficients in mixtures A and B will be given later.

A dipole can interact with a non-polar molecule by producing in the latter a transient, induced dipole. The induction interaction of a polar molecule A (permanent dipole μ_A) with a neutral molecule B (uncharged and spherically symmetrical) of polarizability α_A (m^3 mol^{-1}) leads to instantaneous dipole formation in molecule B (μ_B is proportional to α_B). The average energy (J) of the induction interaction has the value (P. Debye, 1920)

$$\langle u^{\text{ind}} \rangle = - \frac{2}{(4\pi\varepsilon_0)^2} \frac{\mu_A^2 \alpha_B}{r^6} \qquad (1.230)$$

In contrast to dipole-dipole interaction, this energy does not depend on temperature in this approximation, but is valid only for isotropically polarizable molecules. The polarizability α increases with molar volume. Experimentally it is usually determined from the molar refraction R_∞ at infinite wave-length, $\alpha = 3.96 \times 10^{-31} R_\infty$.

The **interaction of nonpolar molecules** can only be explained by means of quantum mechanics (*W. Heitler, F. London, 1927; F. London, 1930*). Dispersion interactions can be modeled on the basis of an instantaneous dipole moment of one molecule inducing a dipole moment in a nearby molecule, the centers of charges being compensated. In the language of quantum mechanics, the dispersion energy is better explained as the force between locally-excited molecules: with respect to electron exchange between molecules, the wave function corresponds to the exhibition of a relative minimum in comparison with the function for the isolated molecules.

The **dispersion energy** is known as a **London potential**

$$u^{disp} = -\frac{3}{2(4\pi\varepsilon_0)^2} \frac{I_A I_B}{I_A I_B} \frac{\alpha_A \alpha_B}{r^6} \tag{1.231}$$

where α refers to the polarizabilities of the molecules and I to their ionization potentials. The ionization potentials of common molecules do not differ greatly, falling in the range 880-1100 kJ mol^{-1} (8.8-11 eV per molecule) and an average value of 950 kJ mol^{-1} (9.5 eV per molecule) can be considered. This means that the dispersion interaction will be determined mostly by polarization constants. These, however, depend on the character of the chemical bonds in the molecule and, being strongly anisotropic, they should be averaged, e.g. by rotational ellipsoids of the bonds.

Interaction energies play different roles in various types of molecules: for polar molecules, dipole interactions are dominant (up to 65% of the interaction energy) while for nonpolar ones, dispersion interactions are most significant. Induction forces can, as a rule, be neglected if the molecules do not contain multiple bonds. Further, because of the high degree of mutual compensation of dipole-dipole interactions in solids and liquids, and due to the additivity of dispersion forces on the interaction of a number of molecules (the energy of the dispersion interactions of N molecules is proportional to $N(N-1)/2$ pairs of mutual contacts), the dispersion interactions are usually the most important in condensed systems. Therefore, the additivity of all the forces discussed, $u = u^{rep} + u^{dip} + u^{ind} + u^{disp}$ is valid only for relatively high distances between the molecules, in the gaseous phase.

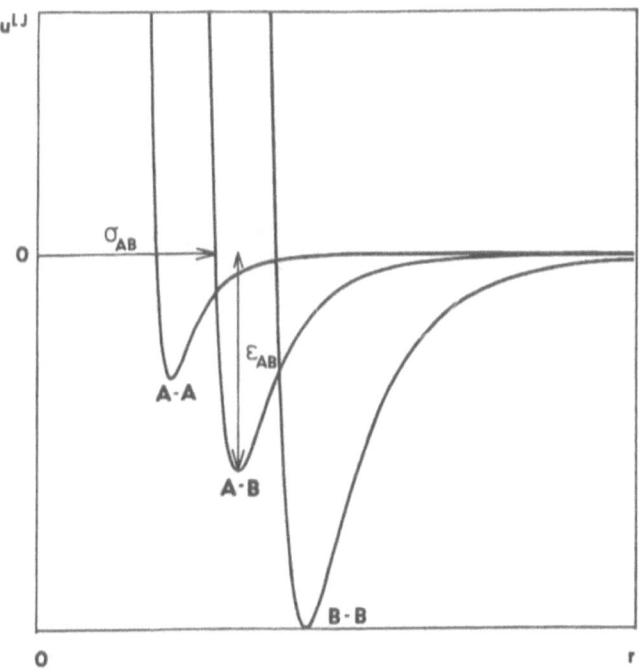

Figure 15. Lennard-Jones (6-12) potential of pairs of molecules A ... A, B...B and A ... B.

To express intermolecular forces at a short distances, various semi empirical formulae have been proposed. For non-polar molecules, the **Lennard-Jones potential** (*J.E. Lennard-Jones, 1924, 1931*) is widely used:

$$u^{LJ} = 4\varepsilon \left[\left(\frac{\sigma}{r}\right)^{12} - \left(\frac{\sigma}{r}\right)^{6} \right] \quad (1.232)$$

where ε is a constant (J), i.e. the depth of potential energy minimum (Figure 15) and σ is the effective (collision) radius of the molecule. On the interaction of different molecules A and B, the constants comprise the geometric mean:

$$\varepsilon_{AB} = \sqrt{\varepsilon_A \varepsilon_B} \tag{1.233a}$$

and arithmetic mean:

$$\sigma_{AB} = \frac{\sigma_A + \sigma_B}{2} \tag{1.233b}$$

It is seen that for attractive forces the typical power dependence $1/r^6$ is preserved, while the repulsive forces are modelled by the $1/r^{12}$ function. This is the reason why the Lennard-Jones potential is often called the "**6-12 potential**". It contains only two adjustable parameters, ε and σ, making it convenient for calculations: collision integrals, diffusion coefficients and thermo diffusion coefficients in real mixtures can be calculated (*J. Hirschfelder et al., 1954; R.C. Reid, T.K. Sherwood, 1958*). Some data are illustrated as follows (*T.K. Sherwood et al., 1975*):

molecule	ε/kJ mol^{-1}	σ/nm
helium	0.0850	0.2551
argon	0.7757	0.3542
methane	1.236	0.3758
ethane	1.793	0.4443
n-hexane	3.320	0.5949
benzene	3.428	0.5349
chloroform	2.829	0.5389
ethanol	3.015	0.4530
methanol	4.006	0.3626
mercury	6.236	0.2969
water	6.727	0.2641

The Lennard-Jones potential depends strongly on the geometry of the molecule. For instance, if the diameter of the hydrocarbon moiety $-CH_2-$ which is 0.09 nm, is increased by substitution with a methyl or ethyl group to 0.11–0.13 nm, the intermolecular attraction is supposed to decrease 10-30 times.

The LJ potential is unsatisfactory for polar molecules and in this case the Stockmayer potential can be used:

$$u^S = u^{LJ} - \frac{\mu_A \mu_B}{r^3} \Phi \tag{1.234}$$

which extends the LJ potential u^{LJ} by a term expressing dipolar interaction (inversely proportional to r^3); Φ is a function of the mutual dipole orientation (for linear orientation and antiparallel dipoles, $\Phi = 2$ and the minimum value of $\Phi = -2$).

Obviously, force fields in nonelectrolyte solutions are less precisely defined than for Coulombic forces in ionic systems. The Debye-Hückel relation between the ionic activity coefficient and the energy of ionic interactions includes the ionic strength as a single parameter representing the composition of solution. However, in nonelectrolyte solutions a structural model of the mutual contacts of the molecules should be made to use any of the formulae discussed concerning intermolecular forces, when the composition of the mixtures varies. The task for separation chemistry is to predict the thermodynamic properties of components in non-ideal mixtures when the pure-component parameters are known (*E.A. Guggenheim, 1952; M.I. Shakhparonov, 1956*).

Pioneering works on the structural theory of solutions were done using quasi-chemical equilibria (*F. Dolezalek, 1908*) and by the real-gas approximation (*I.I. Van Laar, 1925*). Then it was developed (*J.H. Hildebrand, G. Scatchard, 1930; V.A. Kireev, 1942; E.A. Guggenheim, 1944; J. Barker, 1952, 1953*) to the quasi-lattice structure of solutions.

The total change of energy on formation of a solution from its components A and B can be calculated from the number of molecular contacts (z) and the energy of interaction (u_{AB}).

To a first approximation, if the molecules A and B have the same volume, i.e. so-called isomegetic solutions, the change of interaction energy can be derived from the following assumptions. In a liquid consisting of molecules A, every molecule interacts with z molecules A with an energy u_{AA}. The same is true in respect of liquid B. If one molecule in liquid A is replaced by molecule B (Figure 16) or vice versa, the change of energy in the quasichemical reaction

$$A \cdots A + B \cdots B \rightleftarrows 2A \cdots B \qquad (1.235)$$

is

$$u_{A-B} = -u_{AA} - u_{BB} + 2u_{AB} \qquad (1.236)$$

and the interchange energy Δu (J mol^{-1}) corresponding to the introduction of one "alien" molecule B is one-half of this value,

$$\Delta u = u_{AA} - \frac{1}{2}(u_{AA} + u_{BB}) \qquad (1.237)$$

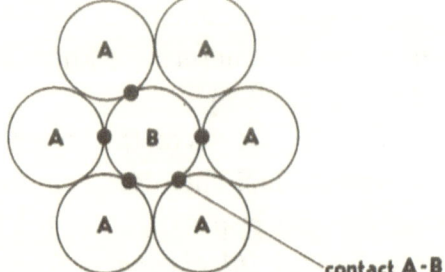

Figure 16. Model of contacts in solutions.

For instance, if the energy can be expressed by the dipole-dipole interaction of A and ($r_A = r_B = r$), according to 1.229 we have

$$\Delta u^{\text{dip}} = \frac{2N_A}{3RT(4\pi\varepsilon_0)^2 r^6}[\mu_A^2 \mu_B^2 - \tfrac{1}{2}(\mu_A^4 + \mu_B^4)] = \frac{2N_A}{3RT(4\pi\varepsilon_0)^2}\left(\frac{\mu_A^2 + \mu_B^2}{r^3}\right)^2 \quad (1.238)$$

For molecules of different molar volumes, $\overline{V}$ and $\overline{V}$ (m^3 mol^{-1}) the contact number should be proportional to the volume, and the volume interaction energy u_V (J m^{-3}) is

$$\Delta u_V = \frac{u_{AB}}{\sqrt{\overline{V_A V_B}}} - \frac{1}{2}\left(\frac{u_{AA}}{\overline{V_A}} + \frac{u_{BB}}{\overline{V_B}}\right) \quad (1.239)$$

The interaction energy u_{AB} is roughly approximated by the geometric mean

$$u_{AB} = \sqrt{u_{AA} \cdot u_{BB}} \quad (1.240)$$

though in reality u_{AB} has a lower value. Now Δu_V can be expressed through the parameters of the pure components,

$$\Delta u_V = \left(\sqrt{\frac{u_{AA}}{\overline{V_A}}} + \sqrt{\frac{u_{BB}}{\overline{V_B}}}\right) \quad (1.241)$$

The square root of the volume energy was termed (J.H. Hildebrand, S.E. Wood, 1933) the **solubility parameter** (J$^{1/2}$m$^{-3/2}$)

$$\delta = \sqrt{\frac{u}{v}} \quad (1.242)$$

Because many of the solubility parameter values are tabulated in older units, i.e. cal$^{1/2}$ cm$^{-3/2}$, (J.H. Hildebrand, R.L. Scott, 1950) the transformation δ[cal cm$^{-3/2}$] = 0.49δ [kJ$^{1/2}$ dm$^{3/2}$] = $490\,\delta$ [J$^{1/2}$ m$^{-3/2}$] should be used. For instance, hexane has $\delta = 7.3$ cal$^{1/2}$dm$^{-3/2}$ = 14.9 kJ$^{1/2}$ dm$^{-3/2}$ = 0.0149 J$^{1/2}$ m$^{-3/2}$.

Solubility parameters represent the density of cohesion energy of molecules. Using δ's, 1.241 can be given as

$$\Delta u_V = (\delta_A - \delta_B)^2 \quad (1.243)$$

This difference serves as a criterion for the mutual solubility of compounds. Mixing proceeds when $\Delta G_{mix} < 0$ (see equation 1.11–1.13) i.e. for the endothermic process when $\Delta H_{mix} < T\Delta S_{mix}$. The molar excess enthalpy (see equation 1.68) for nonpolar components A and B and $\overline{V}^E = 0$ was derived (J.H. Hildebrand, S.E. Wood, 1933) as

$$\Delta \overline{H} = \overline{H}^E = v_A v_B (\overline{V_A} + \overline{V_B})(\delta_A - \delta_B)^2 \qquad (1.244)$$

where v's are volume fractions (equation 1.49). The condition of mutual solubility is approximated by

$$(\overline{V_A} + \overline{V_B})(\delta_A - \delta_B)^2 < 2RT \qquad (1.245)$$

For a difference of about $|\delta_A - \delta_B| > 5 \div 6$ kJ$^{1/2}$ dm$^{-3/2}$ (strong positive non-ideality), the components may form separate phases. However, the miscibility of lower alcohols ($\delta \approx 25 - 30$) with water ($\delta = 48$) shows that the rule is inapplicable to polar components.

If the additivity 1.232 of various terms of the VdW interactions is valid, which is substantiated by the proportionality $u \propto \delta^2$, then the solubility parameter can also be split into its components as

$$\delta^2 = (\delta^{\text{dip}})^2 + (\delta^{\text{ind}})^2 + (\delta^{\text{dup}})^2 \qquad (1.246)$$

Solubility parameters are widely applied in separation science. They can be found experimentally from the molar heat of evaporation ($L = \Delta H - RT$) and molar volume of substances. For instance, the evaporation enthalpy of ethanol is $\Delta H = 39.4$ kJ mol^{-1} (15°C) and its molar volume $\overline{V}$ is 0.058 dm^3 mol^{-1}. Hence, the cohesion energy is $39.4 - 2.4 = 37$ kJ mol^{-1} and $\delta = \sqrt{37/0.058} = 25.3$ kJ$^{1/2}$ dm$^{-3/2}$. Theoretical calculations according to equation 1.244 indicate that in this case the contribution of the dispersion forces is about 37%, that of the dipole interactions about 11% and hydrogen bond interactions between C_2H_5OH molecules are responsible for the rest.

Solubility parameters can also be derived from coefficients of thermal expansion (α) and pressure expansions (β), see equation 1.70, as

$$\delta = \sqrt{T\frac{\alpha}{\beta}} \qquad (1.247)$$

(*J.H. Hildebrand, S.E. Wood, 1933*). For involatile compounds (polymers etc.), the solubility parameters are estimated indirectly from solubilities, swelling and gas-chromatographic elution data (*H.C. Van Ness, M.M. Abolt, 1982; P.J. Schoenmakers, H.A.H. Billet, Degalan, L. 1982; A.E. Nesterov, 1988*) or from the size of molecular balls in solvents (*A.S. Kertes, 1965*).

Instead of solubility parameters, related constants are used. Thus, the **Small's molar attraction constants** are defined (*P.A. Small, 1953*) as

$$F_S = \delta\overline{V} = \sqrt{u\overline{V}} \qquad (1.248)$$

where $\overline{V}$ is the molar volume ($m^3 mol^{-1}$), and they have dimensions of $J^{1/2}\, m^{3/2}\, mol^{-1}$.

In calculations, the additivity of cohesion energies between polyfunctional molecules is assumed (*I. Langmuir, 1925*)

$$u = \sum v_i u_i \qquad (1.249)$$

where u_i is the interaction energy and v_i the number of the i-th group constituting the molecule. Some of the values used are as follows:

group	u_i / kJ mol^{-1}
-CH$_2$	4.19
-CH$_3$	9.64
-CH(CH$_3$)-	10.0
-CH=CH-	10.2
CF$_3$	7.83
-CH$_2$Cl	17.2
-C$_6$H$_5$	40.6
-CO-NH-	60.8

In surface chemistry and in chemical practice (e.g. *B.V. Hassas, F. Karakas, M.S. Celik, 2014*), a simple characteristic of molecular polarity and the contribution of dipole interactions uses a **hydrophilic-lipophilic balance (HLB) of a** molecule defined in a 20-degree scale (*W.C. Griffin, 1954; R. Heusch, 1970*) as

$$\text{HLB} = 20 \frac{M_{\text{hydrophil}}}{M} \qquad (1.250)$$

where $M_{\text{hydrophil}}$ and M are molecular masses of the hydrophilic group and the whole molecule respectively.

Empirical determination of solvent polarity in binary mixtures is based on solvatochromic changes (*Z.B. Maksimovic, C. Reichardl, A. Spizic, 1974*).

Sophisticated calculation methods based on the **quasichemical theory** (*E.A. Guggenheim, 1952*) generalized (*G.M. Wilson, 1964; J.M. Prausnitz, 1969, 1981; J. Novak et al., 1974*) and applied to interactions of functional groups within molecules (*A.S. Kertes, F. Grauer, 1973; A. Fredenslund, et al., 1975,*

1977) need more detailed parameters. They require use of an area fraction of component *i*

$$\theta_i = \frac{a_i x_i}{\sum a_i x_i} \quad (1.251)$$

where x_i is the mole fraction of component *i* and a_i is the area parameter of the pure component, and the segment fraction of component *i*

$$\Phi_i = \frac{v_i x_i}{\sum v_i x_i} \quad (1.252)$$

where v_i is the volume parameter of the pure component. The parameters a_i and v_i are obtained by summation of the surface areas (A_j) and VdW volumes (V_j) of groups in the molecule:

$$a_i = \sum v_j A_j \quad (1.253)$$

$$v_i = \sum v_j V_j \quad (1.254)$$

Binary interaction energies between the groups are determined from vapor-liquid data (*A. Fredenslund, R.L. Jones, J.M. Prausnitz, 1975; A. Fredenslund et al., 1977*).

By employing the principal expression 1.244 for molar excess enthalpy, it is possible to obtain absolute activity coefficients in regular solutions. Taking into account eqn. 1.89 and definition 1.9, in the absence of the enthalpic term ($\overline{G} = \overline{H}$)

$$RT \ln \gamma_i = \mu_i^E = \frac{\partial \overline{H}^E}{\partial n_i} \quad (1.255)$$

Because in the two-component regular system, the volume fraction

$$v_A = \frac{n_A \overline{V}_A}{n_A \overline{V}_A + n_B \overline{V}_B} \quad (1.256)$$

we have also

$$dn_A \frac{n_B \bar{V}_B}{\bar{V}_A} - \frac{dv_A}{(1-v_A)^2} = \frac{n_B \bar{V}}{\bar{V}} - \frac{dv_A}{v_B^2} \qquad (1.257)$$

Combining equations 1.255–1.257 with respect to 1.244, results in

$$RT \ln \gamma_A = \Delta u_v \bar{V}_A v_B^2 \qquad (1.258a)$$

$$RT \ln \gamma_B = \Delta u_v \bar{V}_B v_A^2 \qquad (1.258b)$$

or, in isomegetic solutions ($\bar{V}_A = \bar{V}_B = \bar{V}$; $\Delta u_v \bar{V} = \Delta u$)

$$RT \ln \gamma_A = \Delta u x_B^2 \qquad (1.259a)$$

$$RT \ln \gamma_B = \Delta u x_A^2 \qquad (1.259b)$$

where $x_A + x_B = 1$. Considering that $\gamma_A = \gamma_A(0)$ at $x_A = 0$ (then $v_B = 1 = x_B$), see equation 1.97, it can be seen that Δu has the meaning $\Delta u = (RT/\bar{V}_A) \ln \gamma_A(0) = (RT/\bar{V}_B) \ln \gamma_B(0)$ and therefore in regular solution $\bar{V}_B \gamma_A(0) = \bar{V}_A \gamma_B(0)$.

The trend in the activity coefficients (both absolute and concentration) corresponds, for a positive non-ideality ($\Delta u > 0$) to the curves on Figure 9, for negative non-ideality ($\Delta u < 0$) they are turned over symmetrically.

The single parameter Δu, describing the behaviour of regular solutions, can be obtained from pure component data, the solubility parameters, according to 1.243. The advantage of the regular solution model is its simplicity and symmetry.

For k-component regular solutions, the activity coefficients include $k(k-1)/2$ unknown binary mixtures parameters u:

$$RT \ln \gamma_i = \sum_{j=0}^{k} u_{ij} x_j^2 + \sum_{j=1}^{k} \sum_{m=1}^{k} (u_{ij} + u_{im} - u_{jm}) x_j x_m \qquad (1.260)$$

where both $m \neq i$ and $j \neq i$. A three-component system, for instance, is described by three parameters (u_{AB}, u_{AC} and u_{BC}) but for four-component systems 6 parameters are necessary.

Regular solution theory has been applied mostly for vapor-liquid but also liquid-liquid distribution separation systems (*e.g. T. Wakayashi, et al., 1964; A.S. Kertes, 1965; H. Freiser, 1969; A.M. Rozen et al., 1972; D.E. Noel, C.F. Meloan, 1972*).

As illustrated above, further development of solution theory is associated with qualitative differentiation of intermolecular contacts at individual segments of molecules and its goal is to achieve this with the minimum number of adjustable parameters, the number of which is always greater than in regular solution theory. The resulting number, however, may be much lower. For instance in the **UNIFAC** ("universal quasichemical functional group activity coefficients") method there is generally, in addition to 7 k geometric (size and shape) molecular parameters — see equations 1.251-1.254 — $n(n + 1)$ group interaction parameters (n is the number of types of group in the molecule) in the combinatorial term of the activity coefficient (*T. Magnusson, P. Rasmunssen, A. Fredenslund, 1981; J. Gmehling, P. Rasmussen, F. Fredenslund, 1982*). Nevertheless, very often only two parameters are finally needed to represent the behavior of binary mixtures (*A. Fredenslund et al., 1977*). Software package DDBSP for UNIFAC calculations is commercially available [1].

1.2.6 Specifics of microconcentrations

All previous considerations of mixture properties and component behavior were based on the presence of the large sets of molecules to be separated. In practice (radiochemistry, electronic materials) the concentration of micro components drops to 10^{-6} mol dm^{-3}. The content of impurities in semiconducting germanium should be less than 10^{-8} % and for silicon even less, at 10^{-11} %. Glass for optical fibres should contain less than 10^{-8} % of impurities of transition metals. The "omnipresent" concentration of even such rare elements as platinum, osmium and rhenium is about

---

[1] http://www.ddbst.com/unifac-calculation.html

10^{-8} %. Radioactive indicator concentrations generally may be of the order of $10^{-9} - 10^{-18}$ %. The natural concentration of radon in air is about 2 atoms per 1 cm³, i.e. less than 10^{-17} % (*P. Benes, V. Majer, 1980*).

Fluctuations in concentration and energy, which should be considered as probable in the existence of a certain number of molecules with a certain energy in a given volume element, may become significant (*I. Prigogine, 1950; R. Haase, A. Muenster, 1950; A.Yu. Zakgeym, 1966; I.P. Krichevskii, 1975*). The **probability of fluctuations**, when the actual entropy S' of the system deviates from the equilibrium entropy S ($S' < S$) is, according to equation 1.23,

$$W = \exp\left(-\frac{S'-S}{R}\right) \qquad (1.261)$$

and from this it can be derived that the mean square deviation of a thermodynamic parameter X is proportional to the square root of the mean number of particles N_i,

$$\langle \Delta x^2 \rangle \propto \sqrt{N_i} \qquad (1.262)$$

For instance, if the mean number of atoms in a total set of N_S atoms is expressed by the mole fraction x_j, this value fluctuates with a statistical deviation

$$|\Delta x_i| = \sqrt{\frac{x_i}{kTN_S}} \qquad (1.263)$$

($x_i/N_S \approx N_i/N_S^2$). On this basis, it can be assumed (*A.A. Benedetti-Pichler, J.R. Rachele, 1940*) that quantitative reproducibility of analytical evidence (for the criterion $\Delta x < 1\%$) occurs in a set of at least 10^4 molecules and qualitatively ($\Delta x < 10\%$) in a set no smaller than 100 molecules. However, these postulates cannot be true, because successful and reproducible separation of rare transuranium elements (lawrencium, kurchatovium, mendelevium etc.) was performed with samples numbering just 10-100 atoms (*A. Ghiorso et al., 1955; I. Zvara et al., 1966, 1970; A. Ghiorso, 1967; R.J. Silva et al. 1970*).

For chemistry on this scale, the statistical sets independent of time should be replaced by statistical behavior in time (*R. Guillaumont, G. Boussiéres, 1972; D. Trubert, C. Le Naour, 2003*). Then it is necessary to formulate the entropic term of the chemical potential in statistical thermodynamic terms, i.e.

$$\mu_i = \mu_i^0 + RT \ln W_1 \tag{1.264}$$

where W_1 is the probability of occurrence of the i-th atom in a particular chemical state or phase. It can be related for example to the distribution constant — section 1.3.1. — or the retardation factor — section 1.4.3. — and found by repeated experiments.

Ultra dilute solutions can be, from a thermodynamic point of view, unstable, or **metastable**. The thermodynamic criterion of ultra-dilute solutions is formulated so that collisions between micro component molecules become very improbable and their interaction does not affect activity coefficients and Henry's law 1.79 is valid. If the probability of mutual collisions 1:1000 is sufficiently low, the corresponding concentration is below $x_A < 0.03$ if A is a non-electrolyte (about 10^{-2} M solutions) and $x_A < 2 \times 10^{-5}$ (less than 10^{-3} M solutions) if A is an electrolyte. Further, if the interaction of solvent (S) with solute (A) is stronger than the interaction between the solvent molecules, i.e. $u_{AS} > u_{SS}$ (positive non-ideality) then the Brownian motion does not interfere with the general tendency of solvates to A$\cdots$S formation. However, if $u_{AS} < u_{SS}$ (negative non-ideality), the mixture can pass through metastable states. The relaxation time $\tau(s)$ for the change from the metastable to stable state can be determined (*G.G. Devyatykh, S.M. Vlasov, 1966*) from the **Smoluchowski equation** for particle coagulation:

$$\tau = 1/4\pi Dr N_i \tag{1.265}$$

where D is the diffusion coefficient $(m^2 s^{-1})$, r is the molecular radius (m) and N_i the number of molecules in a volume element (m^{-3}). Its value gives the time necessary for the mutual collision of half of the molecules present $(N_i/2)$ and even for $10^{-10} \div 10^{-9}$ M solutions it is rather low ($\tau \approx 1s$).

Fluctuations in homogeneous systems occur only under special circumstances (*R. Rigler, 1976*). Theoretically, for instance, in cell organelles having a very small volume ($V \approx 10^{-20}$ m^3), the acidity at pH 6 corresponds to the presence of 10^{-23} moles of H$^+$ molecules, i.e. according to equation 1.263, 6 ± 2.5 H$_3$O$^+$ molecules may be found in the organelle (the pH would fluctuate over the range 6 ± 0.4). The fluctuations of two-dimensional structures, such as molecular films, are of interest for membrane separation processes; the question is: how many molecules are required for a functional unit of a molecular assembly? (*H. Kuhn, 1957, 1983*).

1.3 HETEROGENEOUS EQUILIBRIA

Most separation systems are heterogeneous and, as shown in section 1.1.1, phase separation is the final stage for the majority of separation processes. Gas-liquid, liquid-solid and liquid-liquid systems play a special role in chemical separation. Gas-liquid (vapor-liquid) and liquid-solid systems are used in classical distillation and crystallization processes, and for the most part belong to those physical separation techniques that proceed without change of chemical composition. The first part of this chapter is devoted to phase equilibria important for physical separation systems and the second part to biphasic systems where the separation is based on chemical composition and interactions (see e.g. *S. Peter, 1979; J.M. Prausnitz, 1981; V.M. Glazov, L.M. Pavlova, 1988*).

Although multiphase systems can be described by processes in isolated phases, phase contact should be noted as a factor in their mutual influence. The separation system as a whole may be a **closed system**, thus a boundary detaches the system from its surroundings and there is no transfer of any particle with non-zero rest mass (atoms, molecules, electrons, ions), but energy transfer is possible. However, the interphase boundary between phases of the separation system should allow a flux of separated components, i.e. these parts (phases) of the separation system are **open systems (subsystems)**. From a thermodynamic point of view, there is no great difference between closed and open systems. It was J.W. Gibbs who overcame problems in the thermodynamic description of complicated closed systems by analyzing them as **phase systems** consisting of a number of open phase subsystems (see e.g. *S.M. Walas, 1985*).

1.3.1 Phase equilibria

The existence of a boundary surface between phases makes separation easier, see section 1.1.1., but it is not a thermodynamic condition. The phase boundary is only an *apparent* attribute of a phase and even an abstract one if it is a surface in the geometrical sense, without considering its molecular structure and continuum of physic-chemical properties. Thus, the **phase** is a multitude of particles, the thermodynamic properties of which are described by the same thermodynamic intensive properties (temperature, pressure, concentration), i.e. by some of the fundamental Gibbs equations for open systems:

$$G = U + pV - TS + \sum \mu_i n_i \tag{1.266}$$

or

$$dG = SdT - Vdp + \sum \mu_i dn_i \tag{1.267}$$

—cf. equations 1.10 and 1.17. This means that the phase is homogeneous and has macroscopic dimensions (i.e. fluctuations do not play a role).

For equilibria between phase I and phase II, the equality of the chemical potential 1.38 is valid, in the general form of the equality of the total differential

$$d\mu_i^I = d\mu_i^{II} \qquad (1.268)$$

The total differential of the chemical potential for component i, $\mu_i = \mu_i(T, p, x_1 \ldots x_i \ldots x_k)$ will be for binary mixtures (components A and B):

$$d\mu_A = \left(\frac{\partial \mu_A}{\partial T}\right) dT + \left(\frac{\partial \mu_A}{\partial p}\right) dp + \left(\frac{\partial \mu_A}{\partial x_A}\right) dx_A \qquad (1.269a)$$

Considering equations 1.9 and 1.110, we find

$$\frac{\partial \mu_A}{\partial T} = \frac{\partial}{\partial T}\left(\frac{\partial G}{\partial \mu_A}\right) = \frac{\partial}{\partial x_A}\left(\frac{\partial G}{\partial T}\right) = \frac{\partial S_A}{\partial x_A} = \bar{S}_A, \qquad (1.269b)$$

$$\frac{\partial \mu_A}{\partial p} = \bar{V}_A \qquad (1.269c)$$

and for an ideal mixture, when equation 1.37 is valid,

$$\frac{\partial \mu_A}{\partial x_A} = \frac{RT}{x_A} \qquad (1.269d)$$

This means that

$$d\mu_A = -\bar{S}_A dT + \bar{V}_A dp + \frac{RT}{x_A} dx_A \qquad (1.270a)$$

and by analogy

$$d\mu_B = -\bar{S}_B dT + \bar{V}_B dp + \frac{RT}{x_B} dx_B \qquad (1.270b)$$

where $\bar{S}$'s and $\bar{V}$'s are, respectively, molar entropy and molar volume.

It can be seen that in two-phase systems the equilibrium may depend on various parameters. For separation processes, it is important to know how the intensive parameters of the system (T, p, c_i) can be used for the change of phase equilibria. In the k-component system with f phases there are $k(f - 1)$ equilibrium relations 1.268 and $f(k - 1)$ concentration balance equations 1.46, and two more universal variables, temperature and pressures, to define the system completely. The smallest number v of independent variables necessary to define the state of a system is given by the **phase rule** (J. W. Gibbs, 1876):

$$v = f(k - 1) + 2 - k(f - 1) \quad (1.271a)$$

or

$$v = k - f + 2 \quad (1.271b)$$

where v is called the **variance** (number of **degrees of freedom**) of the system. The number of independent components k is given by the number of substances (s) diminished by the number of independent reactions between them (r) and number of independent relationships between the substance concentrations (m):

$$k = s - r - m \quad (1.272)$$

In the two-phase system of ideal binary mixtures ($f = 2, k = 2$) the variance is $v = 2$ (T and p, or T and x_A, or p and x_A). For instance, if the system consists of three substances ($s = 3$), A_1, A_2 and A_3 connected via reaction 1.126 ($r = 1$), the number of independent components is generally $k = 3 - 1 = 2$ and in the one phase system the variance is $v = 3$. However, if the system is formed with an equivalence of reactants (see equation 1.136), $m = 1$ and $k = 3 - 1 - 1 = 1$ and the variance decreases to unity. Further, if two actual initial concentrations of interacting components are estimated, according to mass balances 1.130 and 1.130b ($m = 2$) the system behaves as an univariant phase ($k = 0, v = 1$), where either temperature or pressure shifts the chemical equilibrium between the interacting species (e.g. *J.E. Ricci, 1951*).

Number 2 on the right hand side of the phase rule equation 1.271 is connected with two universal parameters, temperature and pressure. This number increases with increasing parameters (external fields, surface tension etc.) which exert an influence on the system. Conversely, in condensed systems, pressure plays a minor role on phase equilibria and the practical **variance of condensed systems** is

$$v_c = k - f + 1 \qquad (1.273)$$

In separation processes the mass balances are evidently controlling factors and the total variance (V) includes the masses of individual f phases as extensive parameters, so that

$$V = v + f = k + 2 \qquad (1.274)$$

For example, for two miscible liquids ($k = 2$) the liquid-vapor equilibrium ($f = 2$) can be controlled by $v = 2 - 2 + 2 = 2$ parameters (temperature and pressure) and the total variance $V = 2 + 2 = 4$ includes the masses of both liquid and vapor phases. The maximum number of phases for the two-component system is, at zero variance, $f = 2 + 2 = 4$ (e.g. gas, liquid and two solids).

According to the number of degrees of freedom, there can be distinguished **invariant** ($v = 0$), **univariant** ($v = 1$), **bivariant** ($v = 2$) or **trivariant** ($v = 3$) **systems**. The greater the number of components, the greater is the variance of the system. The components which arise due to interactions of other components do not change the variance, as is seen from equation 1.269a. Because biphasic systems of binary mixtures are bivariant, temperature and pressure are the universal and sole controlling parameters in **physical separation processes** (distillation, crystallization) (*T.P. Carney, 1949; C.S. Robinson, E.R. Gilliland, 1950; R.S. Tipson, 1950; S.A. Bagaturov, 1961; J.W. Mullin, 1972; H.Z. Kister, 1992; K. Sattler,*

H.J. Feindt, 1995). If the separation is to be controlled by chemical composition, a third component should be added, see section 1.3.2. For instance, in the presence of the third component, whether inert ($k = 3$) or interacting with the two other components, when two new compounds, e.g. solvates, or complexes appear (the number of independent components is still $k = 5 - 2 = 3$) the systems become trivariant and the third component also plays its controlling role. This is the principle of **chemical separation processes**, e.g. extractive distillation, precipitation, etc. (*E.J. Henley, J.D. Seader, 1981; S. Kulprathipanja, 2002*).

The phase rule 1.271 provides the most universal picture of separation processes, and this is both its strength and weakness at the same time. Usually, for the two phases used in most of separation systems, the rule is reduced to the **equivalence of variance and number of independent components** in biphasic separation systems:

$$v^l = k \qquad (1.275)$$

i.e. the concentrations of the components in one phase determine the concentrations in the second phase. The total variance is higher by two units, which are the masses of the two phases.

The actual behavior of multiphase systems is described both analytically or by **phase diagrams** expressing the equilibrium concentrations and other variables of the systems in two-or three-dimensional (3D) plots.

The simplest phase equilibrium in separation systems is the vapor-liquid (*E. Hala et al., 1967*) or solid-liquid (*R. Haase, H. Schonert, 1969*) equilibrium. Let us consider first the binary mixtures, which consist of one volatile or crystallizing solvent (S) and the second component (A) exists only as dissolved in the solution (e.g. salt in water). If the liquid phase is phase I, this means that $x_A^{II} = 0$ and $x_S^{II} = 1$. At constant pressure $dp = 0$ and, with respect to equation 1.270, the equilibrium condition 1.268 for the solvent is

$$-S_S^I dT + \frac{RT}{x_S^I} dx_S^I = -S_S^{II} dT \qquad (1.276)$$

At equilibrium ($\Delta G = 0$) the molar entropy difference is equal to the latent energy of solvent phase change (evaporation, melting),

$$(S_S^{II} - S_S^I)T = \Delta \overline{H}_S \qquad (1.277)$$

and, because for dilute solutions ($x_A \ll 1$) $dx_S/x_S = d \ln x_S = d \ln (1 - x_A) \cong x_A$, the equation 1.276 is

$$dT = \frac{RT^2}{\Delta \overline{H}_S} dx_A^I \qquad (1.278)$$

or, in integral form (when $T_1 T_2 \cong T^2$) it gives the change in temperature of the phase equilibrium in the presence of solute A:

$$\Delta T = \frac{RT^2}{\Delta \overline{H}_S} \Delta x_A^I \qquad (1.279)$$

For vapor-liquid equilibrium, T is the boiling-point ($T = T_b$) and $\Delta \overline{H}_S$ is the molar enthalpy of vaporization of solvent. Because the latter value is positive, the difference ΔT represents the elevation of the boiling point at a concentration of the involatile component, x_A^I. For solid-liquid equilibria, T is the melting (freezing) point temperature ($T = T_m$) and $\Delta \overline{H}$ is the molar enthalpy of freezing for the solvent. Now the difference ΔT gives the lowering of solvent freezing point by a concentration of the solute x_A.

As regards to freezing points, the product $RT^2/\Delta \overline{H}_S$ is mostly under 30 K and for the boiling point it falls in the range 50-100 K. For ideal solutions, these shifts are equal for any solute, and the activity coefficients of non-volatile compounds can be derived if this is not so (C.A. Eckert et al., 1981). This situation is illustrated by the **temperature-pressure phase diagram** (Figure 17). The diagram will be of a similar character for one-or quasi-one-component (S) systems. The presence of an involatile and/or non-crystallizing component (A) is expressed by the shift of the line

b, giving the pressure dependence of the melting point, and line c that is the vapor pressure above the solution. The application of these relations in separation science comes into consideration on **freezing-out** the solute or **matrix evaporation**.

If the solute is volatile or crystallizes together with the matrix, **isobaric temperature-concentration diagrams** are constructed in the coordinates: boiling temperature versus mole fraction (*E. Hala et al., 1967; M. Hirata, S. Ohe, K. Nagahama, 1976*).

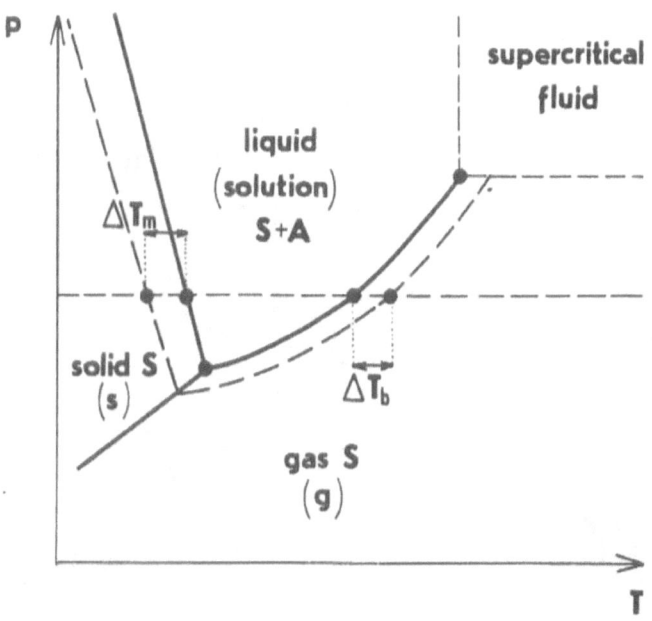

Figure 17. Temperature-pressure $(T-p)$ phase diagram of pure matrix S and solution (dotted lines) of non-volatile and non-crystalizing solute A in S.

A simple temperature-concentration diagram is presented in Figure 18. It corresponds to the coexistence of two homogeneous phases (liquid-vapor or liquid-solid). Lines I and II give the composition of phases I and II respectively at a given temperature T. For vapor-liquid equilibrium the lines are called the **liquid** and **vapor curves** respectively. and for solid-liquid equilibrium their names are **solidus** and **liquidus**.

Assuming the vapor phase to be ideal and the deviations from Raoult's law 1.78 are included in the activity coefficients of the liquid phase, the equilibrium composition is given by the equality:

$$x_i^{II} p = p_i = \gamma_i^I x_i^I p_i^0 \qquad (1.280)$$

where p is the total pressure. For binary system we have the ratio

$$\frac{x_A^{II}}{x_B^{II}} = \frac{x_A^I p_A^0 \gamma_A^I}{x_B^I p_B^0 \gamma_B^I} \qquad (1.281)$$

The ratio p_A^0 / p_B^0 is a constant, the **relative volatility** of the pure components (T. Boublik, V. Fried, E. Hala, 1973).

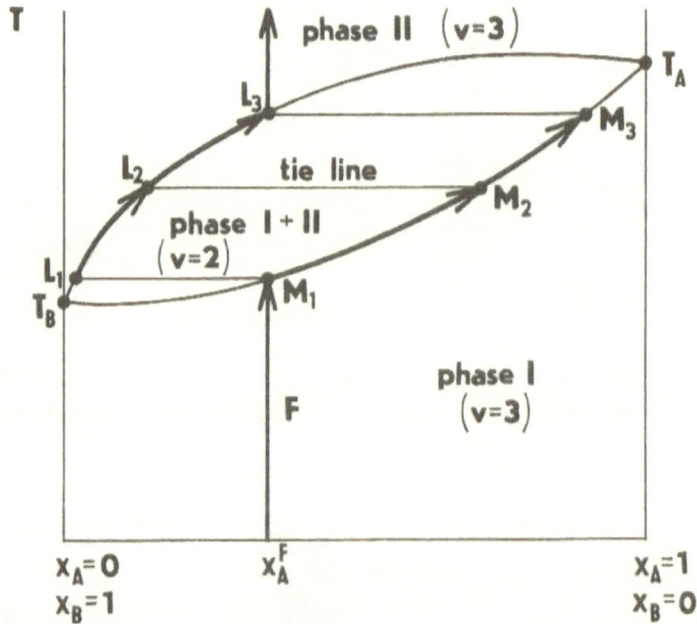

Figure 18. Temperature-concentration diagram for two-phase (I,II) system. The variance is either 3 or 2 (combining T and x at given p). Separation starts at M_1 and ends at L_3, or vice versa.

The line connecting the points L-M of equilibrium phases is called a **tie line** or **conode** and the coexisting equilibrium phases are called **conjugate phases**. This particular diagram shows that phase II is always richer in component B which has a lower boiling (melting) point ($T_B < T_A$). When the feed mixture has the composition indicated by line F, its intensive separation starts by evaporation at the boiling temperature, T_1. (Actually, separation occurs at any temperature, the liquid being in equilibrium with the gaseous phase saturated by its vapor; however, then the air or other gas should be considered as the next component present, or a new diagram, corresponding to the partial vapor pressure should be constructed.). If it is the melting process, the separation begins at the melting temperature T_1. At this point only evaporation will be discussed, fusion being completely analogous if there are not several solid phases.

Thus, at temperature T_1 an infinitesimal amount of mixture is evaporated, the vapor phase having the composition given by point L_1 and the liquid the composition $M_1 \cong K$. The liquid becomes slightly enriched by the less-volatile component A, its composition shifts to the right and a new equilibrium is established. At any moment of evaporation the mass balance for the total molar amount of mixture is

$$n^F = n^I + n^{II} \tag{1.282a}$$

and the same is valid for any component, e.g.

$$x_F^A n^F = x_A^I n^I + x_A^I n^{II} \tag{1.282b}$$

From equations 1.282 it can be shown that

$$\frac{n^I}{n^{II}} = \frac{x_A^I - x_A^F}{x_A^F - x_A^{II}} \tag{1.283}$$

which is called the **lever rule** for the estimation of the amount of phase from the segments $x_A^I - x_A^{II}$ and $x_A^F - x_A^{II}$ of the tie line at the given temperature. If evaporation is performed without removal of vapor, the process is finished at the temperature T_3 when the composition of the vapor is the same as that of the original mixture $(x_A^{II} = x_A^F, n^I = 0)$. Further heating leads only to an increase of temperature. It is seen for this closed system that only poor separation, completely limited by the relative volatility p_A^0/p_B^0 of the components, can be achieved. **Distillation** in which the vapor is removed continuously will be described later in section 2.1.2.

The actual diagrams of **solid-liquid equilibria** (*R.M. Garrels, C.L. Christ, 1965; R. Haase, H. Schonert, 1969; J. Nyvlt, 1977*) are usually more complicated than those given in Figure 18, because the components in the solid phase are rarely fully isomorphous and do not form solid solutions over the whole concentration range. Often pure solids freeze out from solution (e.g. the crystallization of individual salts from aqueous solutions).

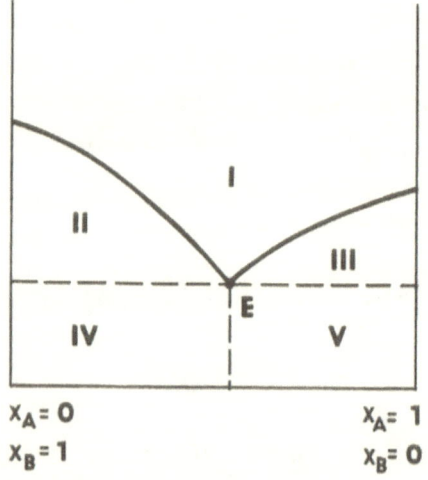

Figure 19. Typical solid-liquid diagram; I -liquid system, II, III — solid-liquid systems, IV, V — solid systems, E — eutectic.

Sometimes the diagram resembles that in Figure 19. This is the region of the liquid mixture A + B(I) and at the composition given by line E, has the same composition as the solid phase: the mixture of this composition is referred to as a **eutectic**. Region

II corresponds to the equilibrium of the liquid with component B; area III corresponds to the coexistence of the liquid phase with the pure component A. The line separating area I from II and III is the liquidus. Region IV belongs to the coexistence in the solid phase of component B and the eutectic, and in region V we have the eutectic with component A in excess.

Conversely, if there is a compound $A_{v_1}B_{v_2}$ existing both in the solid and liquid phases, the melting diagram has a maximum melting temperature at a given composition. On the mole fraction scale such maxima appear at x corresponding to the ratio of integers $v_1 : v_2$, e.g. for the mixture of Mg and Sb, at the ratio 3:2, and for benzoyl chloride — aluminium chloride at a figure 1:1.

If the components A and B are isomorphous (*E. Mitscherlich, 1818; D.I. Mendeleev, 1856; H. Grimm, 1924; V.M. Goldschmidt, 1927*) i.e. they can form common crystal lattices over a wide range of concentrations, e.g. K-Rb, albit $NaAlSi_3O_8$-anortite $CaAl_2Si_2O_8$, $BaSO_4 - KMnO_4$, $LiCl - MgCl_2$, $CH_3COOH-CHCl_3$, etc., the melting lines split to solidus and liquidus, because the conjugate phases, the liquid and solid phases have different compositions.

Special problems arise in the separation of mixtures that behave like ideal solutions: the eutectics, for instance, have the same compositions in the liquid and solid phase and no separation can be achieved on their melting or freezing. This is also true of vapor-liquid equilibria, where the mixtures are called **azeotropes**. Their composition (*R.H. Ewell, J.M. Harrison, L. Berg, 1944; L.H. Horsley, 1952; J. Gmehling, 1994; A.K. Frolkova, 2010*) on distillation can be changed only at a different pressure or by the addition of a third component (section 2.2). Examples are ethanol-water (at 4.4 wt % of water) and nitric acid- water (at 31.6 wt % and 11.9 wt % of water). If the pressure is decreased, e.g., to one-tenth of atmospheric pressure, the HNO_3-H_2O azeotrope has a slightly higher water content, 12.6 wt % and, in principle, a small part of the concentrated nitric acid can be separated from the original azeotropic mixture.

A particular case of liquid-vapor equilibrium is the **two-liquid-vapor equilibrium**. The vapor pressure above two immiscible liquids is the sum of the vapor pressure of the two pure liquids. Applying eqn. 1.281 for the case $x_A^I = x_B^I = 1$ and $\gamma_A^I = \gamma_B^I = 1$ the ratio of mole fractions in vapour phase is kept constant,

$$\frac{x_A^{II}}{x_B^{II}} = \frac{p_{II}^0}{p_B^0} \qquad (1.284)$$

Thus, in **condistillation** it is possible to evaporate and distil a substance of low volatility in the presence of an immiscible high-volatile component (water, inert gas) in the ratio given by the relative volatility, 1.284 (independent of the composition of the condensed phase). **Steam distillation** proceeds by blowing steam through a separated liquid, the vapors of which form a separate phase in the condensate. If further volatile compounds are present, their relative content in the distillate changes as in normal distillation, i.e. steam distillation is equivalent to distillation at low pressure (vacuum distillation). For example, in mixtures of aniline (A) and water (B), the pressure ratio at 100°C is $p_A^0/p_B^0 \approx 6.5/100$ (in kPa) and in the distillate there appears as much as 0.06 mole of aniline per mole of water.

The formation of a **third liquid phase** is typical for highly polar ammonium salt solutions in aqueous-organic extraction systems and it is suppressed by addition of higher alcohols (*A. V. Ochkin, 1980*).

To represent **three-component phase diagrams** both spatial (3D) diagrams or plane rectangular and triangular coordinates are used (*D.R.F. West, 1982*). In rectangular coordinates the mole fractions of the two components at constant temperature and pressure are plotted on x and y axes, because the content of the third component is obtained from the balance $x_A + x_B + x_C = 1$. However, if the content of the component is very low, other coordinates are used. The coordinates $Y = \frac{x_C^{II}}{x_B^{II}}$

vs. $X = \frac{x_C^I}{x_A^I}$ represent the **distribution isotherm** of component C between the matrix A in phase I and matrix B in phase II (see the next section). Analogous coordinates in the bilogarithmic scale, i.e. log Y vs. log X, have also been proposed (*D.B. Hand, 1930*). To present the phase diagram mixture A + C, **Jänecke coordinates** (*E. Janecke, 1906*), $Y = x_B/(x_A + x_C)$ vs. $X = x_C/(x_A + x_C)$ are widely used because they are as instructive as triangular coordinates but easier to construct.

A symmetrical in-plane representation for a three component phase diagram is given by equilateral **triangular coordinates** (*J. W. Gibbs, H.W.B. Roozenboom, 1892*). At constant pressure and temperature, the distance from the side opposite to the apex representing a pure component (A, B, C) gives the mole fraction of the component in the mixture. This parameter can also be projected on one of the non-perpendicular sides of the triangle (Figure 20).

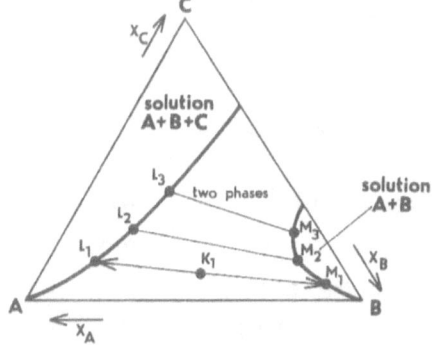

Figure 20. Triangular diagram of partially miscible (pair B + C) or immiscible (pair A + B) liquids.

The diagram in Figure 20 represents an example of a system of two pairs of partially-miscible liquids (A + B and +C), one pair (A + B) being immiscible in the absence of the third component. The region between the lines L and M corresponds to the exfoliation into two conjugate phases, the saturated solutions, the amounts of which are given (for an initial composition given by point K) by the lever rule mentioned above, 1.283. Usually the tie lines are not parallel and have to be constructed experimentally and by extrapolation. Sometimes the conodes have a pole, i.e. they intercept at one point which indicates structural regularity e.g. compound formation (*A. V. Nikolayev, I.I. Yakovlev, Yu.A. Dyadin, 1967*). For the composition corresponding to complete miscibility of all three components, the tie line shrinks to a single **critical point**, the "**plait point**", when the interface disappears.

The graphical presentation of conjugate phases is an important tool for engineering calculations of multifold separation when the compositions of the phases change during the process (section 1.3.2 and 2.2.1).

1.3.2 Chemical equilibria and distribution in biphasic systems

A particular type of phase diagram is the activity diagram which represents the composition of a system of chemically interacting components and their products as a function of their thermodynamic activity. For instance, for equilibrium 1.126, the linear dependence,

$$\log a_3 = \log K^0 + v_1 \log a_1 + v_2 \log a_2 \qquad (1.285)$$

corresponds to the coordinates $\log a_3$ vs. $\log a_1$ or $\log a_2$, the slope being the stoichiometric coefficients v_1 or v_2. When the product of reaction (A_3) is a pure substance, e.g. solid crystals, its activity is unity and $\log a_3 = 0$. These diagrams are useful for precipitation and solid-solution equilibria (*R.H. Garrels, C.L. Christ, 1965; E. Rolia, 1974; J. Celeda, S. Skramovsky, J. Zilkova, 1984*). This type of diagram is recommended when the activities are directly measurable. Thus, for oxidation-reduction processes combined with protonation,

$$v_1 Ox_A + v_2 Red_B + v_H H^+ + v_e e^- \rightleftarrows v_3 Red_A + v_4 Ox_B \qquad (1.286)$$

where Ox and Red are the oxidized and reduced forms of substances A and B respectively, both the activity of hydrogen ions and electrons can be obtained from measurements of pH = $-\log a_{H+}$ (e.g. by the glass electrode) and electromotive force with an inert electrode (section 1.2.2.). According to the Nernst equation 1.122, the potential of the solution is

$$E = E^0 - \frac{RT}{v_e F} \ln \frac{(a_{RedA})^{v_3}(a_{OxB})^{v_4}}{(a_{OxA})^{v_1}(a_{RedB})^{v_2}(a_{H+})^{v_e}} \quad (1.287)$$

If the potential is measured at $25°C$ ($RT \ln 10/F = 0.0591\ V$) then

$$E = E^0 + \frac{0.059}{v_e}[\log \frac{(a_{OxA})^{v_1}(a_{RedB})^{v_2}}{(a_{RedA})^{v_3}(a_{OxB})^{v_4}} - v_H pH] \quad (1.288)$$

Construction of the diagrams was proposed for metal corrosion systems in the coordinates E vs. pH (*M.G.N. Pourbaix, 1963*). **Pourbaix diagrams** represent the area of coexistence of several components in oxidation- reduction-hydrolysis equilibria (Figure 21).

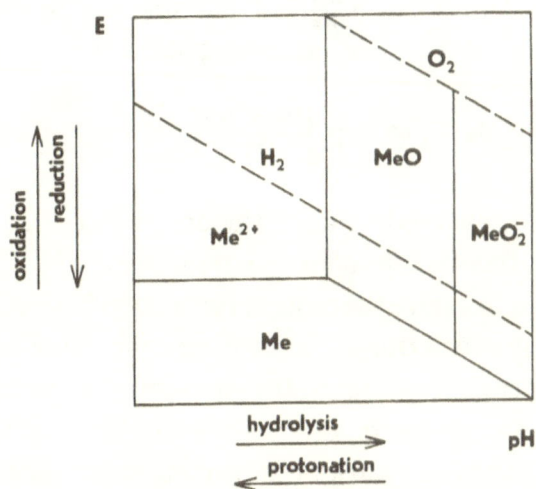

Figure 21. Schematic Pourbaix diagram.

The horizontal lines ($v_H = 0$) correspond to equal activities of the oxidized and reduced forms: in the area above the line the activity of the oxidized form prevails and at more negative potentials the reduced form is dominant. If the oxidation is accompanied by hydrolysis, as in the reaction Me $+H_2O - 2H^+ + 2e^- \rightarrow$ MeO, the line has the slope v_H (in the latter case $v_H = -2$). The vertical lines ($v_e = 0$) indicate the protonation (hydrolysis) equilibria when the activities of the protonated and dissociated forms are equal. In aqueous solutions at sufficiently low potentials, gaseous hydrogen starts to develop and at high potentials oxygen is generated. Hence, the lines of H_2 and O_2 evolution give the practical limits of the processes at atmospheric pressure. The position of the oxidation-reduction equilibrium line depends on the concentration of oxidized form in solution; at lower concentrations it shifts to a more negative potential. According to 1.288, the decrease of concentration for one order is equivalent to a shift for 59 mV for one-electron or of 29 mV for two-electron reduction processes. Pourbaix diagrams are used in hydrometallurgical processes: a drop in the emf of about $80 - 100$ mV is an indication of good recovery of a metal ion (a chemically-bound or reduced form) from solution (*H.T.S. Britton, 1928; F. Habashi, 1969; G.M. Ritcey, A. W. Ashbrook, 1979*). For example, on the electrolytical reduction $Me^{z+} + ze \rightarrow Me$, according to 1.288,

$$[Me^{z+}] \propto \exp \left[\frac{(E-E^0)zF}{RT}\right] \qquad (1.289)$$

the reversed value of the activity coefficient γ of the ion being the coefficient of proportionality.

To derive the real concentration from potentiometric data, the **Gran functions**, substantiated for redox, acid-base, precipitation or complex-ometric titrations are well established (*G. Gran, 1952, D. Dyrssen, et al. 1968*). They combine both equilibrium data (pH, E) and the total amounts of interacting components (volumes of solutions). The total amounts of system components are often more easily (analytically) available than the equilibrium data. In this way the diagrams for **precipitation reactions**, i.e. the

solubility curves or **surfaces** are constructed (*N.I. Stepanov, 1936; J. Klas, 1970*). If the precipitation reaction is

$$mM + nL \rightleftarrows (M_mL_n)_S \qquad (1.290)$$

the condition of equilibrium is

$$\mu_S = m(\mu_M^0 + RT \ln a_M) + n(\mu_L^0 + RT \ln a_L) \qquad (1.291)$$

Because the chemical potentials of solid components, μ_S, μ_M^0 and μ_L^0 are constant, the product

$$(a_M)^m (a_L)^n = K_S^0 \qquad (1.292)$$

is the thermodynamic constant, the **solubility product**. Because

$$K_S^0 = f_M^m f_L^n [M]^m [L]^n \qquad (1.293a)$$

or, if M_mL_n is an electrolyte, then

$$K_S^0 = f_\pm^{(m+n)} [M]^m [L]^n \qquad (1.1293b)$$

(see eqn. 1.102) and the **concentration solubility product** (*S. Lewin, 1960*)

$$K_S = [M]^m [L]^n \qquad (1.294)$$

corresponds to the definition of an effective constant 1.165. In accordance with equations 1.129 and 1.104, it should include the degree of dissociation of M_mL_n and the activity coefficients of M and L (*M.I. Shakhporonov, 1951; V. V. Rachinskii, L.A. Zhukova, 1978; N.P. Komar, 1978*).

The solubility products of many metal hydroxides, sulphides and phosphates are very small and their values should be considered critically (*M. Haissinski, 1969; Yu.P. Davydov, 1978*).

Typical solubility diagrams are given in Figure 22. The real process of **precipitation** in the vicinity of the solubility curve may be rather slow, and super saturation of the solution may occur at high solubilities, but larger crystals are formed (A.S. Myerson, 2001. The insoluble compounds are readily separated by filtration or by inhibited diffusion in the gel media used for micro amounts of **bioprecipitates** (*H. Heidelberger, F.E. Kendall, 1935; J. Oudin, 1946; G. Mancini, A.O. Carbonare, J.T. Heremans, 1965*). The solubility curves are limited with asymptotes when dissociation in solution is incomplete (Figure 22, curves 1, 2), otherwise each concentration can be reduced to any extent by an increase of the second interacting component (curve 3), the limits being their own solubilities. However, the solubility may increase when some side reaction takes place that produces more soluble products (the upper part of curve 2 corresponds to formation of soluble ML_2 as a

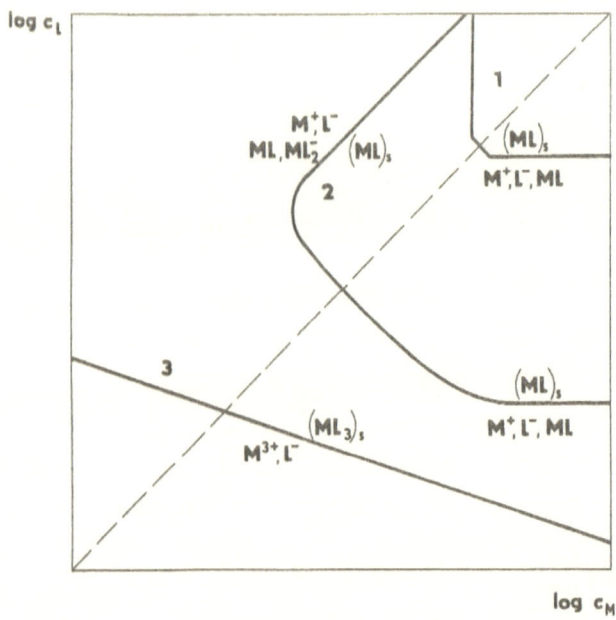

Figure 22. Solubility curves (broken line corresponds to solutions of pure ML_n compounds, when $mc_L = nc_M$) for different K_S values, $K_S(1) > K_S(2) > K_S(3)$. Solid phase region starts from the upper right corner of the diagram.

product). Such soluble complex formation in an excess of one reactant is typical both for inorganic and organic systems, charged metal complexes or biospecific complexes with higher antigen-antibody ratios being examples.

The precipitation isotherm in the case of a small solubility product value K_S or high values of m and n is rather steep (*V. V. Rachinskii, A.A. Lurie, 1963*). Equilibrium 1.290 can be applied to the formation of ionic micelles (*R. Zsigmondi, 1922*) from polyelectrolytes (ionic detergents, phospholipids, etc.) (*S.A. Rice, M. Nagasawa, 1961; A.S. Kertes, G. Markovits, 1968*). Because typical values of m and n exceed 50-100 in this case, the **micellization curves** (analogous to precipitation curves) are practically linear on the bilogarithmic scale; the **critical micelle concentration** (CMC) is defined as the characteristic in this case (*P. Mukerjee, K.J. Mysels, 1971*). It is supposed that no intermediate species (oligomers) arise during M_mL_n formation. For small micelles the law of mass action is considered,

$$K^m = \frac{[M_mL_n]_{mic}}{[M]^m[L]^n} \tag{1.295}$$

Because of large m-values, the change in micelle activity is insignificant (large micelles represent a separate phase of constant activity) and the dependence of the CMC for the solution of L in the presence of M can be derived as

$$\log \text{CMC} = \text{const} - \frac{m}{n} \log c_M \tag{1.296}$$

The CMC is the solubility limit for a monomer, above which the micelles appear and grow slowly, changing their shape from spherical to rod-like; finally, the micro heterogeneous solution separates into two phases, a true liquid phase and liquid crystal or hydrated solid. The solubilizing action of micelles (incorporation of lipophilic molecules into the hydrocarbon core of micelles) may be both a positive or negative factor in separation processes, e.g. by the interaction of biopolymers with detergents (*A.S. Jones,*

1953; D. Guerritore, L. Bellelli, 1959; F. Sebba, 1982; K.E. Goklen, T.A. Hatton, 1985).

More complex, step-wise equilibria will be described in further detail and also in section 2.2, because the solubility diagrams do not provide exact information on the yield of product formation.

Precipitation can be also achieved by a change of solvent properties, e.g. due to a decrease in dielectric permittivity in mixed water-organic solvents. The solubility of non-electrolytes diminishes in the presence of electrolyte and the salting-out effect (see equation 1.218) is described by the empirical **Setchenow equation** (*J. Setchenow, 1889*) relating the activity coefficient of the solute γ_L, or solubility S, and electrolyte concentration, c_M:

$$\log \gamma_L = \log \frac{S^0}{S} = k_S c_M \qquad (1.297)$$

where k_S is the salting constant ($k_S \approx 0.1 \div 1$ dm^3 mol^{-1}) which is proportional to the hydration number of the electrolyte and the molar volume of the non-electrolyte (*M. Dixon, E.C. Webb, 1961*). Ammonium sulphate or sodium sulphate ($1 \div 10$ M) is usually applied for the precipitation of proteins. For example, for the salting out of bovine serum albumins with ammonium sulphate we have $k_S = 7.65$ mol^{-1} dm^3 and because $\ln S_0 = -21.6$, $\ln S = -21.6 - 7.65$ $c_{(NH_4)_2SO_4}$, i.e. one-molar ammonium sulphate decreases its solubility by 7 orders of magnitude.

Hitherto, a system has been considered in which the interacting components form individual phases on **electrodeposition** or **precipitation**. Very often the separated components are in such small amounts that no macroscopic phase arises and the solids remain in colloidal solution (*P. Benes, V. Majer, 1980*), or as a film along the liquid-liquid interface (*S. Winitzer, 1973; S. V. Chizhevskaya, et al., 1977; G.A. Yagodin, V. V. Tarasov, S. Yu. Ivakhno, 1982*), or at the liquid-solid interface, which is well known from immunoassay precipitation reactions (*S.A. Berson, R.S. Yalow, 1959; R.S. Yalow, S.A. Berson, 1960*). Beside micellar solutions (*A.S. Kertes, H. Gutmann, 1975*), such micro heterogeneous systems are obvious in practice but in principle final mechanical separation is possible by ultracentrifugation and ultrafiltration —see Sections 1.1.1 and 1.4.3.

In many modern separation techniques, however, the separated component is not assumed to form an individual phase. In such methods as **adsorption** and **chemisorption** (*e.g. O. Hahn, 1926; R. Kaiser, 1963; S.J. Gregg, K.S.W. Sing, 1967; E.J. Fuller, 1973; T. Braun, J.D. Navratil, A. Farag, 1985*), **cocrystallization** (*e.g. E. Rolia, 1974; I. V. Melikhov, M.S. Merkulova, 1975*), **ion exchange** (*e.g. Y. Marcus, A.S. Kertes, 1969; F. Helfferich, 1962; O. Samuelson, 1963; W. Rieman III, H.F. Walton, 1970; J.X. Khym, 1974; K. Dorfner, 1991*) and **solvent extraction** (*e.g. R.E. Treybal, 1951; G.H. Morrison, H. Freiser, 1957; Z. Ziolkowski, 1963; J. Stary, 1964; C. Hanson, 1968; G.M. Ritcey, A.W. Ashbrook, 1979; G.A. Yagodin, S.Z. Kagan, V.V. Tarasov, 1981; J.Rydberg, C.Musikas, G.R.Choppin, 1992*) two-phase systems are formed by the macro components playing the role of matrix (carrier), weakly-interacting with the components to be separated. Separation in this case is based on the distribution of the minor components between the matrices of the two phases that serve as physical carriers in the step of mechanical separation. Chemical reactions can be used to enhance the distribution due to formation of a species with specific properties with respect to affinity towards one specific phase (i.e. with a certain molecular size, geometry, conformation, charge, dipole moment, hydrophobicity, volatility, etc.).

The condition of equilibrium distribution between two phases was formulated in general in sections 1.1.2. and 1.3.1. by the equality of chemical potentials — equations 1.38 and 1.268. If a pure substance is chosen as the standard state for each phase (Convention I), then

$$\mu_i^{I0} = \mu_i^{II0} = \mu_i^*, \tag{1.298}$$

(compare with eqn. 1.88) and absolute activities are also equal at interphase equilibrium

$$a_i^I = a_i^{II} \tag{1.299}$$

For ideal solutions ($\gamma_i = 1$) it becomes

$$x_i^I = x_i^{II} \tag{1.300}$$

which means an entirely random distribution of the i-th component in all parts (phases) of the system occurs. In the ideal system the entropic term of the chemical potential is the moving force for each component and no mutual separation of the components can be obtained.

For real solutions

$$\frac{x_i^I}{x_i^{II}} = \frac{\gamma_i^{II}}{\gamma_i^I} \tag{1.301}$$

and this ratio is generally not constant. When infinitely-dilute solutions are chosen as standard states (Convention II), the equality 1.38 gives

$$\dot{\mu}_i^I + RT \ln \dot{a}_i^I = \dot{\mu}_i^{II} + RT \ln \dot{a}_i^{II} \tag{1.302}$$

and

$$\frac{\dot{a}_i^{II}}{\dot{a}_i^I} = \exp\left(\frac{\dot{\mu}_i^I - \dot{\mu}_i^{II}}{RT}\right) \tag{1.303}$$

The right-hand side of the equation is a thermodynamic constant which is called the **distribution constant**:

$$K_i^D \equiv \exp\left(\frac{\dot{\mu}_i^I - \dot{\mu}_i^{II}}{RT}\right) \tag{1.304}$$

Now the ratio, eqn. 1.303, gives

$$\frac{x_i^{II}}{x_i^{I}} = \frac{\gamma_i^{I}}{\gamma_i^{II}} K_i^D \tag{1.305}$$

Dividing eqns. 1.305 by 1.301 and considering relation 1.97,

$$K_i^D = \frac{\gamma(0)_i^{I}}{\gamma(0)_i^{II}} \tag{1.306}$$

i.e. the distribution constant is the ratio of absolute activity coefficients at infinitely low concentrations and it is the greater, the stronger the interaction of the distributed substance (or distribuend) with matrix II (the smaller is $\gamma(0)^{II}$) and vice versa (R. Collander, 1949). In such solutions where Henry's law is obeyed ($\dot{\gamma} = 1$) for the distributed species in each phase, eqn. 1.305 may be written as

$$\frac{x_i^{II}}{x_i^{I}} = K_i^D \tag{1.307}$$

or, on the molar scale

$$\frac{c_i^{II}}{c_i^{I}} = \frac{\bar{V}_I}{\bar{V}_{II}} K_i^D \tag{1.308}$$

The ratios 1.307 and 1.308 differ by the ratio of the molar volumes of matrices I and II, $\bar{V}_I$ and $\bar{V}_{II}$ when no other component is present — see equation 1.55.

Usually the distribution constants given in the literature refer to the molar scale and they implicitly include the ratio $\frac{\bar{V}_I}{\bar{V}_{II}}$. However, when the data for distribution between various matrices need to be compared, the mole fraction scale 1.307 is preferred. For instance, if the distribution of a substance between water ($\bar{V}_I = 0.018$ dm^{3}mol^{-1}) and benzene ($\bar{V}_{II} = 0.078$) is on the mole fraction scale, $K^D = 1$ (ideal or pseudo-ideal solution system) and it is the same for the distribution in the water-dodecane ($\bar{V}_{II} = 0.17$) system, these will be $K^D = 0.23$ and 0.106

in molar concentration scale respectively; however, this does not prove stronger interactions in phase I.

The distribution according to 1.308, as found theoretically and experimentally (*M. Berthelot, J. Jungfleisch, 1872*) and was substantiated by W.Nernst (*1904*), is called the Nernst or **Berthelot-Nernst** distribution law: the distribution constant in dilute solutions does not depend on the concentration of the distribuend, the equilibrium concentration in one phase changing linearly with the concentration in the second phase. The distribution coefficient can be predicted theoretically (*R. Zahradnik, P. Hobza, Z. Slanina, 1973*) and linked to solubility parameters (*H. Buchowski, 1962; D.F. Noel, C.E. Mellon, 1972*), molecular surface data (*E. V. Komarov, 1970; G.L. Amidon, et al. 1975*), Hammett substituent constants (*C. Hansch, et al., 1963, 1964*), electronegativity (*A.M. Rozen, Z.I. Nikololova, 1964*) and solvent polarity (*Yu.G. Frolov, V.V. Sergievskii, 1971, 1972*).

The ratio of distribuend concentrations is the most important characteristic for the separation process and is defined as the **distribution ratio**

$$D_i = \frac{c_i^{II}}{c_i^{I}} \tag{1.309a}$$

or

$$D_i = \frac{\bar{V}_I}{\bar{V}_{II}} \frac{x_i^{II}}{x_i^{I}} \tag{1.309b}$$

The distribution ratio, in contrast to the distribution constant K^D, depends not only on the phase composition but also on the concentration of the distribuend and, according to eqns. 1.305 and 1.55, includes activity coefficients and densities (molar volumes) as other characteristics of the solutions. The aim of the theory is to predict the distribution ratio as a function of solution composition, which is the basis for the chemical control of separation (section 2.2.).

The plot of c^{II} as a function of c^I is the **separation isotherm** or **equilibrium line** (Figure 23). Its deviation from linearity can be expressed by activity coefficients, particularly due to mass balance such as 1.139. In the initial part the isotherm is linear (region of validity of Henry's law). If it decreases with concentration, the decrease in the $\dot{\gamma}_i^I/\dot{\gamma}_i^{II}$ ratio is responsible due to positive non-ideality in phase II (or, more precisely, due to the increase of $\dot{\gamma}$ II). The opposite tendency usually occurs for positive non-ideality in phase I.

If the distribution proceeds between a bulk phase (liquid, gas) and the surface of the other phase (solid or liquid) the distribution, which is known as **positive** or **negative adsorption** (D.M. Ruthven, 1984), is expressed in another way. The amount of substance is related to the mass of the solid phase II (mol kg^{-1}),

$$N_i = \frac{n_i}{G} \tag{1.310}$$

or to the **surface concentration** (mol m^{-2}),

$$\Gamma_i = \frac{n_i}{A} \tag{1.311}$$

It should be stressed that N_i does not have the meaning of molality of a solid solution, though formally it is the same as 1.53a (the i-th component is localized on the surface

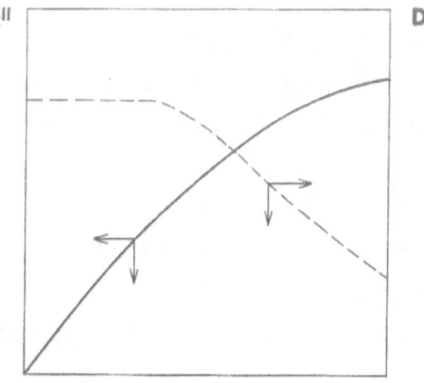

Figure 23. Separation isotherm (equilibrium line), c^{II} vs. c^I function, or distribution ratio (D) as a c^I function.

of the bulk phase) and it is still used for simplicity in presenting experimental data. The ratio

$$a = G/A \qquad (1.312)$$

is the **specific surface area** ($m^2\ kg^{-1}$) of the carrier phase which is usually the dispersed solid. Non-porous dispersed materials have a specific surface area $10 \div 10^4\ m^2\ kg^{-1}$ (*S.J. Gregg, K.S. Sing, 1967*) while for porous materials a is in the range 10^4 to $10^6\ m^2\ kg^{-1}$. On compression of the surface molecules at the water-air interface, the polar macromolecules can form **organized monolayers** (*I. Langmuir, 1925*); each linear macromolecule occupies an area about $2 \div 3 \times 10^{-20}\ m^2$ and therefore the limiting Γ ranges between 40 and 70 micromoles m^{-2}. The maximal surface concentration (Γ_∞) for low-molecular substances is also of the order $10^{-6}\ mol\ m^{-2}$ (*J.J. Bikerman, 1958; J.J. Kipling, 1965; D. Valenzuela, A. Myers, 1989*).

The solid-liquid distribution ratio can be expressed as

$$D_i^S = \frac{N_i}{c_i} = \frac{\Gamma_i a}{c_i} \qquad (1.313)$$

and has the dimensions $dm^3\ kg^{-1}$ (when c_i is a molarity).

In principle, the interaction of a neutral molecule with a non-polar surface is described by the Lennard-Jones potential 1.233. However, integrated force vectors in the neighborhood of the interacting molecule make the attractive potential dependence proportional to $1/r^3$ (instead of $1/r^6$ for isotropic molecular interactions) and therefore, the molecule-surface interaction is much stronger than molecule-molecule interaction at the same distance. **Physical sorption** is characterized with a binding energy of about $5 - 50\ kJ\ mol^{-1}$ and **chemisorption** of about $150 - 500\ kJ\ mol^{-1}$. The total interaction energy is the sum of Coulombic, induction and dispersion interactions and, in general, many-body effects are encountered (*W.A. Steele, 1974*). For example, for the adsorption of gases on zeolites the dispersion component of the interaction energy dominates for non-polar molecules (N_2, CO_2, alkanes) but dipolar interactions dominate for H_2O molecules induction interaction become important for unsaturated species such as benzene (*R.V. Golovnya, T.A. Misharina, 1980; S. Miertus, E. Scrocco, J. Tomasi, 1981; S. Miertus, I. Miertusova, 1984*). On **ionic adsorption** on crystals, Coulombic interactions and the existence of the electric double layer strongly prevail (*O. Stern, 1924; A.P. Ratner, 1933; E.J.W. Verway, 1935; R.M. Barker, R.P Townsend, 1973*) even by adsorption on neutral polymers (*P. Benes, M. Paulenova, 1973*). **Bioaffinity**, i.e. the chemical coupling of peptides and proteins, is achieved by binding active groups as cyanogen halides, to solid carriers (*R. Axen, J. Porath, S. Ernback, 1967; R. Axen, S. Ernback, 1971; F. Farron-Furstenhal, J.R. Lightholder, 1977; S. Ahuja, 1979; G. Johansson, G. Kopperschlager, P.A. Albertsson, 1983; G. Johansson, M. Andersson, 1984*). Solid phase antibodies or antigen –haptens become a predominant tool in immunoassay (*R.A. Masseyeff, W.H. Albert, N.A. Staines, 1992; J.P. Gosling, L.V. Basso, 1994*).

The chemical potential of an adsorbed molecule (ion) and its distribution coefficient is proportional to the interaction potentials. The distribution law 1.303 can be applied for the distribution:

$$\Gamma_i = c_i \frac{\gamma_i^I}{\gamma_i^{II}} \exp\left(\frac{\mu_i^I - \mu_i^{II}}{RT}\right) \qquad (1.314)$$

When the distribution of ions is considered, the difference of electrochemical potentials between the solution and solid surface enters into the exponent. As will be shown, the main problem in the practical use of such an equation is that, because of the non-homogeneous surface, both γ and μ depend on the adsorption site. The standard state μ^{II} corresponds to the interaction between a molecule and a given site for the hypothetical adsorption of 1 mole per square metre ($\Gamma = 1, a^{II} = 1$). To the first approximation, γ^{II} ($m^2 mol^{-1}$) can be considered as inversely proportional to the fraction of free surface (not occupied by adsorbed molecules), i.e. $1/\dot{\gamma}^{II} \propto -\Gamma/\Gamma_\infty$ (compare with equation 1.139) and increases with the amounts adsorbed unless the formation of the second adsorption layer is energetically more favorable.

The interaction energy has the same meaning as the **surface tension** σ (J m^{-2}) or **surface Gibbs energy**. The equation which relates the surface concentration of component i at the surface of macro component 1 was derived by J.W. Gibbs (*1876*). The excess Gibbs energy (J) appearing on contact of the two phases, which is due to a molar amount change n_i^S (mol) at the interface with total area A(m^2) is

$$G^E = \sum \mu_i n_i^S + A\sigma \qquad (1.315)$$

where σ is the interfacial tension (J m^{-2}), μ's are the chemical potentials (J mol^{-1}) of the components and

$$n_i^S = A\Gamma_i. \qquad (1.316)$$

(see e.g. *A.A. Abramzon, Ye.D. Shchukin, 1984*). By analogy with how the Gibbs-Duhem equilibrium condition 1.20 was obtained, the derivation of equation 1.315 gives

$$dG^E = \sum A \Gamma_i d\mu_i + Ad\sigma = 0 \qquad (1.317)$$

and the **Gibbs adsorption isotherm** is

$$d\sigma = -\sum \Gamma_i d\mu_i \qquad (1.318)$$

In a two-component system, when the macro component 1 does not dissolve component 2 (μ_1 = const, μ_1 = 0), the latter remains at the interface. Knowing that $d\mu_i = RTd \ln a_i$, integration of 1.318 for the two-component system gives

$$\Gamma_2 = -\frac{1}{RT}\frac{d\sigma}{d \ln a_2} \qquad (1.319)$$

If the solution of component 2 is ideal, its adsorption isotherm is

$$\Gamma_2 = -\frac{c_2}{RT}\frac{d\sigma}{dc_2} \qquad (1.320)$$

i.e. the surface concentration on Gibbs adsorption of component 2 is proportional to its concentration c_2 (mol m^{-3}) in the bulk phase and the **surface activity coefficient** $d\sigma/dc_2$ (J m mol^{-1}) (*G. Aniansson, 1951; D.K. Chattaraj, K.S. Birdi, 1984*). When the latter is negative (surface-active component 2), there is positive adsorption ($\Gamma > 0$) and vice versa. It can be used, for example, for removal of hydrophobic species and surfactants from aqueous solution when a solid phase with a large interface, as in polyurethane foam, is used (*H.J.M. Bowen, 1970; T. Braun, A.B. Farag, 1978; C.M. Smith, J.D. Navratil, 1979; T. Braun, 1983; Palagyi, S., T.Braun, 1993*). Many other separation processes are surfactant-based (*R.Lemlich, 1972; A.N. Clarke, D.J. Wilson, 1983; J.F. Scamehorn, J.H. Harwell, 1989*). If $d\sigma/dc_2 = $ const, the adsorption isotherm is linear in Γ_2 vs. c_2 coordinates. The Gibbs isotherm can be used for the establishment of interfacial concentrations based on measurements of interfacial tension (*M. Cox, M. Elizalde, J. Castresana, 1983; D. Bauer et al., 1986*). Interfacial tension isotherms can indicate reactions near organic/water interfaces (*J. Szymanowski, K. Prochaska, 1989*).

The ideal distribution between the solid and gas, or the solid and liquid phases with a linear distribution isotherm occurs rarely. The principal reasons for the deviations are:
(i) geometrical and chemical non-uniformity of the surface of the solid phase,
(ii) gradual saturation of the interfacial layer and multilayer adsorption. The first results in non-linearity of the adsorption isotherm even at small concentrations when Henry's law would be applicable.

Empirical data can be modeled by the **Freundlich isotherm** (*H. Freundlich, 1922*)

$$\Gamma = ac^b \tag{1.321a}$$

linearized in coordinates

$$\log \Gamma = \log a + b \log c \tag{1.321b}$$

where a and b are constants ($b < 1$). For gas-solid equilibria, instead of c the gas pressure is used. The isotherm reflects the properties of the surface with several classes of site (*E. Cremer, S. Flugge, 1938*) and occurs frequently for the adsorption of both non-electrolytes and ions (*Yu.B. Zeldovich, 1934; Ya.D. Zelvenskii, V.A. Shalygin, Yu.V. Golubkov, 1962; V.S. Sotnikov, A.S. Belanovskii, 1966*).

Gradual saturation of the monomolecular adsorption layer is expressed by the **Langmuir isotherm** (*I. Langmuir, 1916*),

$$\Gamma = k_1 \frac{k_2 c}{1+k_2 c} \tag{1.322c}$$

This hyperbolic isotherm can be linearized in coordinates $1/\Gamma\,(\text{m}^2\text{mol}^{-1})$ vs. $1/c\,(\text{m}^3\text{mol}^{-1})$,

$$\frac{1}{\Gamma} = \frac{1}{k_1} + \frac{1}{k_1 k_2}\frac{1}{c} \tag{1.322d}$$

which is applicable also to ion exchange (*G.E. Boyd, J. Schubert, A.W. Adam, 1947; N. Misak, 1993, 1995*). The constant product $k_1 k_2 (\text{m})$ is equivalent to the reciprocal constant in Henry's law 1.79 ($c = k_A \Gamma$) and the constant k_1 (mol m^{-2}) is identical with the maximal adsorption capacity as $c \to \infty\,(k_1 = \Gamma_\infty)$. Thus, **Henry's law** also represents a linear-type isotherm.

Modified three-parameters Freundlich isotherms generalized for multi-site adsorption were derived by **R. Sips** (1950)

$$\Gamma = k_1 \left(\frac{c}{k_2 + c}\right)^{k_3} \tag{1.323}$$

and **J.Toth** (1971):

$$\Gamma = k_1 \frac{c}{(k_2 + c^{k_3})^{1/k_3}} \tag{1.324}$$

At low concentrations the first one turns to Freundlich and the second one to Henry's isotherm respectively.

Semi empirical adsorption rules can be also derived by the law of mass action (*Yu.V. Yegorov, 1971*). For multicomponent systems (*F. Helferrich, G. Klein, 1970*) the partial value of the adsorption will be given by

$$\Gamma_i = k_1 \frac{k_{2i} c_i}{1 + \sum k_{2i} c_i} \tag{1.325}$$

Multilayer formation on the adsorption of vapor and capillary condensation was considered in the derivation of the **Brunauer-Emmett-Teller (BET) isotherm** (*P.H. Emmett, S. Brunauer, 1937; S. Brunauer, P.H. Emmett, E. Teller, 1938*):

$$\Gamma = \frac{k_1 k_2 p}{(1 + k_2 p - p/p_0)(1 - p/p_0)} \tag{1.326}$$

where p is the pressure of the vapor and p_0 is the vapor pressure over the pure liquid. The BET isotherm can be linearized in the form

$$\frac{1}{\Gamma} \frac{p}{p_0 - p} = \frac{k_2}{k_1 p_0} + \left(\frac{1}{k_1 p_0} - \frac{k_2}{k_1}\right) p \tag{1.327}$$

Because p has the same meaning as the c used previously in the Langmuir isotherm, the BET isotherm transforms to it when $\ll p_0 (p/p_0 \ll 1)$, and at $p = p_0$ bulk condensation of vapors occurs.

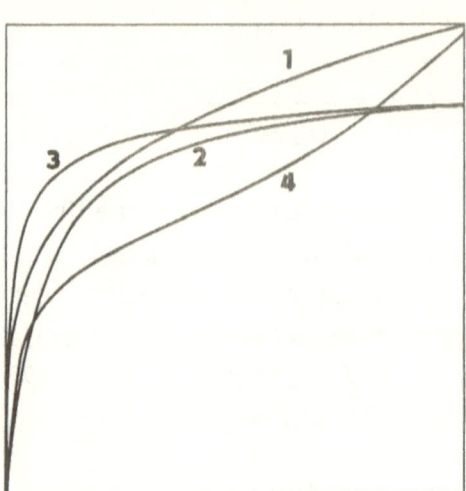

Figure 24. Types of isotherm:
1 — Freundlich,
2 — Langmuir,
3 — Langmuir-Freundlich,
4 — BET.

The isotherms are illustrated by Figure 24.

It is too difficult to use non-linear isotherms for multicomponent adsorption. For two-component systems the equilibrium approach is still possible (*C.M.Yon, P.H.Thurnock, 1971; W. Fritz, W. Merk, E.V. Schleunder, 1981*). The **Fritz-Schleunder isotherm**,

$$\Gamma_i = \frac{a_{i0}(c_i)^{b_{i0}}}{c_1 + \sum_{i=0}^{k} a_{ij}(c_j)^{b_{ij}}} \qquad (1.328)$$

was derived for a k-component system and a and b are coefficients of individual ($j = 0$) and mixed absorption isotherms.

The Langmuir isotherm is a special case of 1.321 when $b_{i0} = b_{ij} = c_i = 1$. For a single-solute system ($k = 1$), eq. 1.328a gives the **Freundlich-Langmuir isotherm**:

$$\Gamma = k_1 \frac{k_2 c^b}{1 + k_2 c^b} \qquad (1.329)$$

For three- or more component systems, the diffusion kinetic model of adsorption (*A.R. Mansour, 1986*) is easier to calculate. The linking of the sets of partial differential equations of the first order is useful for the modeling (*H.K. Ree, A. Rutherford, N.R. Amundson, 1989*).

The adsorption of ions often proceeds as the heterogeneous **ion-exchange** reaction,

$$\nu_1(A^{z_1})_I + \nu_2(B^{z_2})_{II} \rightleftarrows \nu_1(A^{z_1})_{II} + \nu_2(B^{z_2})_I$$

in which the ions A^{z_1} in phase I (solution) replace the ions B^{z_2} in phase II (ion exchanger), i.e. the equality

$$\nu_1 z_1 = \nu_2 z_2 \tag{1.330}$$

is valid. The distribution ratio of the ions is considered both in the mole fraction (rational) or molality scale and also in the hybrid scales used generally for solid-liquid distribution (reaction 1. 313). The physical meaning of the latter value (dm^3 kg^{-1}) is different however, because the important feature of ion exchangers, unlike other types of adsorbents is that they operate while swollen by a solvent and the absorbed ions enter the bulk of the exchanger (*B.A. Adams, E.L. Holmes, 1937; O. Samuelson, 1939; K. Dorfner, 1991*). Hence, the exchange capacity is of the order 1-10 mol kg^{-1}, much above the upper limit of adsorbents. Acid-insoluble inorganic exchangers also have remarkable properties in this respect (*K.A. Kraus, T.A. Carlson, J.S. Johnson, 1956; J. Krtil, 1962; V. Kourim, J. Rais, B. Million, 1964; A. Clearfield, 1964, 1982: M. Abe, K. Uno, 1979; O. Mikes, P. Strop, Z. Hostomska, 1984*). The acid-base properties of "inert" sorbents must be also buffered for liquid chromatography purposes (*H. Brockmann, H. Schodder, 1941; H. Brockmann, 1947*).

The inner solution of an ion exchanger always represents a highly concentrated electrolyte solution and cannot be treated without incorporating activity coefficients. Since the solid phase carries non-diffusible charges of exchangeable groups, the Donnan mechanism for co-ion adsorption together with counter ions takes place, see section 1.5.1 (*H.P. Gregor, 1951; F. Helfferich, 1959; J.A. Marinsky, 1976*).

The relation between the thermodynamic constant of ion exchange reaction and the ion distribution ratio can be derived for **ion exchange** (practical activities are used for exchanger phase II and molar activities for the solution, phase I):

$$K^D = \frac{(a_A^{II})^{v_1}(a_B^{I})^{v_2}}{(a_A^{I})^{v_1}(a_B^{II})^{v_2}} = Y\left(\frac{m_A^{II}}{c_A^{I}}\right)^{v_1}\left(\frac{c_B^{I}}{m_B^{II}}\right)^{v_2} \qquad (1.331)$$

i.e.

$$K^0 = Y Q_B^A \qquad (1.332)$$

where Y is the product of the activity coefficients — see 1.98:

$$Y = \left(\frac{\gamma_A^{II}}{y_A^{I}}\right)^{v_1}\left(\frac{y_B^{I}}{\gamma_B^{II}}\right)^{v_2} \qquad (1.333)$$

and

$$Q_B^A = \frac{D_A^{v_1}}{D_B^{v_2}} \qquad (1.334)$$

is called the **selectivity coefficient**; it is always arranged in such a way that, under given conditions, the ion indicated by the superscript (A) displaces the ion with subscript (B) from the exchanger, i.e. $Q > 1$. Changes in selectivity with composition are obviously due to many non-ideality factors (*Y. Marcus, A.S. Kertes, 1969*). If there are several sorption sites even when all are ideal, the isotherm is not linear (*R.M. Barrer, J. Klinowski, 1972*). Selectivity increases with charge and polarizability of the ion A and is inversely proportional to its radius of hydration (*F.W.E. Strelow, R. Rethemeyer, C.J.C. Botha, 1965; M.H. Abraham, J. Liszi, 1981; F.W.E. Strelow, 1984*). The model of ion exchange is also suitable for **co-precipitation reactions** (*A.P. Ratner, 1933; L.A. Zhukova, 1981*).

When a component exists in various chemical forms, the distribution ratio in the complex system is generally given by the sum of the concentrations of all forms in phases I and II, i.e. instead of the simple component ratio 1.309, we have

$$D = \frac{\sum c_i^{II}}{\sum c_i^{I}} \tag{1.335}$$

Such a distribution ratio certainly represents the possibility of the separation processes having preparative or analytical value. For instance, in the purification of semiconducting silicon tetrachloride, deleterious impurities of phosphorus or arsenic may exist in various forms (PCl_3, $POCl_3$ and PCl_5, or $AsCl_3$ and H_3AsO_4). Radioiodine can occur in air not only in molecular form I_2 but also as I^-, IO_3^-, HIO and alkyl iodides. A great variety of species can exist in aqueous solution and the environment due to complexation, solvation and also in redox reactions (*N. Suzuki et al., 1986; N.S. Shvydko, N.P. Ivanova, S.I. Rushonik, 1987; F. Jitoh, H. Imura, N. Suzuki, 1990; M. Lederer, 1992*). Thus iron in chloride solution may form a variety of complexes of the type $Fe_m(OH)_pCl_n$. The formation of various mixed complexes as $TaF_m(HSO_4)_n$ is typical of systems for metal separation in complex media (*R. Caletka, R. Hausbeck, V. Krivan, 1989*) and in some systems a strong **synergistic effect** of matrix components on the separation process occurs (*G.R. Choppin, 1981*).

When chemical equilibrium between coexisting species occurs, the distribution ratio can be calculated and predicted by the universal procedure described in section 1.2.3. (*N.P. Komar, 1983*). The application of one-phase mathematical formalism to two-phase systems will be illustrated using complex formation equilibria (*L.G. Sillen, 1961; F.J. Rossotti, H. Rossotti, 1961; M. T. Beck, 1970*).

Various chemical forms of a component will be indexed as individual components. Then, according to equations 1.309 and 1.335, the distribution ratio becomes

$$D = \frac{\sum D_i c_i^I}{\sum c_i^I} \qquad (1.336)$$

i.e. it is expressed entirely through the concentrations in phase I, c^I. However, because the distribution ratios of the individual species D_i with respect to non-linear distribution isotherms are also functions of the solution composition, the latter equation does not offer any advantage compared with equation 1.329 (even if it is known that many D_i's have zero values when the species occur in one phase only: e.g. ions do not exist practically in gaseous or non-polar phases, ion-exchangers contain either cations or anions, crystals are formed by isomorphous or isodimorphous ions, macro-molecules do not leave the condensed phase, etc.). The practical use of equation 1.336 becomes preferable in dilute solutions where the Berthelot-Nernst law is valid and $D_i = K_i^D$ (equation 1.307). Then

$$D = \frac{\sum_i^k K_i^D c_i^I}{\sum_i^k c_i^I} \qquad (1.337)$$

which obviously corresponds, according to the phase rule 1.275, to the variance of the distribution $v \leq k$. Because the components are not independent, the variance is decreased by each reaction between them — see discussion relating to equation 1.272. If the distributed species are the metal complexes $M_m L_n (OH)_p$, according to equation 1.197 and the following ones, the distribution constants K_i^D refer to the distribution

$$K_{mij}^D = \frac{[M_m L_n OH_j]_{II}}{[M_m L_n OH_j]_I} \qquad (1.338)$$

and the distribution ratio 1.331 for metal M converts to (J. Rydberg, 1950, 1955)

$$D = \frac{\sum K^D_{mij} \, m[M_m L_i OH_j]_{II}}{\sum m[M_m L_i OH_j]_I} \qquad (1.339)$$

The denominator of this expression can be expressed through the complexity function 1.199. The numerator is also transformed to the same form by means of the analogous function, complexity in phase II,

$$X^{II} = \sum_{m=1}^{m} \sum_{i=0}^{n} \sum_{j=0}^{p} K^D_{mij} \, \beta_{mij} [M]^{m-1}[L]^i[OH]^i \qquad (1.340)$$

Now, the constant

$$\beta^H_{mij} \equiv K^D_{mij} \beta_{mij} \qquad (1.341)$$

is a constant of a heterogeneous equilibrium, i.e. the **two-phase stability constant**,

$$\beta^H_{mij} = \frac{[M_m L_i OH_j]_{II}}{[M]_I^m [L]_I^i [OH]_I^j} \qquad (1.342)$$

A typical example of such a heterogeneous constant is the **extraction constant** for the extraction of metal M by a weak, chelating acid HL which forms the organophihc complex:

$$M^{n+} + n\text{HL} \rightleftarrows \text{ML}_n + n\text{H}^+ \qquad (1.343a)$$

It is practical to represent this reaction as heterogeneous (regardless of the real mechanism), relating the species to the phases where they dominate: ionic forms M^{n+} and H^+ in the aqueous phase (I) and neutral, organophilic forms (HL, ML$_1$) in the organic solvent, phase II:

$$(M^{n+})^I + n(HL)^{II} \rightleftarrows (ML_n)^{II} + n(H^+)^I \qquad (1.343b)$$

Therefore, the extraction constant will be (*I.M. Kolthoff, E.B. Sandell, 1941; H. Irving, T.B. Pierce, 1959; J. Stary, 1963*):

$$K_{ex} = \frac{[ML^{n+}]_{II}\,[HL]_I^n}{[M^{n+}]_I\,[HL]_{II}^n} = \frac{\beta_n^H}{\left(K_1^H\right)^n} \qquad (1.344a)$$

where

$$K_1^H = \frac{[HA]_{II}}{[H^+]_I[L^-]_I} \qquad (1.344b)$$

is the two-phase protonation analogue of constant 1.194.

When the species in the organic phase are dissociated, the extraction constant is essentially such as given by eq. 1.331 (*J. Rais, 1971*).

At the same time, the distribution ratio of metal M will be given by these equilibrium parameters:

$$D = \frac{[ML_n]_{II}}{[M^{n+}]_I} = K_{ex}\left(\frac{[HL]_{II}}{[H^+]_I}\right)^n \qquad (1.345)$$

In more general form, 1.339 becomes

$$D = \frac{\sum_{m=1}^m \sum_{i=0}^n \sum_{j=0}^p m\beta_{mij}^H [M]^m [L]^i [OH]^j}{\sum_{m=1}^m \sum_{i=0}^n \sum_{j=0}^p m\beta_{mij} [M]^m [L]^i [OH]^j} \qquad (1.346)$$

or

$$D = \frac{x^{II}(1+\langle m^{II} \rangle)}{x^I(1+\langle m^I \rangle)} \qquad (1.347)$$

where $\langle m^I \rangle$ is the average number of central atoms M in the polynuclear complexes in phase I — function 1.202 — and $\langle m^{II} \rangle$ is the analogous function defined through two-phase stability constants, 1.341.

Now, to achieve the parameters analogous to one-phase systems in biphasic systems at the phase volume ratio,

$$r = \frac{V_{II}}{V_I} \tag{1.348}$$

it is helpful to investigate mass balances. It can be shown (*F. Macasek, 1974*) that if **weighted two-phase stability constants**

$$\beta^W_{mij} = \beta_{mij}(1 + rK^D_{mij}) \tag{1.349}$$

or

$$\beta^W_{mij} = \beta_{mij} + r\beta^H_{mij} \tag{1.350}$$

are used, the balances 1.199–1.200 and functions 1.201–1.204 are isomorphous (of identical mathematical form) with those in two-phase systems. This means that the **total complexity** will be

$$X^t = X^I + rX^{II} \tag{1.351}$$

or

$$X^t = X\left(1 + \frac{R}{1-R}\frac{1+\langle m^{II}\rangle}{1+\langle m^I\rangle}\right) \tag{1.352}$$

and the **total average ligand number** will be given as

$$\langle n^t \rangle = (1 - R)\langle n^I \rangle + R\langle n^{II} \rangle \tag{1.353a}$$

$$\langle m^t \rangle = (1 - R)\langle m^I \rangle + R\langle m^{II} \rangle \tag{1.353b}$$

where R is the fraction of central atom M in the phase which is related to the distribution ratio and the phase ratio as

$$R = \frac{rD}{1+rD} \tag{1.354}$$

This ratio is universally known as the **yield of separation** in biphasic systems (see eqn. 2.13).

The mass balances in this complicated two-phase system are

$$c_M = [M]_I(1 + \langle m^t \rangle) X^t \tag{1.355}$$

and

$$rc_L = \Pi_I^t [L]_I + \frac{\langle n^t \rangle}{1+\langle m^t \rangle} c_M \tag{1.356a}$$

where the initial concentration c_M refers to phase I and c_L to phase II — compare with eqns. 1.199 and 1.200, and Π^t is a function which represents the ratio between the amounts of all free and protonated forms of A in both phases to that of the free form L,

$$\Pi^t = \sum (1 + rK_i^D) K_i^p [H^+]^i \tag{1.356b}$$

(compare with eq. 1.195) (*J. Rydberg, 1950; T. Sekine, Y. Hasegawa, N. Ihara, 1973*).

These functions allow treatment of the biphasic complex equilibria as single-phase ones, basically when the weighted stability constants 1.349 are used.

In practical separations, balances are necessary to calculate yields of separation, equation 1.354 — see also section 2.2.1. Such phase diagrams like the activity diagrams and solubility curves, give only the main possibilities of two-phase formation and only rough information about its extent (the distance of the initial composition coordinate from the equilibrium line corresponds to the amount of elutriated product). Further, "pure" two-reactant systems are more the exception than the rule.

For instance, when complex formation equilibria are to be considered in a mixture of k central atoms, instead of one balance 1.190 there are k balances

$$c_{M_k} = [M_k] X_k \tag{1.357}$$

X_k being the complexity function for the central atom M_k, and the ligand balance 1.191 being extended to

$$c_{M_k} = [L] + \sum_{i=0}^{n} \langle n_k \rangle c_{M_k} \qquad (1.358)$$

where $\langle n_k \rangle$ is the average ligand number in the mixture of complexes of metal M_k with ligand L. From the point of view of separation chemistry, it is enough to consider the mixture as a quasi-two-component system (*F. Macasek, D. Vanco, 1981*). All competition reactions can be included in the ligand activity coefficient, likewise in the relation 1.145,

$$\gamma_L = 1 - \frac{1}{c_L} \sum \langle n_k \rangle c_{M_k} \qquad (1.359)$$

Then the protonation of ligand, expressed through function Π^t (eqn. 1.196) is also a particular case of a competing reaction, when M_k is identical with H^+. Nevertheless detailed calculations should be performed in matrix form as in other multicomponent systems — section 1.2.3.

Systems where side reactions and the formation of an individual **solid phase** of the distribuend occurs are also frequent in separation practice.

After including the amounts of solid phase $M_m L_n$ (total amount S, mole) in mass balances 1.190 and 1.191, these equations become

$$c_M = [M]X + \frac{mS}{V} \qquad (1.360)$$

and

$$c_L = [L] + \langle n \rangle c_M + \frac{nS}{V} \qquad (1.361)$$

because the amounts of M and L in the solid phase $M_m L_n$ are always in the stoichiometric ratio $m:n$. The equilibrium concentrations [M] and [L] should fulfill equation 1.294,

$$K_S = [(c_M - \tfrac{mS}{V})\tfrac{1}{x}]^m (c_L - \tfrac{nS}{V} - \langle n \rangle c_M)^n \qquad (1.362)$$

i.e. the amount of solid phase S can be calculated from this implicit equation. Though algebraically it is possible only in simple cases, iteration procedures also can give results for more complicated situations (section 2.2.1). Using eqns. 1.346 and 1.347, it is also possible to consider three-phase systems (e.g. solid- liquid I-liquid II) (*F. Macasek, 1974*).

1.4 DYNAMICS OF SEPARATION

Hitherto, only separation systems in thermodynamic equilibrium have been considered. However, the separation of substances is a transport process such that, *a priori*, it means the achievement of equilibrium (spontaneous separation) or its destruction (induced separation) in real time. Separation is a process which is connected with vector fluxes of substances and non-equilibrium performance. The phenomenological description and modelling of non-equilibrium processes are at least as important as the equilibria of mixture components.

1.4.1 Non-equilibrium thermodynamics

In the first chapter, in connection with equation 1.41, it was shown that the gradient of the chemical potential results in a spatial displacement of a substance. A temperature gradient causes a flux of heat, an electric potential gradient produces a flux of electric charge, etc. The flux (J s^{-1}, mol s^{-1}, C s^{-1}) is a change of some thermodynamic potential F,

$$J = \tfrac{dF}{dt} \qquad (1.363)$$

Generally, proportionality between the flux J and the conjugate gradient X over a certain interval is linear,

$$J_i = L_i X_i \qquad (1.364a)$$

$$J_j = L_j X_j \qquad (1.364b)$$

In any particular case, the **straight phenomenological (kinetic) coefficient** L of proportionality is related to specific physical parameters (viscosity, thermal conductivity, ohmic resistance, etc.). In such cases transport by the flux is called pure transport due to the action of the specific gradient, the thermodynamic force. It is evident that a pressure or concentration gradient causes mass transport or that the flux of ions is due to electromotive force. It has been observed, however, that each gradient exerts some influence on the flux, though it is much less evident. Examples of such "cross influence" are electrokinetic phenomena (electro-osmosis, streaming potentials) or mass transfer in a temperature gradient (Soret's effect, thermodiffusion).

Generalization of the non-equilibrium behavior of systems is the subject of **non-equilibrium thermodynamics**. It is substantiated on the basic postulate that when diminishing the size of a subsystem of a macroscopic system, the relaxation time necessary to reach equilibrium is also diminished and such a subsystem can be considered in local equilibrium. The parameters of subsystems, however, are functions both of space coordinates and time. The differences are destroyed by the flux of energy, mass and charge, and have spontaneous, irreversible characters, resulting in an increase in entropy. Linear non-equilibrium thermodynamics (*S.R. DeGroot, 1961; L Prigogine, 1967; A. Katchalsky, P.F. Curra, 1967*) deal with the particular situation when the fluxes are linear functions of gradients — relation 1.363. Linearity can always be presumed when the system (subsystems) is not far from equilibrium.

The basic laws of non-equilibrium thermodynamics are as follows.

(I) Any thermodynamic force X_j (gradient of intensity factor, e.g. temperature, pressure, chemical potential) exerts some influence on the flux J_i, adding a partial contribution J_{ij} to the overall flux J_j,

$$J_i \propto L_{ij} X_j \qquad (1.365)$$

where L_{ij} is the **coupling (cross) phenomenological coefficient** ($L_{ii} \equiv L_i$ in relation 1.363). The resulting flux is the **superposition** of all the partial effects (*L. Onsager, 1931*):

$$J_i = \sum_j L_{ij} X_j \qquad (1.366)$$

This means that the flux of mass is given not only by the concentration gradient but is also due to the gradients of electric potential, temperature, etc.

According to the **Curie principle** (*P. Curie, 1908*) in homogeneous (isotropic) systems, there is no mutual effect of scalar and vector factors: that is, their cross coefficients are zero. For instance, chemical affinity is a scalar parameter and it cannot cause a real mass transfer of a vector type quantity (the increase or decrease of concentration of reacting substance is only an imaginary, scalar flux). Exceptions may occur in heterogeneous systems, e.g. in membranes: chemical interactions in membranes, or the interface of the membrane and bulk phase may have an effect on the flux of the reactant through the membrane and vice versa.

(II) Between phenomenological cross coefficients there is the **Onsager reciprocity relation**

$$L_{ij} = L_{ji} \qquad (1.367)$$

(or, under the influence of outer fields, $L_{ij} = -L_{ji}$), i.e. the cross coefficients of flux-force pairs form a symmetrical matrix. This relation is often called the **fourth law of thermodynamics**. It was derived (*L. Onsager, 1931*) from the statistical theory of fluctuations.

E.g., Maxwell-Stefan diffusivity of the *i*-th component in the *j*-th matrix is the same as that of j-th species in *i*-th matrix, $D_{ij}=D_{ji}$ (P.W.Rutten, 1992).

(III) The power, **energy dissipation** ($J\ s^{-1}$ = W) necessary for performance of the irreversible process is given through the **entropy production** which is expressed by the sum of the products of the conjugated fluxes and forces,

$$P = T\frac{d\Delta S}{dt} = \sum J_i X_i \quad (1.368)$$

Now the relations can be illustrated by phenomena connected with the transport of electric charge (flux J_1) and liquid (flux J_2) for an electric potential difference $\Delta\phi$ and pressure Δp. The conjugate pairs are

$$J_1 = L_1 \Delta p \quad (1.369a)$$

$$J_2 = L_2 \Delta p \quad (1.369b)$$

where L_1 corresponds to electric conductivity (reciprocal value to electric resistance in Ohm's law) and L_2 is proportional to the cross section of the opening and inversely proportional to the viscosity of the liquid (for laminar flow — see part 1.4.2.). However, the generalization given by non-equilibrium thermodynamics does not concern a particular physical model. The cross phenomena will result in superpositions

$$J_1 = L_{11}\Delta\varphi + L_{12}\Delta p \quad (1.370a)$$

$$J_2 = L_{21}\Delta\varphi + L_{22}\Delta p \quad (1.370b)$$

and the energy dissipation will be

$$P = J_1 \Delta\varphi + J_2 \Delta p \quad (1.371)$$

Because

$$L_{12} = L_{21} \tag{1.372}$$

from eqns. 1.370 and 1.371, the values of the fluxes are eliminated and

$$P = L_{11}\Delta\varphi^2 + 2L_{12}\Delta\varphi\Delta p + L_{22}\Delta p^2 \tag{1.373}$$

Minimal energy dissipation on varying the pressure difference Δp will follow from the condition

$$\frac{\partial P}{\partial \Delta p} = 2L_{11}\Delta\varphi + 2L_{22}\Delta p = 2J_2 = 0 \tag{1.374}$$

This means that there is zero flux of liquid (for a non-zero flux of electricity, J_1). This is the condition of the **steady state** which is generally formulated as

$$J_i = 0 \tag{1.375a}$$

for

$$\partial P/\partial X_i = 0 \tag{1.375b}$$

(P is minimal, but not necessarily zero). It must be recognized that this condition differs from that of **thermodynamic equilibrium** at which $\Delta G = 0$, i.e. in an isolated system

$$P = 0 \tag{1.376}$$

The meaning of the steady state is that there is no particular mass transfer, even in the presence of thermodynamic forces, due to the constant dissipation of energy.

In our case, the steady state J_2 is maintained by the electric current J_1. Under this condition, from eqn. 1.370b we obtain

$$\left(\frac{\Delta\varphi}{\Delta p}\right)_{J_2=0} = -\frac{L_{22}}{L_{21}} \tag{1.377}$$

and for $\Delta E = 0$, from eqns 1.370 and 1.372,

$$-\frac{L_{22}}{L_{21}} = \left(\frac{J_2}{J_1}\right)_{\Delta\varphi=0} = 0 \tag{1.378}$$

In this way

$$\left(\frac{\Delta\varphi}{\Delta p}\right)_{J_2=0} = \left(\frac{J_2}{J_1}\right)_{\Delta\varphi=0} = 0 \tag{1.379}$$

and, by analogy,

$$\left(\frac{\Delta\varphi}{\Delta p}\right)_{J_1=0} = \left(\frac{J_2}{J_1}\right)_{\Delta p=0} = 0 \tag{1.380}$$

These relations can be proved to be identical with **Saxen's relations** (*U. Saxen, 1892*) which connect the ratio of liquid and electricity transfer (J_2/J_1) at zero potential difference, i.e. the electric current under the action of pressure with the ratio of potential and pressure difference ($\Delta\varphi/\Delta p$) as for steady liquid transfer and vice versa. However, these **electrokinetic phenomena** were derived without a particular physical model. To relate these phenomenological coefficients with physical parameters of the systems, a model of their behavior must be proposed (Figure 25). The model of transfer of liquid under the influence of an electric field, **electro-osmosis**, is based on the existence of diffusion charged layer at the solid — liquid interface (*O. Stern, 1924*) in a slot (or capillary) filled by electrolyte solution. As a result, a certain part of the charges remain fixed in a layer of thickness δ and others can move freely with the liquid. The electric potential difference between the inner part of the electric double layer and the solution is called the **electrokinetic potential (zeta potential, ζ –potential)**:

$$\zeta \approx \varphi_d \tag{1.381}$$

and it plays a substantial role in electrokinetic phenomena. Usually the zeta-potential reaches 40 – 120mV (*H. Freundlich, 1922; P. Benes, M. Paulenova, 1973*). Any movement of liquid under such circumstances is the analogue of movement of one layer of an electric condenser. There are two types of force between the layers of liquid. The first is friction arising from weak intermolecular interactions and causing the inner pressure (N m^{-2} = Pa); this is a force per unit area of two adjacent molecular planes by moving of one plane relative to the other, according to *I. S. Newton (1686)*:

$$\vec{P}_{visc} = \eta \frac{dw}{dz} \tag{1.382}$$

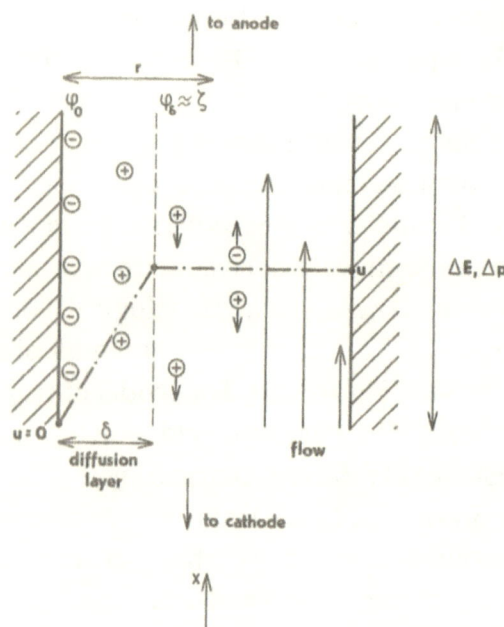

Figure 25. Scheme of electro-osmosis (streaming potential).

where η is the dynamic viscosity (N s m^{-2} = Pa s) and dw/dz is the radial gradient of the velocity vector (s^{-1}) which can be at $\delta \ll r$ linearly approximated by

$$\frac{dw}{dz} = \frac{w}{z} = \frac{w}{\delta} \tag{1.383}$$

(H.L.F. Helmholtz, 1879).

The second force in liquids is due to Coulombic interactions between the ions in the flow and the stationary diffusional layer, which produce an additional friction in the opposite direction,

$$\vec{P}_{el} = \sigma \frac{d\varphi}{dx} \tag{1.384a}$$

where σ is the charge density (C m^{-2}) in the layer. The value of σ can be obtained from the theoretical capacity of the condenser, and the electrokinetic potential, eqn. 1.381, becomes

$$\vec{P}_{el} = \frac{\varepsilon_0 \varepsilon_r}{\delta} \zeta \frac{d\varphi}{dx} \tag{1.384b}$$

In the steady state the forces $\overleftarrow{P}_{visc}$ and $\vec{P}_{el}$ are equilibrated and the relation between the velocity of liquid and the gradient along the slot (capillary) is obtained from 1.382, 1.383 and 1.384.

$$w = \frac{\varepsilon_0 \varepsilon_r \zeta}{\eta} \frac{d\varphi}{dx} \tag{1.385}$$

Now, returning to the general equations for electric current and liquid stream in the absence of a pressure drop ($\Delta p = 0$) and dividing 1.370 by the cross sectional area (m^2), the electric current density i (A m^{-2}) is

$$i = l_{11} \frac{d\phi}{dx} \tag{1.386a}$$

and the liquid stream density, which is identical with the linear velocity (m^3 s^{-1} m = m s^{-1}):

$$w = l_{21} \frac{d\varphi}{dx} \tag{1.386b}$$

where $l_{11} = L_{11}/A$ and $l_{21} = L_{21}/A$. Comparing 1.386b and 1.385 it can be seen that

$$l_{21} = \frac{\varepsilon_0 \varepsilon_r \zeta}{\eta} \tag{1.387}$$

(m V^{-1} s^{-1} = A m N^{-1}). Note that, according to the reciprocity relation 1.367, the coefficient in the relation $i = l_{12} dp/dx$ (electric current induction by the action of pressure) will have the same value and dimensions. According to Ohm's law, l_{11} should be identical with the specific conductivity κ (S m^{-1})

$$l_{11} = 1/R = \kappa \tag{1.388}$$

This means that

$$\frac{w}{i} = \frac{\varepsilon_0 \varepsilon_r \zeta}{\kappa \eta} \tag{1.389}$$

which expresses the efficiency of liquid movement in the electric field (liquid stream to consumption of electric current) in electro-osmotic separation.

Considering the laminar flow of liquid — see section 1.4.2. — we have

$$w = l_{22} \frac{dp}{dx} \tag{1.390a}$$

where for a capillary of radius r

$$l_{22} = \frac{r^2}{8\eta} \tag{1.390b}$$

and w can be replaced in equation 1.385 by the parameter of the cross-sectional area (πr^2); therefore

$$\frac{r^2}{8\eta} \frac{dp}{dx} = \frac{\varepsilon_0 \varepsilon_r \zeta}{\eta} \frac{d\varphi}{dx} \tag{1.391}$$

or

$$\left(\frac{\Delta p}{\Delta E}\right)_{v=0} = \frac{8\varepsilon_0\varepsilon_r\zeta}{r^2} \tag{1.392}$$

giving a further expression for the **streaming potential** ΔE which maintains a steady zero-liquid-flow ($w = 0$) for a pressure drop Δp. In the particular case of ion exchangeable membranes containing ions with charge z (elementary charge units) and concentration c, the electrokinetic potential of the Helmholtz double layer is $\zeta \approx zc$ and

$$\frac{\Delta \varphi}{\Delta p} = \frac{zcF}{\kappa\rho} \tag{1.393}$$

(G. Schmid, 1952; R. Schloegl, 1955) where F is Faraday's constant and

$$\rho \propto \eta/r^2 \tag{1.394}$$

is the resistance to liquid flow. Usually $\Delta\varphi$ is several millivolts for $\Delta p = 0.1$ MPa (F. Helfferich, 1959).

For instance, for the electrokinetic potential at the water-glass interface, $\zeta = 50$ mV, according to equation 1.387, $\varepsilon_0 = 8.85 \times 10^{-12}$ C^2 J^{-1} m and for water $\varepsilon_r = 78.6$, thus $l_{21} = \frac{(8.85\times10^{-12})(78.6)(0.050)}{10^{-3}}$ m^2 V^{-1} s^{-1} = 3.48×10^{-8} m^2 V^{-1} s^{-1}. At the potential gradient $\frac{d\varphi}{dx} = 1000$ V m^{-1}, the velocity of water flow will be $w = 3.48 \times 10^{-5}$ m s^{-1} i.e. 12.5 cm per hour, and it is oriented against the ion current (towards the cathode in this case, because the glass surface is negatively charged). If the specific conductivity of liquid (aqueous solution of electrolyte) is = 100 mS m^{-1}, according to equation 1.386a and 1.388 there is an electric current density $i = (100 \times 10^{-3})10^3 = 100$ A m^{-2}. Through a capillary of diameter 0.05 mm will flow an electric current $I = i\pi r^2 = 7.85 \times 10^{-7}$ A = 0.8 µA and a water flow

$\dot{V} = w\pi r^2 = 2.7 \times 10^{-13}$ m³ s⁻¹ $= 0.97$ µl h⁻¹. In comparison, such a flow will occur at the pressure gradient — see equation 1.459 —

$$dp/dx = \frac{8\eta w}{r^2} = 2.78 \times 10^{-4}/(5 \times 10^{-5}) \text{ Pa m}^{-1} = 0.11 \text{ MPa m}^{-1}$$

On the basis of non-equilibrium thermodynamics, some missing data can be obtained empirically, as demonstrated by Saxen's relations 1.379–1.380, without detailed physical substantiation.

In separation science, electro-osmosis plays either a positive or negative role.

Proposals have been made to realize separation methods in which the electro-osmotic flow serves to reduce analysis time and improve resolution as **electrochromatography** (*J.H. Knox, I.H. Grant, 1987*) or **electrokinetic chromatography** (*S. Terabe et al., 1984; S. Terabe, 1989*). Its principle of separation is based on the chromatographic principle of distribution between a charged carrier and solvent matrix in which a relative countercurrent flow is driven by electrophoresis (see below), usually in fused silica capillaries; ionic micelles (*T. Nakagawa, 1981*) or polymeric ions can be employed as the carriers. A solute moves the faster the higher the distribution ratio between carrier and solvent is (*S. Terabe et al., 1984*).

In electrodialysis (*R.L. Lacey, 1972*) or capillary electrophoresis (*F.M. Everaerts, J. L. Beckers, T.P.E.M. Verheggen, 1976*), electro-osmosis is an undesirable phenomenon. This can be minimized by the addition of neutral polymeric compounds, such as poly(vinyl alcohol), which suppress the zeta-potential at the liquid-liquid interface and increase the viscosity. The electroendo-osmotic properties of gels used in electrophoretic separations also depend on the presence of bound charged groups (e.g. sulphate groups in agarose) and in this respect polyacrylamide is preferred (*V. Piljac, G. Piljac, 1986*).

The equations of non-equilibrium thermodynamics also appear useful for other "cross" phenomena, such as the diffusion of matter in a thermal gradient, or **thermodiffusion** (S. Chapman, T.G. Cawling, 1939; W. Jost, 1960).

In the concentration gradient dc/dz accompanied by a temperature gradient $d \log T/dz$ there is a flux of substance (mol m^{-2}s^{-1}) as a superposition of diffusional flux (from higher concentration to lower) and thermo-diffusional flux (from warmer to colder regions),

$$j_1 = l_{11}\frac{dc_1}{dz} + l_{12}\frac{d \log T}{dz} \qquad (1.395a)$$

where the constants in a two-component (A and B) mixture are identical with the mutual diffusion coefficients of components A and B, D_{AB}(m^2 s^{-1}) — see equation 1.404 — and the **thermodiffusion coefficient** D_{AB}^t (mol m^{-1} s^{-1}), respectively:

$$l_{11} \equiv -D_{AB} \qquad (1.395b)$$

$$l_{12} \equiv D_{AB}^t \qquad (1.395c)$$

In the steady state ($j_1 = 0$) the gradient of A on the mole fraction scale is obtained from 1.395 as

$$\frac{dx_A}{dz} = -\beta x_A x_B \frac{d \log T}{dz} \qquad (1.396)$$

where β is called the **thermodiffusion factor**

$$\beta = \frac{D_{AB}^t}{D_{AB} x_A x_B} \qquad (1.397)$$

It was shown (W.H. Furry et al., 1939) from the kinetic model for ideal elastic spheres that

$$\beta = \frac{105}{118} \frac{M_B - M_A}{M_B + M_A} \qquad (1.398)$$

($M_A < M_B$), i.e. more light molecules A prevail in the colder region of the temperature gradient. The thermodiffusion factor of real mixtures A and B is 2-10 times smaller due to the van der Waals interactions. Using the Lennard-Jones potential 1.233, thermodiffusion coefficients for real mixtures can be calculated (*J. Hirschfelder, et al., 1948*).

Thermodiffusion was successfully applied to the sophisticated separation of stable isotopic mixtures (*K. Clusius, G. Dickel, 1939; R.C. Jones, W.H. Furry, 1946; A.M. Rozen, 1960*).

Generally, the universal moving force is a gradient of chemical potential μ, as ilustrated by equation 1.41. When a point with coordinate x and concentration of substance c moves with a velocity w_x, the flux (mol m s^{-1}) through the plane perpendicular to the vector of velocity is

$$j = cw_x = -\frac{c}{f}\frac{\partial \mu}{\partial x} \tag{1.399}$$

and from the definition of activity a, equation 1.81, it follows that

$$j = -\frac{RTc}{f}\frac{\partial \ln a}{\partial x} \tag{1.400}$$

With respect to 1.98b,

$$c\frac{\partial \ln a}{\partial x} = \frac{\partial \ln a}{\partial \ln c}\frac{\partial c}{\partial x} = \left(1 + \frac{\partial \ln \gamma}{\partial \ln c}\right)\frac{\partial c}{\partial x} \tag{1.401}$$

and the general equation for the **flux of substance** (mol m^{-2}s^{-1}) in a **concentration gradient** in a **non-ideal solutions**

$$j = -\frac{RT}{f}\left(1 + \frac{\partial \ln \gamma}{\partial \ln c}\right)\frac{\partial c}{\partial x} \tag{1.402}$$

For an **ideal solution** ($\partial \ln \gamma / \partial \ln c = 0$), we have

$$j = -\frac{RT}{f}\frac{\partial c}{\partial x} \tag{1.403}$$

This means that for ideal solutions the flux is proportional only to the entropic term of the chemical potential gradient and the latter is equivalent to the concentration gradient.

1.4.2 Convective transport

As shown in section 1.4.1, the transport of substances results from gradients of various thermodynamic forces.

Material transport has several forms:

(I) At the molecular level:
  (i) Random replacement of molecules by their mutual collisions —**molecular diffusion**.
  (ii) Directed motion of ions in electric fields — **electrophoresis**.

The oriented motion of molecules in other fields (gravitational, centrifugal) is usually negligible, except for the sedimentation of large, polymer molecules in strong centrifugal fields. Conversely, random motion also occurs for larger particles (e.g. colloids) as **Brownian motion** (*R. Brown, 1828*).

(II) At the level of phase conglomerates:
  (i) Transport with small conglomerates (globules) of fluid matrix which change their velocity chaotically and the substance follows their stochastic movement — **eddy (turbulent) diffusion (dispersion)**.
  (ii) Transport with the bulk of moving fluid in which the inner distribution (local concentration) of the substance is preserved — **convective concentration change**.

Fick's first law (*A. Fick, 1855*) for molecular diffusion corresponds to the phenomenological expression 1.404.

$$j = -D \frac{\partial c}{\partial x} \qquad (1.404a)$$

where D is the **diffusion coefficient (diffusivity)** (m² s⁻¹) and gives the expression for the random transport of substance in the concentration gradient $\partial c/\partial x$.

For multicomponent diffusion,

$$j_i = -\sum_{j=1}^{n-1} D_{ij} \frac{\partial c_i}{\partial x} \tag{1.404b}$$

(R.L. Baldwin, P.J. Dunlop, L.J. Gosting, 1955).

Like any statistical process, this transport is slow and its activation energy corresponds to the level of weak (van der Waals) interactions. Comparing 1.404a with 1.403, yields the relation

$$D = \frac{RT}{f} \tag{1.405}$$

obtained by **A. Einstein** (1905).

Because the frictional coefficient f depends on the molecular size, see eq. 1.42, free diffusion in viscous media can be used for the separation of macromolecules from low-molecular components (D. Freifelder, M. Better, 1982).

For diffusion coefficients D_i of ions M^{z_i}, W. Nernst (1888) showed that, at infinitesimal concentration

$$D_i = \frac{RT}{z_i F} u_i \tag{1.406a}$$

The ionic mobility u_i (m² s⁻² V⁻¹) is connected with the limiting molar (equivalent) conductance of the ion Λ_i^0 (S m² mol⁻¹) via

$$u_i = \frac{\Lambda_i^0}{z_i F} \tag{1.406b}$$

The sign of u_i is conventionally the same as the charge of the ion z_i and both D_i and A_i have positive values.

The diffusion coefficients in gases are of the order 10^{-5} m² s⁻¹ (0.1 cm² s⁻¹), in liquids 10^{-9} m² s⁻¹ (10^{-5} cm² s⁻¹) and in solids and gels $10^{-14} - 10^{-10}$ m² s⁻¹ ($10^{-10} - 10^{-6}$ cm² s⁻¹).

The coefficient of mutual diffusion in ideal mixtures of A and B is

$$D_{AB} = x_A D_A + x_B D_B \tag{1.407}$$

(*L.S. Darken, 1948*) and the mean inter-diffusion coefficient of ions A^{z_A} and B^{z_B} is

$$D_{AB} = \frac{D_A D_B (z_A^2 c_A + z_B^2 c_B)}{D_A z_A^2 c_A + D_B z_B^2 c_B} \tag{1.408}$$

—see equation 1.650. Both 1.407 and 1.408 are mole fraction-or charge fraction-weighted diffusion coefficients. The dependence of diffusion co-efficients on the composition of a real solution is expressed according to equation 1.403 as

$$D = D^0 \left(1 + \frac{d \ln \gamma}{d \ln c}\right) \tag{1.409}$$

(*R.E. Powell, W.E. Roseware, E. Eyring, 1941; L.S. Darken, 1967*), or empirically (*N.N. Li, R.B. Long, 1969; V. Sanchez, M. Clifton, 1977*) usually as an exponential function on the diffusant concentration c,

$$D = D^0 \exp(kc) \tag{1.410}$$

where D^0 is the diffusion coefficient at zero concentration of diffusant - **Maxwell-Stefan diffusivity** - and the constant k can be fitted for a given medium and concentration range. The relation 1.405 is more suitable for the establishment of the Stoke's radius of molecules than vice versa, especially in the case of ions, the radii of which differ greatly from crystallographic data (see section 1.2.4).

For the diffusion of A in liquid matrix B, the semiempirical **Wilke-Chang equation** (*C.R. Wilke, P.C. Chang, 1955*) can be applied:

$$D_{AB} = 7.4 \times 10^{-12} \frac{(\Phi_B M_B)^{1/2} T}{\eta_B \overline{V}_A^{0.6}} \qquad (1.411)$$

where D_{AB} is given in m²s⁻¹, M_B is the molecular mass of B, Φ_B is the parameter of association of the matrix molecules ($\Phi \cong 1$ for non-polar solvents, $\Phi \cong 1.5$ for ethanol and $\Phi \cong 2.6$ for water), T is the temperature, η is the dynamic viscosity of the matrix (mPa s) and $\overline{V}$ is the molar volume of A at the boiling point (cm³mol⁻¹).

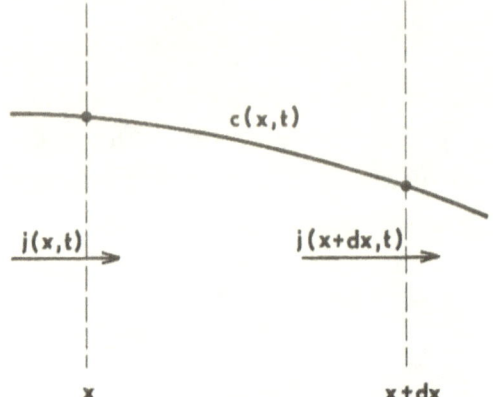

Figure 26. Diffusion in non-linear concentration gradient.

For many practical purposes it is necessary to consider concentration changes due to **diffusion in a non-linear concentration gradient** i.e. in a medium where the flux varies from coordinate x to coordinate $x + dx$ (Figure 26). A flux $j(x)$ enters the element and a flux $j(x + dx)$ leaves it. This means that in a volume element Adx (A is the cross-sectional area of the element), the net gain in molar quantity of diffusing substance is

$$\frac{dc}{dt} = \frac{A[j(x) - j(x+dx)]}{Adx} \qquad (1.412)$$

The value of $j(x + dx)$ can be obtained from the first term of the Taylor series,

$$j(x + dx) = j(x) + \frac{\partial j}{\partial x} dx \qquad (1.413)$$

and the difference-differential eqn. 1.412 becomes a partial differential equation

$$\frac{dc}{dt} = -\frac{\partial j}{\partial x} \qquad (1.414)$$

which is referred to as the **equation of continuity**. Substitution of eqn. 1.404 yields

$$\frac{dc}{dt} = \frac{\partial}{\partial x}\left(D\frac{\partial c}{\partial x}\right) \qquad (1.415)$$

or

$$\frac{dc}{dt} = D\frac{\partial^2 c}{\partial x^2} + \frac{\partial D}{\partial c}\left(\frac{\partial c}{\partial x}\right)^2 \qquad (1.416)$$

If the diffusion coefficient is independent of the concentration ($\partial D/\partial c = 0$), then we obtain

$$\frac{dc}{dt} = D\frac{\partial^2 c}{\partial x^2} \qquad (1.417)$$

i.e. **Fick's second law** of molecular diffusion.

If the convection flux with constant velocity w is included, according to eqns. 1.399 and 1.414

$$\frac{\partial j}{\partial x} = w_x \frac{\partial c}{\partial x} \qquad (1.418)$$

In respect of eqn. 1.416,

$$\frac{dc}{dt} = -w_x \frac{\partial c}{\partial x} + D\frac{\partial^2 c}{\partial x^2} \qquad (1.419)$$

which is known as the equation of **one-dimensional convective diffusion**. For **three-dimensional diffusion** it can be shown that

$$\frac{dc}{dt} = -(\vec{w}_x \frac{\partial c}{\partial x} + \vec{w}_y \frac{\partial c}{\partial y} + \vec{w}_z \frac{\partial c}{\partial z}) + D(\frac{\partial^2 c}{\partial x^2} + \frac{\partial^2 \varepsilon}{\partial y^2} + \frac{\partial^2 c}{\partial z^2}) \quad (1.420)$$

or, more concisely, through operators

$$\frac{dc}{dt} = -\vec{w}\,\text{grad}\,c + D\,\text{div grad}\,c \quad (1.421a)$$

or

$$\frac{dc}{dt} = -\vec{w}\nabla c + D\nabla^2 c \quad (1.421b)$$

where

$$\text{grad} \equiv \nabla \equiv i\frac{\partial}{\partial x} + j\frac{\partial}{\partial y} + k\frac{\partial}{\partial z} \quad (1.422)$$

$$\text{div} \equiv \frac{\partial}{\partial x} + \frac{\theta}{\partial y} + \frac{\theta}{\partial z} \quad (1.423)$$

$$\text{div grad} \equiv \nabla^2 \text{ (Laplacian)} \equiv \frac{\partial^2}{\partial x^2} + \frac{\partial^2}{\partial y^2} + \frac{\partial^2}{\partial z^2} \quad (1.424)$$

and i, j, k are respectively the unit vectors (concentration is the scalar function and the concentration gradient is the vector function). In the case of ion transport in an electric field (L. Onsager, R.M. Fuoss, 1932) equation 1.420 is extended by the term

$$\frac{dc}{dt} = \cdots + D_i z_i c_i \frac{F}{RT}\,\text{grad}\,\varphi \quad (1.425)$$

(compare with expression 1.406).

For **diffusion in spherical particles**, e.g. single drops (C. Hanson, M.A. Hughes, R.J. Whewell, 1978), polar coordinates are suitable. In an isotropic medium (with the equal properties in each direction from the center) we have

$$\frac{dc}{dt} = D\left(\frac{\partial^2 c}{\partial r^2} + \frac{2}{r}\frac{\partial c}{\partial r}\right) \quad (1.426)$$

where r is the distance from a center. By substitution of $c' = rc$, the equation is easily transformed to type 1.417.

The one-dimensional equation is useful in normalized, dimensionless time and length coordinates which reduce the number of independent parameters of transport. Let there be a characteristic length d of a system (e.g. d can be the thickness of the motionless layer where diffusion is the only factor in transport of the substance) and coordinate x is replaced by the **dimensionless length**

$$\xi = \frac{x}{d} \tag{1.427}$$

Subsequently, $\partial \xi = \partial x/d$ and $\partial^2 \xi = \partial^2 x/d^2$. Dividing equation 1.419 by D/d^2 and considering

$$\tau = \frac{Dt}{d^2} \tag{1.428}$$

to be a **dimensionless time**, the equation is transformed to

$$\frac{dc}{d\tau} = -\frac{w_x d}{D}\frac{\partial c}{\partial \xi} + \frac{\partial^2 c}{\partial \xi^2} \tag{1.429}$$

i.e. instead of two parameters, w_x and D in eqn. 1.419 there is only one dimensionless parameter which can be referred to as the **diffusion Péclet number** (Pe_D or simply Pe),

$$Pe = \frac{w_x d}{D} \tag{1.430}$$

The Péclet number means **dimensionless transport velocity**. It is sufficient to characterize one-dimensional convective diffusion: the concentration is expressed either as the function $c(t, x, w, D, d)$ or as $c(\tau, \xi, Pe)$ at a fixed value of d. For instance, if the convective transport of material with a diffusion coefficient $D = 1.5 \times 10^{-10}$ m^2 s^{-1} is to be described in a fluid with flow velocity $w = 3 \times 10^{-7}$ m s^{-1} (about 1 mm per hour), it may be done for each coordinate x and boundary coordinate d according to eqn. 1.419. However, such a solution can also be obtained by eqn. 1.429 and is valid for all systems having the same Péclet number, for an arbitrary characteristic dimension d. For example, at $d = 1\,\text{cm} = 0.01\,\text{m}$, $Pé = (3 \times 10^{-7})(0.01)/1.5 \times 10^{-10} = 20$. The dimensionless length is obtained by dividing the Cartesian coordinate by the value of d and the dimensionless time is $\tau = (3 \times 10^{-7}/10^{-4})t = 0.003\,t$ (t is given in seconds). Furthermore, this means that convective diffusion in any other system with various D's and w, or size, but equal Pe values will be modelled in the same dimensionless coordinates (see further the similitude theory).

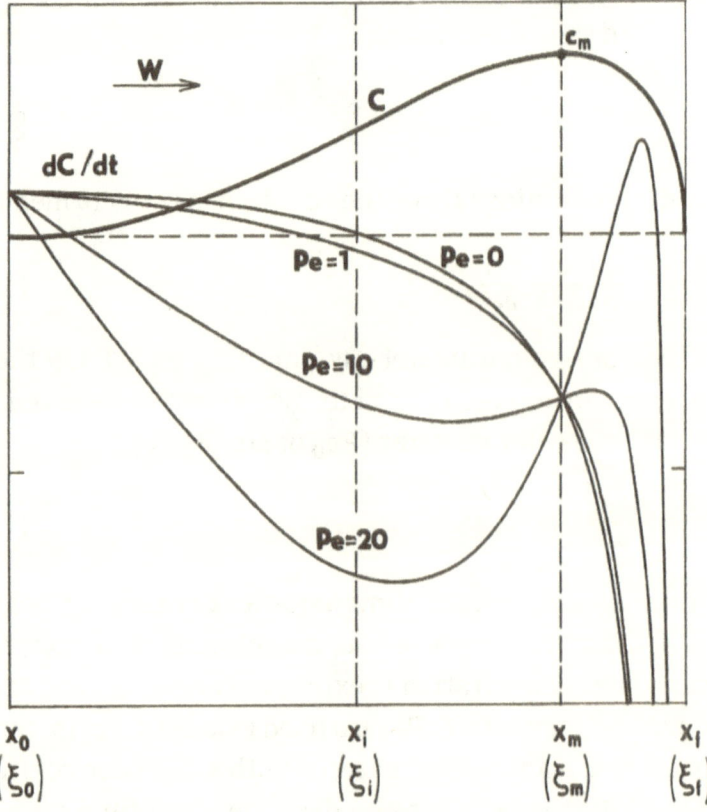

Figure 27. Concentration change (dc/dt) of a concentration wave (c) at various Péclet numbers.

The influence of the two terms on the right hand side of eqn. 1.429 can be illustrated graphically (Figure 27). The curves dc/dt correspond to various Péclet numbers when the illustrative concentration profile c exists in the matrix which flows with velocity w_x in the direction from x_0 to the front of the concentration profile x_f. Several distinguishing features can be observed. The concentration profile is simulated such a way that c drops to zero both at x_0 and x_f but in the tail $(x = x_0)$ the gradient becomes $dc/dx = 0$, and at the leading edge $(x = x_f)$ the gradient turns to $dc/dx = -\infty$. Hence, at x_0 there always will be some increase of concentration $(dc/dt > 0)$ due to backward diffusion from the profiles of higher concentrations, but with an increasing forward flow w (increase of Péclet number) the contribution of diffusion becomes negligible (even $Pe = 20$ corresponds to a very low flow velocity w in liquids). Right from the point of inflection x_i the concentration always decreases $(dc/dt < 0)$ even at $x = 0$, i.e. at zero flow rate, because the substance diffuses in opposite directions from x_i or it is carried away by the bulk of fluid. Obviously, the faster the flow (the greater is Pe), the larger part of the moving profile belongs to the region of decreasing concentration. At some value of Pe number ($Pe = 20$ in the figure) there occurs an increase of concentration in the front of the moving substance, because of rapid convective transport from the maximum concentration profile (x_m) to the front (x_f), which prevails strongly over the diffusion flux. It is remarkable that at maximal concentration (x_m) dc/dx always equals zero and the rate of decrease of concentration corresponds entirely to the diffusion component: diffusion spreading of the maximum is the same as in the motionless matrix $(w = 0 \text{ or } Pe = 0)$. This is equivalent to Fick's second law 1.417,

$$\frac{\partial c(\xi_m, \tau)}{\partial \tau} = \frac{\partial^2 c(\xi_m, \tau)}{\partial \xi^2} \tag{1.431}$$

The solution of the linear parabolic differential equation 1.431 can best be found by an **operational method**, eliminating the real variable r using Laplace operators (L) (*R. V. Churchill, 1950; V.A. Ditkin, A.P. Prudnikov, 1965*). If $f(\xi,s)$ is a function of a complex variable s and $c(\xi,\tau)$ is real function then the operator gives the linear relation

$$f = Lc \tag{1.432a}$$

which is obtained when

$$f(s) = Lc(\tau) = \int_0^\infty e^{-s\tau} c(\tau)\, d\tau \tag{1.432b}$$

Using this substitution (and that is the point) equation 1.431 transforms to the linear differential equation

$$sf(s,\xi) = \frac{d^2 f(s,\xi)}{d\xi^2} \tag{1.433}$$

Its solution is, in the first step, an imaginary function

$$f(s) = c(\tau)\exp(-\xi\sqrt{s}) \tag{1.434}$$

After having obtained this, the initial and boundary conditions should always be formulated.

For an infinitesimally thin initial layer with planar concentration c_S (mol m^{-2}), they are as follows:

$c_S = c_m$ at $\tau = 0$ and $\xi = 0$ and

$c_S = 0$ at $\tau = 0$ and $\xi < 0$ or $\xi > 0$.

According to tables of Laplace operations and transforms, the real function c can be obtained as the planar concentration (mol m^{-2}) function:

$$c_S = c_m \left[1 - \mathrm{erf}\left(\frac{\xi}{2\sqrt{\tau}}\right)\right] \tag{1.435a}$$

The "erf" is the Gaussian error function (the integral of the normal probability curve)

$$\text{erf}(x) = \frac{2}{\sqrt{\pi}} \int_0^x e^{-x^2} \, dx \tag{1.435b}$$

(Figure 28)[2]

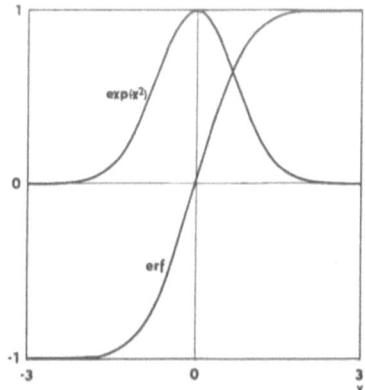

Figure 28. The normal probability and ert (Gaussian error) functions.

The finite planar concentration c_m is an abstraction, so called the **Dirac impulse**, not realizable physically, because actually any real amount of diffusant must give $c_m \to \infty$ at layer thickness $d \to 0$. After the next integration of the planar concentration,

---

[2] The integral cannot be presented with the help of elemental functions but as the series

$$\int_0^x e^{-x^2} \, dx = \sum_{n=0}^{\infty} (-1)^n \frac{1}{n!} \frac{x^{2n+1}}{2n+1}$$

Hence, we have

erf $(-x)= -$erf(x), erf$(0)=0$, and erf$(\infty)=1$.

The function erf should not be misunderstood as the complementary function, erfc=1-erf. Both are often used in calculations of diffusion processes. To obtain reliable data, at least 20-30 terms of the sum should be calculated, because the sum does not converge very fast.

$$c_0 d = \int_{-\infty}^{+\infty} c_S \, d\xi \qquad (1.436a)$$

(the definition of the fundamental **characteristic dimension** d as the real thickness of the initial thin layer) the eqn. 1.435a transforms

$$c = \frac{c_0}{4\pi\tau} \exp\left(-\frac{\xi^2}{4\tau}\right) \qquad (1.436b)$$

(Fig. 29).

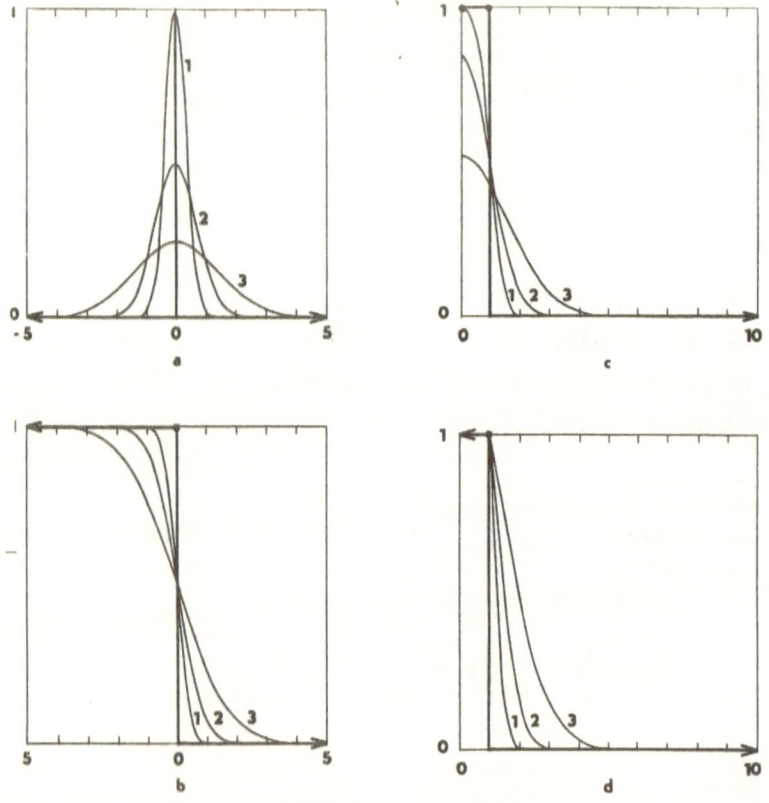

Figure 29. Concentration profiles at various initial and boundary conditions at dimensionless time $\tau = 0.0625, 0.25$ and 1.0 (curves 1, 2 and 3 correspondingly: (a) diffusion from a thin layer to infinity, (b) diffusion from an infinite layer to another infinite layer, (c)

diffusion from a thick layer to infinity, (d) diffusion from a constant concentration boundary to infinity.

According to the properties of the Gaussian function, the half-width σ of the probability curve exp $(-x^2)$ at its inflection point has the meaning (*O.L. Davies, 1949; L.M. Batuner, M.Ye. Pozin, 1960*)

$$\sigma = \langle \Delta\xi \rangle = \sqrt{2\tau} \qquad (1.437a)$$

or

$$\langle \Delta x \rangle = \sqrt{2Dt} \qquad (1.437b)$$

which is the **Einstein-Smoluchowski equation** for the **mean diffusion path** of molecules, the mean displacement in the x-direction (*A. Einstein, 1905*). At typical values in liquids, $D = 10^{-10} - 10^{-9}$ m^2 s^{-1}, the diffusion distance has a mean value $10 - 50$ μm within one second and, being proportional to the square root of time, only 60-times more after one hour (about 0.6–3 mm).

The solution of 1.431 at the boundary conditions above corresponds to diffusion of the substance from an infinitesimally thin layer to unlimited space (e.g. the behavior of a narrow chromatographic band —see section 1.4.5). Some other solutions of eqn. 1.431 under other obvious initial and boundary conditions are as follows (see e.g. *J. Crunk, 1956; N.I. Nikolaev, 1980*).

(i) **Diffusion from a thick layer** (its thickness being the characteristic length d) to an unlimited layer, e.g. a broad chromatographic band, diffusion on contact of a large bulk of the other phase, etc. The initial and boundary conditions are

$c = c_0$ at $\tau = 0$ and $\xi \leq 1$ (it spreads from $\xi = 0$ to $\xi = 1$)

$c = 0$ at $\tau = 0$ and $\xi > 1$ or $\tau = \tau$ and $\xi = \infty$, and the solution is

$$c = \frac{c_0}{2}\left[\operatorname{erf}\left(\frac{1+\xi}{2\sqrt{\tau}}\right) + \operatorname{erf}\left(\frac{1-\xi}{2\sqrt{\tau}}\right)\right] \quad (1.438)$$

(Fig. 29b). This solution is symmetrical for diffusion in the opposite direction, i.e. for diffusion from the band $\xi \in (-1,1)$.

(ii) Diffusion from an unlimited layer to another unlimited layer, e.g. on contact with a large bulk of phases at arbitrary d, which are contacted at coordinate $\xi = 0$, i.e. the initial and boundary conditions being

$c = c_0$ at $\tau = 0$ and $\xi \leq 0$ or at $\tau = r$ and $\xi = -\infty$,

$c = 0$ at $\tau = 0$ and $\xi > 0$ or at $\tau = \tau$ and $\xi = \infty$.

Then the solution of eqn. 1.431 gives (Fig.29c)

$$c = \frac{c_0}{2}\left[1 - \operatorname{erf}\left(\frac{\xi}{2\sqrt{\tau}}\right)\right] \quad (1.439a)$$

(iii) The analogous solution is obtained by **diffusion from a surface with a constant (maximal) concentration**, e.g. the dissolution of a salt in a large layer of unmixed solvent, sorption from a gas flow of constant concentration, etc.,

$c = c_0$ at $\tau = \tau$ and $\xi \leq 0$, $c = 0$ at $r = 0$ and $\xi > 0$,

when

$$c = c_0\left[1 - \operatorname{erf}\left(\frac{\xi}{2\sqrt{\tau}}\right)\right] = c_0 \operatorname{erfc}\left(\frac{\xi}{2\sqrt{\tau}}\right) \quad (1.439b)$$

which is formally the same as eqns. 1.435a and 1.439a (Fig.29d).

(iv) **Diffusion from a thick layer** (thickness d) **to unlimited space** that keeps the outer concentration negligible, e.g. diffusion from a capillary of length d, a membrane of thickness

d or a chromatographic band of width d to a large bulk solution. The boundary conditions are now formulated as

$c = c_0$ at $\tau = \tau$ and $\xi \leq 1$ and

$c = 0$ at $\tau = \tau$ and $\xi > 1$ (i.e. c is zero at any time, not only at the beginning).

Double integration of eqn. 1.431 gives the solution for the average concentration in a thick layer in the form of the infinite series

$$\langle c \rangle = c_0 \sum_{n=0}^{\infty} \frac{8}{\pi^2} \frac{1}{(2n+1)^2} \exp\left[-\frac{(2n+1)^2 \pi^2 \tau}{4}\right] \qquad (1.440)$$

However, the second term ($n = 1$) of the above sum is much smaller than the first ($n = 0$) and the latter is an adequate and useful approximation:

$$\langle c \rangle \approx \frac{8c_0}{\pi^2} \exp(-\pi^2 \tau / 4) \qquad (1.441)$$

It is worthwhile noting that the average concentration does not depend on any other parameter of diffusion systems, such as on the radius of the capillary from where the diffusion proceeds, than the depth d, because, as a result of diffusion equation treatment, the density of the diffusion flux (mol m^{-2} s^{-1}) is the determining parameter of the material transport.

In separation systems in which chemical reaction is involved, the boundary conditions are much more difficult to define in obtaining a simple analytical solution (see section 1.4.4.). However, for calculating the flux, it is sufficient to obtain the solution of the diffusion eqn. only in respect of the value of $\partial c/\partial x$.

(v) **Diffusion through a thick layer** (thickness d) **from the region of constant concentration** c' to one constant concentration c'' (e.g. diffusion through a membrane) leads to the boundary conditions

$c = c'$ at $\xi = 0$ for $\tau = \tau$,

$c = c''$ at $\xi = 1$ for $\tau = \tau$,

Now, the solution of diffusion eqn. 1.431 gives

$$c = c' + (c'' - c')\xi + \frac{2}{\pi}\sum_{n=1}^{\infty}\left[\frac{c'' \cos(n\pi) - c'}{n} \sin(n\pi\xi) \exp(-n^2\pi^2\xi)\right] \quad (1.442)$$

where the trigonometric functions are calculated in radians. At high τ (above $\tau = 1$), a linear gradient of the concentration inside the diffusion layer,

$$c = c' + (c'' - c')\xi \quad (1.443)$$

is rapidly approached (Figure 30).

This state can be reached even in thin layers of solids rather quickly. For instance, the relaxation time ($\tau = 1$) for the diffusion of oxygen from water at 20°C through 0.23 mm ($= d$) silicone rubber membrane ($D = 5.1 \times 10^{-10}$ m² s⁻¹) is $t = \frac{d^2}{D} = 104$ s.

It can be said that the expression for the time dependence of diffusion always contains the functional parameter $e^{-\tau}$.

The general solution of diffusion eqn. 1.431 for varying boundary concentrations and in systems featuring chemical reactions is obtainable only by numerical methods of integration (see e.g. W.F. Ames, 1969; G.F. Carey, B.A. Finlanson, 1975; R. Juin et al., 1982; V.M. Aguilella et al., 1986).

As previously indicated, the treatment of all the systems is much easier for **steady diffusion** when the concentration $c(\xi)$ in any point of the diffusion layer is constant, i.e. if (see eqn. 1.414):

$$\frac{dc}{dt} = -\frac{\partial j}{\partial x} = 0 \quad (1.444)$$

According to eqn. 1.419 the concentration gradient is given by the equality

$$\frac{\partial^2 c}{\partial x^2} = \frac{w_x}{D}\frac{\partial c}{\partial x} \qquad (1.445a)$$

or

$$D\frac{\partial^2 c}{\partial x^2} = w_x c \qquad (1.445b)$$

which is equivalent to the equality of the diffusion and convection flux. At zero convection flux ($w_x = 0$) this means that

$$\frac{\partial^2 c}{\partial x^2} = 0 \qquad (1.446a)$$

which occurs, in a planar diffusion layer, when

$$\frac{\partial c}{\partial x} = \text{const} = \frac{\Delta c}{\Delta x} \qquad (1.446b)$$

This is equivalent to a **linear concentration gradient** in planar diffusion layers for a steady convective diffusion. The steady state is reached rapidly when $\tau > 1$ (see Figure 30); it is characterized by a constant flux

$$j = D\frac{\Delta \varepsilon}{\Delta x} \qquad (1.447)$$

It was shown (*T.K. Sherwood, R.L. Pigford, 1952*) that for **eddy (turbulent) diffusion**, or better **dispersion**, the convection, after a sufficiently long time of displacement, obeys Fick's first law, eqn. 1.404, and the **total diffusion convection** is given by

$$j = -(D^M + D^E)\frac{\partial c}{\partial x} \qquad (1.448)$$

where D^M and D^E are the diffusion (dispersion) coefficients of molecular and eddy diffusion (dispersion) respectively.

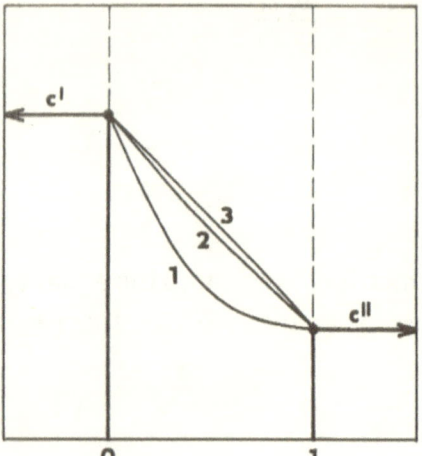

Figure 30. Diffusion profiles in a thick layer (membrane) at dimensionless time $\tau = 0.0625, 0.25,$ and 1.0 (curves 1, 2 and 3 respectively).

Turbulence is evidenced by irregular fluctuations of the instantaneous velocity w about the mean-time value. Mean square values are the components of the intensity of the fluctuations, or the intensity of turbulence (m s^{-1})

$$I = \sqrt{\langle \Delta w^2 \rangle} \qquad (1.449)$$

For example, in a periodic mixer the turbulence is proportional to the amplitude a (m) and frequency f(s^{-1}) of mixing, i.e. $I \approx af$. The **coefficient of eddy diffusion** can be derived as a function

$$D^E = I^2 t_L \qquad (1.450)$$

where t_L is the characteristic "Lagrangian" time, necessary to reach a linear correlation between the mean displacement Δx^2 and time — cf. eqn. 1.437b. After a short time of operation molecular diffusion prevails (*T.K. Sherwood, 1940*).

When a fluid passes through a porous medium of particles of diameter d at a velocity w, we have approximately

$$D^E \approx wd/2 \qquad (1.451)$$

(*R. Aris, N.R. Amundson, 1957*).

For the movement of fluid in smooth tubes and slots there is an exponential relation between D^E, the mean linear velocity $\langle w \rangle$ and the tube diameter (slot thickness) d, $\log D^E \sim \log (\langle w \rangle d)$ and D^E varies from 10^{-5} m² s⁻¹ in liquids (for $wd = 10^{-2}$ m² s⁻¹) to 10^{-2} m²s⁻¹ in gases (for $wd = 10$ m² s⁻¹) (*T.K. Sherwood, R.L. Pigford, C.R. Wilke, 1975*).

For a concentration "piston-pulse" in a flowing liquid, the eddy diffusion coefficient may be several orders of magnitude higher than molecular diffusion. However, when the thickness of the band transported is large enough, its diffusional broadening is negligible.

For practical calculations, steady state diffusion (Figure 31) is often presumed and is defined in the direction of increasing x and decreasing c ($\Delta c = c'' - c', \Delta x = x'' - x'$) as

$$j = K\Delta c \tag{1.452}$$

where K(m s⁻¹) is the **transfer coefficient**

$$K = \frac{D^M + D^E}{\Delta x} \tag{1.453}$$

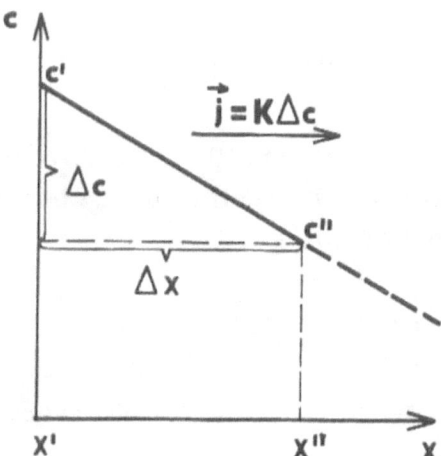

Figure 31. Linear gradient model of mass transfer.

This parameter is convenient for experimental determination for gas-liquid and liquid-liquid interfaces (*W.K. Lewis, W.G. Whitman, 1924; J.B. Lewis, 1954, 1958*). The description of convective transport in turbulent layers, however, remains one of the most difficult tasks of chemical engineering (*J.T. Davies, 1972; V.S. Krylov, 1980*).

In another form of mass transport, effusion, where gas escapes through an opening which is small compared with the mean free path of the molecules, i.e. at low pressures, the collisions of the molecules with the walls of the opening become the determining factor. On the basis of the kinetic theory of gases it was shown (*M. Knudsen, 1909*) for the **effusion molar flux** through a hole of diameter r

$$j = -\frac{2}{3} r \left(\frac{8RT}{\pi M}\right)^{1/2} \frac{dc}{dx} \qquad (1.454)$$

(mol m^{-3} s^{-1}) where M is the molar mass (kg mol^{-1}) and dc/dx is the concentration gradient along the length of the hole. If the latter is equal to d, the gradient can be approximated by $dc/dx = \Delta c/d$. According to ideal gas law 1.36 ($p/RT = n/V = c$), the expression for **Knudsen diffusion** from a region at pressure p_1 to one with pressure p_2 gives for the flux:

$$j = \frac{8}{3} r \frac{1}{(2\pi MRT)^{1/2}} \frac{p_1 - p_2}{d} = w \qquad (1.455)$$

(m^3 m^{-2} s^{-1} = m s^{-1}). In the extreme case, when the opening is short, $d \ll r$, the effusion flux becomes independent of the hole size (*P. Clausing, 1930*):

$$j = \frac{1}{(2\pi MRT)^{1/2}} (p_1 - p_2) \qquad (1.456)$$

and when $p_2 \ll p_1$ (effusion in vacuo),

$$j = \frac{p_1}{(2\pi MRT)^{1/2}} \qquad (1.457)$$

This equation also gives the **molecular distillation** flux (mol m^{-2} s^{-1}) of the substance with vapor pressure p_1 (*I. Langmuir, 1913*). Because the flux is inversely proportional to $p/M^{1/2}$, it is possible using molecular distillation to separate either according to the vapor pressure difference or the molecular mass difference. In a vacuum $0.01 \div 0.1$ Pa (about $10^{-4} - 10^{-3}$ Torr) the mean free path of the molecules is 20-30 mm and the process of molecular transport by evaporation and condensation is quite efficient (*P.C. Carman, 1948*). For instance, even at a relatively low temperature $= 350\ K(77^\circ C)$, such a weakly-volatile substance as stearic acid ($p_1 \cong 0.1$ Pa at the indicated temperature) can be distilled. According to equation 1.457, while $M = 0.284$ kg mol^{-1}, the molar flux reaches 1.38×10^{-3} mol m^{-2} s^{-1} (3.93×10^{-4} kg m^{-2}s^{-1}). This corresponds to evaporation of a layer about 0.027 mm thick per minute. As a result, heat-sensitive materials can be separated relatively quickly in this way and fully from thin layers (usually 0.05–0.5 mm), especially with wiped films (*J. Cvengros, A. Tkac, 1978; A. Tkac, J. Cvengros, 1978*).

Molecular distillation of ice from frozen water-containing materials is used for mild **freeze drying** (lyophilization) of sensitive biological materials (*K. Kroll, 1959; R.B. Keey, 1972*).

In the Knudsen regime, various components of gaseous mixtures are transferred independently, each of the components behaving according to its own partial pressure difference in relations 1.455–1.457.

At higher pressure the mutual collisions of molecules are more frequent and the Knudsen flux changes to a **viscous flow**. As was indicated above, this flow also transfers substances, but without any separation effect: the mixtures behave as a homogeneous gas of average viscosity. Some chemical peculiarities appear, however, in liquid solutions of macromolecules and for high gradients of velocity vectors — equation 1.382 — when shear-induced degradation of biomacromolecules may occur.

Knudsen's formula 1.455 does not include the dynamic viscosity η (Pa s = kg m^{-1} s^{-1}) as a parameter of flux. For **hydrodynamic viscous flux** the **Hagen-Poiseuille relations** (*G. Hagen, 1839; J.L. Poiseuille, 1844*) are more suitable, either for a gas the volume of which is related to pressure P_0,

$$j = w = \frac{r^2}{8\eta} \frac{p_1^2 - p_2^2}{2 d p_0} \tag{1.458}$$

(m s^{-1}), or for a non-compressible liquid, by the integration of Newton's eqn. 1.382, as follows

$$j = w = \frac{r^2}{8\eta} \frac{p_2 - p_1}{d} \tag{1.459}$$

Both of these equations can be applied when the linear convection velocity w of fluid passing through a hole of diameter r is to be found, e.g. by membrane separation — see section 1.5.1.

It should be stressed that the viscosity in the Hagen-Poiseuille equations 1.458 and 1.459 is the <u>average</u> viscosity of the mixture and does not characterize individual components of the mixtures. There are no simple functions relating the composition and **viscosity of mixtures**. B. Ya. *Fialkov (1963)* proposed empirical formulae for deviations of the real viscosity η from that (η_{AB}) deriving from the additivity rule in mixtures of A and B, ($\eta_A > \eta_B$):

$$\eta_{AB} = \eta_A v_A + \eta_B v_B \tag{1.460}$$

(v's are volume fractions) at $v_A = 0.25, 0.50$ and 0.75 as follows:

$$\eta = L_{AB} \eta_{AB} \tag{1.461a}$$

where L_{AB} is the correction coefficient for the particular mixture,

$$L_{3:1} = \frac{8}{\eta_A/\eta_B + 7} \qquad (at\ v_A = 0.25\ or\ 0.50) \quad (1.461b)$$

and

$$L_{1:3} = L_{3:1} + \frac{\eta_A/\eta_B - 1}{5\eta_A/\eta_B + 20} \qquad (at\ v_A = 0.75) \quad (1.461c)$$

The viscosity of mixtures is then found by interpolation of the 5 points (including pure A and B, $v_A = 1$ and $v_A = 0$). If $\eta_A/\eta_B < 6$, it is possible to use the so-called **Arrhenius equation** for viscosity of mixtures,

$$\ln \eta_{AB} = x_A \ln \eta_A + x_B \ln \eta_B \quad (1.462)$$

where x's are mole fractions (*B.Ya. Fialkov, 1973*), or other semi empirical relations (*R. C. Reid, T. K. Sherwood, 1958*).

Flow through porous media, or **permeation** is also described by means of a viscosity parameter and pore diameter d. **Darcy's equation** (*H.P. G. Darcy, 1856*)

$$j = w = \frac{K}{\eta} \frac{\Delta p}{d} \quad (1.463)$$

also includes the empirical parameter $K(m^2)$, called the **permeability coefficient** of the medium (the effective area of permeation). When mean diameter of pores is the fundamental dimension then Darcy's low is valid for $1 \le Re \le 10$ (*J. Bear, 1972*).

The unit velocity of permeation, $w = 1$ cm s^{-1} was in use as the "Darcy" unit. It was found later that for pores of irregular form, the permeability coefficient is

$$K = K' \frac{\varepsilon^3}{a^2} \quad (1.464a)$$

(*J. Kozeny, 1927; P.C. Carman, 1937; S. Ergun, 1952*) where ε is the porosity, i.e. the ratio of free volume to the total volume of medium (dimensionless), a is the specific surface area per unit volume (m^{-1}) and K' is the empirical dimensionless constant. In real porous media, the channels are not straight and therefore the flux 1.463 is decreased by the **obstruction (tortuosity) factor** = 2 – 10, on average 3-times that in globular structures.

For spherical or quasi-spherical particles with radius r — cf. eqn. 1.684a, the specific surface area can be calculated as follows,

$$a = \frac{3}{r}(1 - \varepsilon) \tag{1.464b}$$

and K' is about 0.24 — cf. equation 1.8. For instance, in a chromatographic column with a length = 0.6 m, filled by an adsorbent of mean diameter r = 0.038 mm and porosity of the bed ε = 0.4, we have a = 3 × (1 – 0.4)/(3.8 × 10^{-5}) = 4.74 × 10^4 m^{-1}. If the column is filled with water 20 cm above the sorbent bed (h = 0.6 + 0.2 m), the hydrostatic pressure will be Δp = $\rho g h$ = 10^3 × 9.81 × (0.6+0.8) kg m^{-1} s^{-2} = 7.85 kPa. According to equation 1.464a, K = 0.24 × 0.4^3/(4.74 × 10^4)^2 = 6.84 × 10^{-12} m^2 and the hydrostatic pressure will promote the flow of water (viscosity η = 1 mPa s) w = 6.84 × 10^{-12} × 7.85 × 10^3/(10^{-3} × 0.6 m s^{-1} = 8.95 × 10^{-5} m s^{-1} (0.32 m per hour). For a substance with a diffusion coefficient = 2 × 10^{-9} m^2 s^{-1}, the Péclet diffusion number with respect to column length will be (eqn. 1.430) Pe = (8.95 × 10^{-5})(0.6)/10^{-9} ≈ 27000. This means that the diffusion spreading of a chromatographic band is not serious (compare with Figure 27); the eddy diffusion would be still negligible due to low values of D^E — see eqn. 1.451. The diffusion broadening on replacement of the concentration band diminishes with increasing Péclet number — Figure 32.

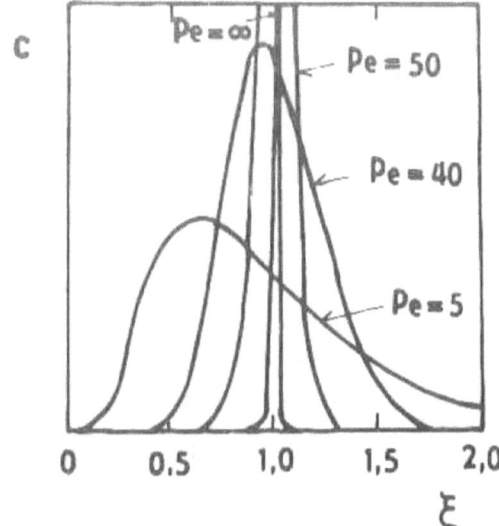

Figure 32. Spreading of concentration band by convective transport at various Péclet numbers.

$Pe = \infty$ corresponds to ideal piston flow (undisturbed zone movement) and $Pe = 0$ refers to ideal (prompt and full) mixing –see also section 1.4.5, eqn. 1.617.

Transfer in the regime intermediate between Knudsen and Darcy flow is given by a linear combination of eqns. 1.455 and 1.463 (*H. Adzumi, 1937*).

The molecules sorbed on the surface of pores may move faster than by 3D molecular diffusion. This **surface migration** is important in sorption phenomena and transfer through micropores (*M. Volmer, J. Esterman, 1921; A. Gosman, S. Linkkonen, P. Passiniemi, 1986*). The effective diffusion coefficient is

$$D^{ef} = D^M + kD^S \tag{1.465}$$

where D^S is the coefficient of **surface (Volmer) diffusion** (m s⁻¹) and k (m) is the absorption coefficient corresponding to product $k_1 k_2$ in the Langmuir adsorption isotherm, eqn. 1.323, or to $1/k_A$ in Henry's isotherm, eqn. 1.79. To give the surface diffusion coefficient the same dimensions as for D^M (m²s⁻¹), k should be the dimensionless distribution coefficient K_1^D defined by eqn. 1.307 (*G. Damkohler, 1936*).

The preceding discussion illustrates the general utility of the dimensionless Péclet number, eqn. 1.430, and dimensionless time, eqn. 1.428, for the description of mass transfer phenomena. There are other dimensionless groups convenient to characterize transport processes. They derive from the general **theory of similitude** *(J. Bertrand, 1848; V.L. Kirpichev, 1874; D. Rayleigh, 1892, 1915; A. Federman, 1911; E. Buckingham, 1914; M.V. Kirpichev, 1953)*.

We have seen that similar systems are described by the same dimensionless initial and boundary conditions. In connection with eqns. 1.419 and 1.429, it was shown that on the introduction of dimensionless parameters, the number of independent parameters decreases by the number of characteristic (fundamental) dimensions introduced: by linking the path and time of diffusion with the characteristic length of the system and diffusion coefficient of the transported substance; instead of five parameters (x, t, w, d and D) the system is described by three dimensionless ones (ξ, τ and Pe). Generally, if the equation of state is given by n parameters x_i:

$$f(x_1, x_2, \ldots, x_i, \ldots, x_n) = 0 \tag{1.466}$$

it can be transformed to equation

$$\varphi(\pi_1, \pi_2, \ldots \pi_i, \ldots, \pi_r) = 0 \tag{1.467}$$

where each of the r dimensionless parameters π_j can be expressed as the product

$$\pi_j = x_1^{\alpha_{1j}} x_2^{\alpha_{2j}} \ldots x_n^{\alpha_{nj}} = \prod_{i=1}^{n} x_i^{\alpha_{ij}} \tag{1.468}$$

and related to p fundamental dimensions so that

$$r = n - p \tag{1.469}$$

This is the so-called **π-theorem** of dimensional analysis (*E. Buckingham, 1914*). For example, the dimensionless time τ introduced by eqn. 1.428 is the π-parameter derived from $x_1 \equiv x$, $x_2 \equiv t$, $x_3 \equiv D$, $x_4 \equiv w$ and $x_5 \equiv d$ ($r = 5$) using fundamental dimensions D and $d(p = 2)$:

$$\tau = x^0 t^1 D^1 w^0 d^{-2} \qquad (1.470)$$

i.e. $\alpha_1 = 0, \alpha_2 = 1, \alpha_3 = 1, \alpha_4 = 0$ and $\alpha_5 = -2$.

Similitude theory says that the necessary and sufficient condition for **similarity of processes** consists of the equality of all dimensionless π-parameters, including those defining the initial and boundary conditions. π-parameters derived from differential equations are the same as those obtained after integration (*T.A. Erenferst-Afanassjewa, 1925; P. W. Bridgman, 1950; H.E. Huntley, 1967; H. Gortler, 1975*).

The reduction of the number of the parameters which are necessary for the determination of a process is of great practical importance. For instance, reducing the number of parameters of the separation process from $n = 5$ to $r = 3$, by introducing two fundamental dimensions, means that from a study of the process at four different values of each of the parameters, in the original case there should be $5^4 = 625$ experiments but in the reduced set only $3^4 = 81$ experiments, i.e. nearly 8-fold less (the same remains valid in the case of process modelling and optimization — see section 2.2.3).

In separation processes, the initial concentration can serve as a fundamental dimension and such dimensionless invariants as the dimensionless distribution constants, eqn. 1.306, or distribution ratio, eqn. 1.309, can be used regardless of the conditions under which they are obtained: the performance of the separation process is the same in general engineering theory — see sections 2.1.1 and 2.4.1.

Though the number of dimensionless parameters Π obtained according to eqn. 1.468 may be very great, few of them are used in practice. In systems with complicated hydrodynamics, the following similitude criteria (similitude numbers) are widely used in convective transport processes, beside the Péclet number 1.430:

Diffusion Fourier number

$$Fo_D = \frac{d^2}{D^M t} \qquad (1.471)$$

is identical with the reciprocal value of dimensionless time, eqn. 1.428, and used for characterization of the deviation from steady-state transport conditions that occur at low values of Fo_D ($Fo_D < 1$).

Reynolds number

$$Re = \frac{dw\rho}{\eta} = \frac{dw}{\nu} \qquad (1.472)$$

where d is the fundamental length (m), w is the flow velocity (m s^{-1}), ρ is the density (kg m^{-3}), η is the dynamic viscosity (Pa s = kg m^{-1}s^{-1}) and ν is the kinematic viscosity (m^2 s^{-1}). The Reynolds number characterizes the similitude of weak (Van der Waals) forces in viscous fluids and particularly the change of viscous (streamline) flow to an eddy (turbulent) one. In practice, we choose as the fundamental length the radius of the tube, the particle radius or the length of the mixer paddle. In the latter case, Re can be calculated as

$$Re = \frac{d^2 \omega \rho}{\eta} \qquad (1.473)$$

where ω is angular speed (s^{-1}) of mixer blades with length d. During flow in round pipes, turbulence occurs at $Re > 30 - 100$ in liquids ($Re \leq 2000$ in case of capillaries), and $Re > 1000$ in gases, and laminar flow is destroyed. E.g. in capillary of internal diameter d=0.53 mm, w=0.15 ml/min, liquid viscosity $\eta > 0.2$ Pa s and density $\rho < 1500$ kg m^{-3}, $Re < 45$ and laminarity is preserved. By flowing gas through adsorbent particles laminar flow is preserved up to Re 4 – 6 ($P.C.\ Carman,\ 1956$).

The Reynolds criterion for characterization of sedimentation processes, regardless of whether the Stokes or Newton regimes of sedimentation occur, can be found from eqns. 1.3 and 1.4. Similitude of the sedimentation process would occur if the ratio of velocities were constant,

$$C = \frac{24\eta}{d\rho_2 w} \tag{1.474}$$

i.e. introducing Re from 1.472

$$C = \frac{24}{Re} \tag{1.475}$$

It was established experimentally that if the drag coefficient C is greater than 12–24 ($Re < 1 - 2$) sedimentation proceeds under the condition of laminar flow, i.e. according to Stokes law, eq. 1.3, and vice versa.

Schmidt number (diffusion Prandtl number) is the dimensionless complex

$$Sc \equiv Pr_D = \frac{\eta}{\rho D^M} = \frac{\nu}{D^M} \tag{1.476}$$

where the values are the same as above. It does not depend on hydrodynamic parameters (flow and dimensions) but is entirely a function of the matrix viscosity and solute diffusion coefficient. The ratio of viscosity and diffusivity indicates that the Schmidt number gives the relative speed of momentum transfer to diffusive transport of mass. The thickness of the laminar hydrodynamic layer in which a gradient of velocity exists (δ_{hydro}) and the diffusion layer where a gradient of concentration exists (δ_{diff}) is

$$\frac{\delta_{\text{hydro}}}{\delta_{\text{diff}}} = Sc^{\frac{1}{n}} \tag{1.477}$$

where n is the coefficient of turbulence dumping: in liquid-solid systems $n \approx 3$, in liquid-gas systems $n \approx 2$ (Figure 33).

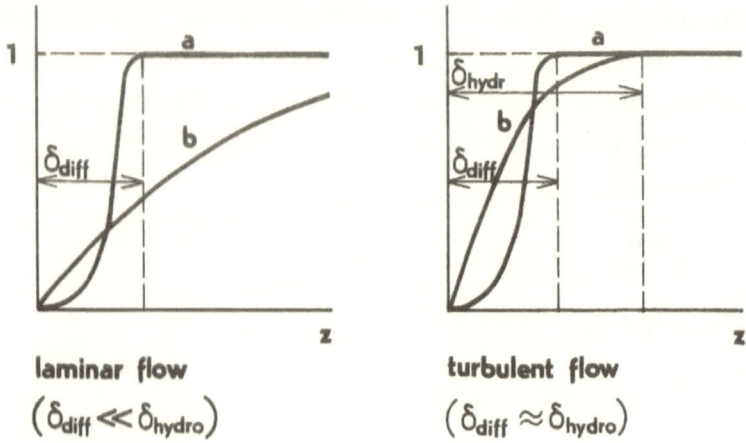

Figure 33. Diffusion (Nernst) and hydrodynamic (Prandtl) layers under laminar and turbulent flow: a — dimensionless concentration, b — dimensionless velocity.

The transfer of momentum (energy) is much more effective than mass transfer and therefore the hydrodynamic layer in liquids exceed the diffusion layer by one order of magnitude: because usually $Sc = 300 - 2300$, (the maximal value of Sc was obtained in saturated $CaCl_2$ solutions, $Sc = 40000$) $\frac{\delta_{hydro}}{\delta_{diff}} \approx 7 - 13$. In gases (vapors) we find $Sc = 0.6 - 2.6$ and the ratio, eqn. 1.447, is in the range 0.8-1.6, i.e. the hydrodynamic and diffusion layers are practically the same size.

Sherwood number (diffusion Nusselt number)

$$Sh \equiv Nu_D = \frac{dK}{D^M} \qquad (1.478)$$

includes the transfer coefficient K (m s $^{-1}$) from equation 1.453. Considering the latter equation, we have approximately

$$D^M \approx \frac{D^M + D^E}{D^M} \qquad (1.479)$$

i.e. it gives the relative importance of eddy diffusion in the transport mechanism.

The Sherwood number is an analogue of the Péclet number and their ratio gives the **diffusion Stanton number**:

$$St_D = \frac{Sh}{Pe_D} = \frac{K}{w} \qquad (1.480)$$

Now, the equations

$$Sh = St_D\, Re\, Sc \qquad (1.481a)$$

or

$$Pe_D = Re\, Sc \qquad (1.481b)$$

give the relative rate of transfer of substance that is caused by convection to that by pure molecular diffusion.

Real systems do not behave exactly according to eqn. 1.480. This serves however as a good approximation and therefore more a general relationship, e.g. for mass transfer, is sought in the form

$$Sh = C\, Re^m\, Sc^n \qquad (1.482)$$

(R. Higbie, 1935) where C, m and n are dimensionless coefficients that should be found empirically. Usually this is done in coordinates $\log Sh$ vs. $\log Re$, for a given composition of the system, where Sc = const, or $\log Sh$ vs. $\log Sc$ for systems of various compositions at constant hydrodynamic conditions when Re = const. The fundamental dimension d_0 is usually excluded and eqn. 1.482 is used in the form

$$Sh = C'(d/d_0)^t\, Re^m\, Sc^n \qquad (1.483)$$

Eqn. 1.482, which extends the validity of the theoretical eqn. 1.478, is a good example of the semi empirical engineering approach to models of complicated systems by means of similitude theory: a great number of hydrodynamic, geometrical and molecular factors are fused into complex dimensionless parameters and empirically adjusted to describe the behavior of real systems.

1.4.3 Transport in external field. Electrophoresis

In a homogeneous solution placed in an external field (gravitational, centrifugal, electric, etc.), a stationary gradient of concentration should occur when the transport of the i-th component eliminates its diffusion according to eqn. 1.445b. Because the flux is most generally expressed through the gradient of the chemical potential μ, the equation can be adjusted for integration by means of eqn. 1.399, resulting in the steady concentration gradient equation for ideal solutions

$$\frac{dc_i}{c_i} = \frac{1}{RT} d\mu_i \qquad (1.484)$$

which can be obtained directly from eqn. 1.81 for the chemical potential. This is because the relative concentration gradient reflects the gradient of the chemical potential caused by external fields (J.C. Giddings, 1969). For a noticeable concentration change, say 1%, $(dc/c \geq 0.01)$ at normal temperature ($T = 300$ K, $RT = 2.5$ kJ mol^{-1}) the chemical potential difference should be $= 0.01 \times 2.5 \times 10^3 = 25$ J mol^{-1}. Let us see how the gradients are created in the fields available.

The contribution of the gravitational field to the molar Gibbs energy of the solute is

$$\mu_i = (M_i - \rho \overline{V}_i)gx \qquad (1.485)$$

where g represents gravitational acceleration (9.81 m s^{-2}) at the standard state at sea level, x is the altitude above sea level, M_i is the molar mass (kg mol^{-1}) of the solute, $\overline{V}$ is molar volume of the solute (m^3 mol^{-1}) and ρ is the density of the buoyant solvent (kg m^{-3}). By analogy, in a centrifugal field

$$\mu_i = (M_1 - \rho\overline{V}_i)\omega^2 x^2 \tag{1.486}$$

where ω is the angular velocity and x is the distance from the rotor center —see eqn. 1.4. The gradient is formed both on sedimentation ($M_i > \rho\overline{V}_i$) or flotation ($M_i < \rho\overline{V}_i$). For ions charged by a net charge $z_i e$ at a potential E, the electrochemical potential 1.109 results in

$$\mu_i = z_i eE \tag{1.487}$$

The steady concentration gradient in the gravitational or centrifugal field can also be described by eqn. 1.484 on substitution from 1.485 or 1.486 for the boundary condition $c_i = (c_i)_0$ at $x = 0$:

$$\ln c_i = -\frac{M_i(1-\rho/\rho_i)}{RT}\omega^2 x^2 \tag{1.488}$$

or

$$\ln \frac{c_i}{(c_i)_0} = \frac{-(M_i - \rho\overline{V}_i)d^2}{RT}\omega^2 x^2 \tag{1.489}$$

(for gravitational field $\omega^2 x^2$ is replaced by gx). The average energy of molecular motion is, at normal temperature, about = 2500 J mol^{-1}. In aqueous solutions ($\rho = 10^3$ kg m^{-3}) such motion can compensate the sedimentation of a solute with a density slightly higher, $\rho_1 = 1020$ kg m^{-3} (typical for bioorganic materials) at a distance $x = 1$ m even if the molar mass of the solute is as high as $M_i = 2500 \times 9.81/(1 - 1000/1020) \approx 1.2 \times 10^6$ kg mol^{-1}, which greatly exceeds the molar mass of high-molecular proteins. Therefore, noticeable concentration gradients can occur only in centrifugal fields (T. Svedberg, K.O. Pedersen, 1940). In strong fields, and especially for small-particle substances, the gradient is very steep and the substance concentrates in a narrow zone (forming a thick suspension, paste, pellet) — section 1.1.

Generally, there is no need to wait for a steady gradient and the substances are separated in conditions, far from the steady state, known as **rate-zonal separation** (*P.C.Wankat, 1944*). In any event, the problem of mechanical separation remains. The latter is easier when a solid deposit (sedimentation pellet or electrodeposit) is to be removed, or if a gel-like medium or capillaries are used for gradient formation. A gel can be cut into 0.5–1 mm slices, each containing a separate solute. Differential pelleting is rather difficult and the pelleting is normally used just for preliminary enrichment of the sedimenting particles. In liquids, the centrifugal gradient zones can be carefully unloaded by upward displacement into 0.05–0.1 cm³ fractions. The rate of transport is given in eqn. 1.3. The sedimenting properties of a substance can be included in the **sedimentation coefficient** (s_i),

$$s_i = \frac{d_i^2(\rho_i - \rho)}{18\eta} = \frac{D_i N_A}{RT} \frac{\pi}{6} d_i^3 (\rho_i - \rho) \tag{1.490}$$

and therefore, under gravity

$$w_i = g s_i \tag{1.491a}$$

and in a centrifugal field (*T. Svedberg, J.P. Nichols, 1923; T. Svedberg, 1926, 1929; T. Svedberg. I. Eriksson, 1933*)

$$w_i = \omega^2 r s_i \tag{1.491b}$$

For example, bacterial DNA (d = 0.08 µm) in CsCl solution has a buoyant density of 1700 kg m⁻³. When centrifuged in a cesium chloride solution with ρ = 1600 kg m⁻³ and η = 1.2 mPa s, the sedimentation coefficient s_i = 2.96 × 10⁻¹¹ s (= 296 Svedbergs). At 50,000 r.p.m. in a bottle ultracentrifuge, sedimentation starts at a radial distance x_1 = 3 cm from the center with the velocity w_1= (50000 × 2π/60)²(0.03)(2.96 × 10⁻¹¹)= 2.43 × 10⁻⁵ ms⁻¹.

The Reynolds number, according to eqn. 1.472, is $Re = (8 \times 10^{-8})$ $(2.43 \times 10^{-5})(1600)/(1.2 \times 10^{-3}) = 2.6 \times 10^{-6}$ and therefore the criterion 1.475 for laminar flow and Stoke's law (eqn. 1.42) holds without question. The time necessary to reach the bottom of the centrifugal tube, which is at a distance $x_2 = 8$ cm from the axis, will be shorter than $(x_2 - x_1)/w_1 = 2.06 \times 10^3$ s, because the sedimentation velocity increases with distance from the axis. To obtain this, eqn. 1.491b is integrated ($w = dx/dt$) which results in

$$\ln \frac{x_2}{x_1} = \frac{d_i^2(\rho_i - \rho)\omega^2}{18\eta} t = s_i \omega^2 t \qquad (1.492)$$

Consequently $t = \ln (8/3) \times 18 \times 0.012/(8 \times 10^{-8})^2 (1700 - 1600)$ $(50000 \times 2\pi/60)^2 = 1.2 \times 10^3$ s, i.e. it takes about 20 minutes for the DNA molecules to settle completely.

Thus, the sedimentation transport rates of molecular species are quite low although the acceleration in centrifuges far exceeds that of gravity and is much greater than that caused by diffusion in the concentration gradient.

The electric field is a superior external field to induce the separation of charged particles (M. Bier, 1959; W. Thorman, D. Arn, E. Schumacher, 1984; W. Thorman, R.A. Mosher, 1985). The transport velocity of ions with mobility u_i (m^2 V^{-1}s^{-1}) — eqn. 1.406 — under an electric field at a field strength $E = d\phi/dx$ (Vm −1) is given as

$$w_i = E u_i \qquad (1.493)$$

which is the basic equation for **electrophoretic separations** (A. Tiselius, 1937) based on the different velocities of ionic solutes in electric fields. The ionic mobility of singly charged, small ions in an aqueous medium is in the range $(5 \div 36) \times 10^{-8}$ m^2 V^{-1}s^{-1} ($5 \div 36 \times 10^{-4}$ cm^2 V^{-1}s^{-1}).

The ionic mobility can be calculated from diffusion coefficients and the Nernst equation, 1.406a. To a first approximation of an ideal spherical molecule ($f = 6\pi N_A \eta r$)

$$u_i = \frac{D_i z_i F}{RT} = \frac{e z_i}{6\pi\eta r_i} \tag{1.494}$$

where e is the elementary charge ($e = F/N_A = 1.60 \times 10^{-19}$ C).

Following the example given above, to move the singly-charged DNA molecule with the same velocity by ultracentrifugation ($w_i = 4 \times 10^{-5}$ m s^{-1}), when its mobility is of the order $u_i = 10^{-6}$ cm^2 V^{-1} s^{-1} = 10^{-10} m^2 V^{-1}s^{-1}, a field strength $E = \frac{w_1}{u_i} = 4 \times 10^5$ V m^{-1} (400 Vcm^{-1}) is required.

In separation practice, the separated component is compared with standard markers and its relative mobility (dimensionless) is defined as

$$v_i = \frac{u_i}{u_s} = \frac{w_i}{w_s} \tag{1.495}$$

where the w_i's are migration velocities determined under given experimental conditions (voltage, carrier and time of separation) from the migration distance l ($w = l/i$):

$$\frac{w_i}{w_s} = \frac{l_i}{l_s} \tag{1.496}$$

An approximate description of mobility as a function of relative molecular mass M_i is given by the equation

$$u_i = a - b \log M_i \tag{1.497}$$

which can be used for a series of similar substances, a and b being constants which should be found empirically under given conditions (G. Piljac, V. Piljac, 1986).

In principle, **electrophoretic separation or electrokinetic separation** (E. Durrum, 1950; M. Bier, 1957; J. Kohn, 1958; J.C. Giddings, 1969; T.M. Jovin, 1973; J. Clausen, 1979; W. Thorman, R.A. Mosher, M. Bier, 1986) can be performed in two ways:

(i) components A and B are transferred independently on the background of a conducting electrolyte. Separation is performed

—on paper wetted by a background electrolyte (*W. Grassmann, K. Hannig, 1953; O.Mikes, 1957; V. Jokl, 1964; J.R. Whitaker, 1967; J. Majer, V.Q. Trinh, I. Valaskova, 1980*).

—in 1-20 mm layers of gels swollen in electrolyte solutions (*R.C.Allen, H.R. Maurer, 1974; C.R. Cantor, C.L. Smith, 1986; M. Burmeister, L.Ulanovsky, 1992; B. Birren, E. Lai, 1993*) or

—in capillaries (*P.D. Grossman, J.C. Colburn, 1992; J.P.Landers, 1994; I. Ali et al., 2007*).

If present in a narrow zone, after time t they pass various distances at **constant E**:

$$l_A = w_A t \qquad (1.498a)$$

$$l_B = w_B t \qquad (1.498b)$$

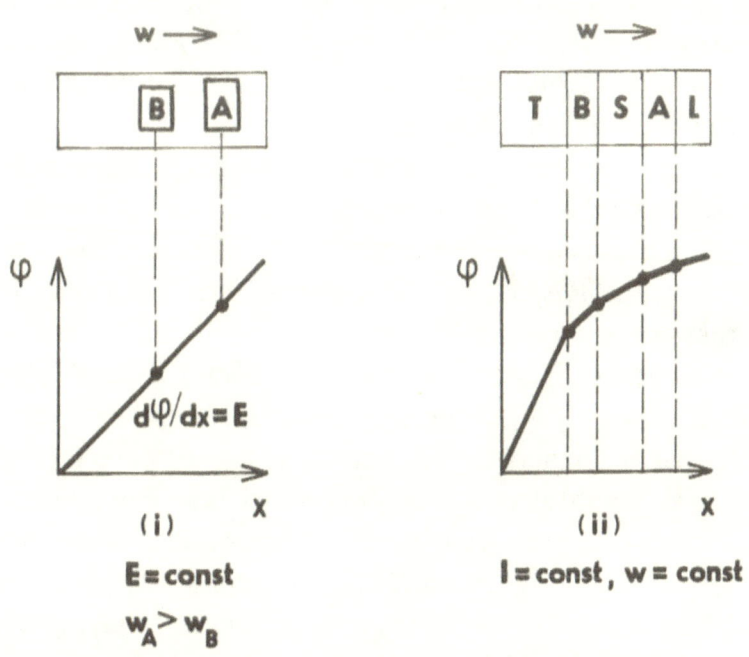

Figure 34. Electrophoretic transport of ionic solutes A and B at constant field strength (E) and free movement or at constant velocity (w) in tight zones, together with other ions (T, S and L).

The interzone distance at time t is, according to eqn. 1.493,

$$l_A - l_B = E(u_A - u_B)t \qquad (1.499)$$

and after some time the zones can be sectioned. For instance, if a mixture of ions with mobilities $u_A=3\times10^{-9}$ and $u_B=5\times10^{-9}$ m^2 V^{-1} s^{-1}, introduced as a 2 cm band into an electric field gradient 500 V m^{-1} is to be separated, it occurs in at least $t = 0.02/500 \times 2 \times 10^{-9} = 20,000$ s (about 6 h).

The separation time is generally reduced by high voltage gradients. However, Joule heating causes the mixing of zones both by diffusion and convection. The resolution is better with proper cooling because of the high Péclet number — eqn. 1.430. Large molecules (with small diffusion coefficients) are better resolved by electrophoresis for long periods of times at a low field strength. High-performance is achieved by **capillary electrophoresis**, HPCE (*R. Weinberger, 1993; P.D. Grossman, J.C. Colburn, 1992; A. Weinberger, 1993; J.P. Landers; 1994; I. Ali, Imran; H.Y. Aboul-Enein, V.K. Gupta, 2009*).

The convective flow of solvent and chromatographic properties can be used either in countercurrent mode to minimize the size of the apparatus down to the maximal distance between zones (*W. Preetz, H.L. Pfeifer, 1967; P.H. O'Farrell, 1985; B.J. McCoy, 1986*) or cross-current, enabling continuous separation by shifting the zones in a perpendicular direction (*J. Barrolier, E. Watzke, H. Gibian, 1958; K. Hannig, 1961; Z. Prusik, 1973*). Unwanted convection is decreased by an increase of electrolyte viscosity and by cooling so as to eliminate electrical heating, which is proportional to the square of the current. A variety of gels are used as viscous media (*A. Gordon, B. Keil, K. Sebesta, 1949; O. Smithies, 1955; M.D. Poulik, 1957; S. Raymond, 1959; S. Raymond, M. Nakamichi, 1962; R.J. Wieme, 1965; J.R. Whitaker, 1967; C.J.O.R. Morris, 1971; H.R. Maurer, 1971; R. DeWachter, W. Furs, 1972; F.M. Jovin, 1973*), polyacrylamide, agarose and starch being the most popular (agarose and starch exhibit more electro-osmotic properties — see section 1.4.1).

Paper, microgranular cellulose and cellulose acetate membranes are used as carriers suitable to hold a large amount of electrolyte (E. Durrum, 1950; W. Grassmann, K. Hannig, 1953; J. Kohn, 1958; O. Mikes, 1957).

(ii) The ionic components A (charge z_A) and B (charge z_B) can be separated in the absence of background electrolyte as the salts of a common counter ion C which conduct the current. To keep the conductivity to a reasonable level, the separation is performed in capillaries and now, instead of the field strength, it is the current which is kept constant. The **constant electric current density** i in the steady state zones is given by Faraday's law

$$i = F(z_A i_A + z_C i_C) = F(z_B i_B + z_C i_C) \qquad (1.500)$$

The current density i ($A\ m^{-2}$) is connected with the specific conductivity κ_i (Sm^{-1}) and the field strengthen in a given field ($V\ m^{-1}$) in the i-th zone, in agreement with Ohm's law, as

$$i = \varphi_i \kappa_i \qquad (1.501)$$

The specific conductivity and molar conductance Λ_i^0 ($S\ m^2 mol^{-1}$) is given by

$$\kappa_i = c_1 \Lambda_1^0 \qquad (1.502)$$

Because the concentration of ions is defined by Kohlrausch's law (F. Kohlrausch et al., 1879) of independent ionic mobilities, we have

$$\Lambda_{AC}^0 = \Lambda_A^0 + \Lambda_C^0 \qquad (1.503a)$$

$$\Lambda_{BC}^0 = \Lambda_B^0 + \Lambda_C^0 \qquad (1.503b)$$

With respect to eqn. 1.406b or to eqn. 1.494, the concentration of the ions in the presence of the common counter-ion C is given as

$$\frac{c_A}{c_B} = \frac{|z_A| t_A}{|z_B| t_B} \tag{1.504a}$$

where t_A and t_B are the dimensionless transference numbers of the ions,

$$t_A = \frac{|z_A| c_A u_A}{|z_A| c_A u_A + |z_C| c_C u_C} \tag{1.504b}$$

$$t_B = \frac{|z_B| c_B u_B}{|z_B| c_B u_B + |z_C| c_C u_C} \tag{1.504c}$$

The result is that the fastest moving zone also contains the highest concentration and vice versa. The ions in this case arrange themselves in order of net mobility in intermediate contact with their neighbors, their velocity being the same — which is why the name **isotachophoresis (ITP)** was given to this **displacement electrophoresis** technique (*R.A. Alberty, E.L. King, 1951; B.P. Konstantinov, O.V. Oshurkova, 1963, 1966; A.J.P. Martin, F.M. Everaerts, 1970; H. Haglund, 1970; F.M. Everaerts, J.L. Beckers, T.P.E.M. Verheggen, 1976; F.M. Everaerts, et al. 1976; F.M. Everaerts, T.P.E.M. Verheggen, E.P. Mikkers, 1979; D. Kaniansky, P. Havasi, 1983; P. Bocek et al., 1985; W. Thormann, R.A.Mosher, M. Bier, 1986*). According to eqn. 1.493, there is a different field strength in the steady state over each zone:

$$w = u_A E_A = u_B E_B \tag{1.505}$$

with a step in the field strength at the zone boundary (Figure 34). In practice, the separated ions move between the continuous zones of two added electrolytes: a high net-mobility ion called the leading electrolyte (L) and a low-mobility ion called the **terminating electrolyte** (T). Usually, in anion separation, chloride and glycinate salts of (hydroxymethyl)- aminomethane (TRIS) are used as leading and terminating electrolytes, respectively, and for cations $L \equiv K^+$ and $T \equiv$ protonated amino acid. To enable resolution of two narrow zones, another electrolyte, the **spacer** (S) having an intermediate mobility

$$u_A > u_S > u_B \qquad (1.506)$$

should be added. For the resolution of n microcomponents, $(n - 1)$ spacer ions are also added.

In separation devices, capillaries of 0.2 − 0.5 mm in diameter or a flat bed of granulated gel are used. Scaling-up of electrophoresis for preparative purposes is a difficult, still unresolved task. One major problem is the free convection caused by electrical heating (minimized in gravity-free space), and gas release at the electrodes, in addition to the manipulation involved in the final mechanical separation of the zones. However, such large-scale electrically-driven separations, such as isoelectric focusing and electrodialysis will be discussed later (section 1.5).

Before discussing isoelectric focusing, the possibilities of control of electric mobility should be understood. This can be achieved either by increasing the size of a molecule by formation of an associate (complex) or by changing the net charge by adding or removing a charged particle. Relaxation of large DNA molecules at higher fields pulse application is used in **pulse-field electrophoresis** (PFG) (*O.J. Lumpkin, P. Dejardin, B.H. Zimm, 1985; C.R. Cantor, C.L. Smith, M.K. Mathew, 1988; J.M. Deutsch, 1987; G.F. Carle, M. Frank, M.V. Olson, 1986; S.D. Levene, B.H. Zimm, 1987; M. Burmeister, L. Ulanovsky, 1992*).

Aggregation is typical for biomacromolecules. (*J.M. Deutsch, 1987*). For instance, RNA often forms specific aggregates which might be interpreted as authentic RNA components. An antibody has two combining sites which will bind specifically with one or several antigenic determinants, to form an "immunoprecipitater" which stops antigen transfer and is used in **immunoelectrophoresis** (*O. Ouchterlony, 1949; M. Kaminski, O. Ouchterlony, 1951; P. Grabar, C.A. Williams, 1953; J. Kohn, 1958; P. Grabar, P. Burtin, 1964; O. Marcini et al., 1965; N.H. Axelsen,*

1975; L.P. Cawley at al., 1976) and by electrophoretic transfer from a gel to another support (W.N. Burnette, W. Neul, 1981). Metal ions easily enter complex-forming reactions with various ligands present in the background electrolyte; addition of strong complexing agents dramatically changes the electric mobility of the ions, both due to an increase in the solvation diameter and a decrease (in the case of anionic ligands) in the positive charge of the central atom (R.A. Alberty, E.L. King, 1951). When complex formation is rapid, the mixture of various complexes has a **mean electric mobility**

$$\langle u \rangle = \frac{\sum u_i c_i}{c} = \frac{\sum u_i c_i}{\sum c_i} \tag{1.507}$$

For the complexation reaction 1.178 and the stability constant 1.185,

$$c_i \equiv [ML_i] = [M]\beta_i[L]^i, \tag{1.508}$$

Corresponding to the average property of a complex mixture — eqn. 1.189 —the expression 1.507 is converted into

$$\langle u \rangle = \frac{\sum u_i \beta_i [L]^i}{\sum \beta_i [L]^i} \tag{1.509}$$

i.e. the effective mobility of the ion depends on the equilibrium ligand concentration in the background electrolyte (V. Jokl, 1964; J. Majer, T. Van Quy, I. Valaskova, 1980; F. Roesch et al., 1989). The mobilities of individual complexes cannot be measured directly. In simple cases, when a small anionic ligand L^- does not significantly change the size of the hydrated ions (on replacing water molecules in the hydration shell) and $r_i \cong r_{i+1}$, then according to eqn. 1.494, the relation between the mobilities and charges can be taken as

$$\frac{u_{i+1}}{u_i} \simeq \frac{z_{i+1}}{z_i} = \frac{z_i - 1}{z_i} \tag{1.510}$$

where z_i is the net charge of complex ML_i (z_0 is the net charge of the central ion). The mobility of the central ion, u_0 can be established experimentaly in the absence of the ligand ($[L] = 0$). For instance, according to eqn. 1.510, the mobilities of the complexes $M^{3+}:ML^{2+}:ML_2^+$ are expected to be in a ratio 1:2/3:1/3 (3:2:1), but the difference is usually smaller because of the actual increase in size of the complex ion.

A remarkable change of charge may occur in proteins. Carboxylic side-chains ionize (at pH 2 − 4) and the amine groups of nucleotide residues are protonated in acid solution (usually below pH 2). The result of these negative and positive charges is a protein with a net charge varying between several positive and negative units (± 7). The change of electrophoretic mobility with pH is characterized by the slope $du/d(pH) \approx dz_i/d(pH)$. Obviously, at some pH the protein has zero net charge; the pH is then called the **isoelectric point** (pI) at which $du/d(pH) = 0$. This is used for electrophoresis in a pH gradient.

In the steady state, the protein stops moving at the position where the *pH* corresponds to its particular *pI* value and is focused there (*H. Svensson, 1961; O. Vesterberg, 1969*). This concentrating effect is a unique feature of **electrofocusing (isoelectric focusing, IEF)** (*E. Schumacher, H.J. Streift, 1958; O. Vesterberg, 1969; H. Haglund, 1971; H. Rilbe, 1973; R.C. Allen, H.R. Maurer, 1974; B. Bjellqkvist et al., 1982; P.G. Righetti, 1983*).

The resolving power of IEF is given (*H. Rilbe, 1973*) at separation along *x* axis as

$$\Delta(pI) = -3\left[\frac{D_i}{E}\frac{d(pH)/dx}{du_i/d(pH)}\right] \quad (1.511)$$

The pH gradient can be produced in ampholyte solutions under an electric field. With a wide pH range, a mixture of different carrier ampholytes is used: aliphatic polyamino-polycarboxylic acids, manufactured under the trade name Ampholine®. Chains with functional groups can be copolymerized within the acrylamide matrix providing a *pH* gradient with unlimited stability, e.g. Immobiline™ systems which can be used even in 5 mm layers due to their very low conductivity (*V. Piljac, G. Piljac, 1985*). In the carrier-ampholyte-base, the resolution corresponding to eqn. 1.511 occurs above $\Delta pI = 0.01$. In Immobiline-type systems, due to the high E and small $d(pH)/dx$ gradients, $\Delta pH = 0.001$ can be achieved (*B. Bjellqvist et al., 1982*).

It has been reported that some crude mixtures can be electrolyzed in the absence of outer ampholyte and that **autofocusing** occurs without buffered solutions (*O. Sova, 1985*).

Electrophoresis can be counterbalanced by an imposed gradient (velocity, chromatographic, and sedimentation) and the components became separated in stationary positions due to the outer field effects (*B.J. McCoy, 1986*).

1.4.4 Interphase transfer. Physico-chemical hydrodynamics

Thus far, we have dealt with transport in homogeneous systems where concentration gradients were induced by external fields and chemical reactions did not play a role, unless a gradient of chemical properties (pH, complexation, etc.) existed in the matrix, in accordance with Curie's postulate in non-equilibrium thermodynamics. Heterogeneous systems are, however, commoner in separation operations and interphase transfer is also often the rate-determining step in chemical separation.

We shall first discuss models of interphase transfer without chemical interactions, following with partial solutions of the diffusion equations, including a chemical rate and electrochemical term, with respect to mass transfer (*J.B. Lewis, 1954, 1958; V.G. Levich, 1959; D.R. Olander, 1960; V.V. Kafarov, 1962; D.A. Frank-Kamenetskii, 1967; M.M. Sharma, P.V.Danckwerts, 1970; K. Sherwood et al., 1975; V.S. Krylov, 1980; T.A Hatton, E.N Lightfoot, 1984; J. Koryta, M. Skalicky, 1989; J.C. Slattery, 1990; J.A. Wesselingh, R. Krishna, 1990; P.C. Wankat, 2012*).

Let us consider the **model of interphase transfer**, where phase I is mobile (moving with velocity w_z in a direction parallel to the interphase plane) and the substance is transferred to the interphase due to diffusion in the perpendicular concentration gradient (Figure 35).

In the two-phase system, the concentration gradient is established because of the non-equilibrium jump in chemical potential at the interphase, due either to zero (or low) concentration in phase II, or the interactions which lead to low activity coefficients in phase II. According to eqn. 1.420, the concentration profile should be calculated from the equation

$$\frac{dc}{dt} = -\vec{w}_z \frac{\partial c}{\partial z} + D \frac{\partial^2 c}{\partial x^2} \tag{1.512}$$

which is valid when longitudinal diffusion (along the z-axis) is negligible compared with the convective transport that follows the flow.

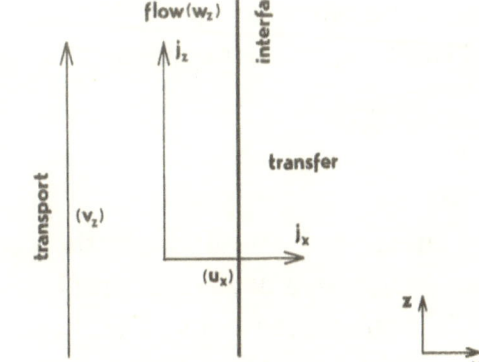

Figure 35. Scheme of transport in phase I combined with transfer to phase II.

The practical task of interphase transfer calculations consists of the calculation of the **interphase flux** which is equal in both phases,

$$j_I = j_{II} = j \tag{1.513}$$

As a result, it is unnecessary to solve eqn. 1.512 to achieve the entire concentration profile. This information will be redundant, because the fluxes are sufficiently determined by derivation at the interphase coordinate $x = 0$,

$$D^I \frac{\partial c^{*I}}{\partial x} = D^{II} \frac{\partial c^{*II}}{\partial x} \tag{1.514}$$

($c^* = c$ at $x = 0$) — see eqn. 1.448. For example, diffusion at a constant concentration at the boundary $c^* = c_0$ to the semi-infinite second phase ($c_\infty = 0$ at any t), using eqn. 1.439b results in

$$j = D^{II} \frac{\partial}{\partial x} \left[c^{*II} \left(1 - \mathrm{erf} \frac{x}{2\sqrt{D^{II}t}} \right) \right] \tag{1.515}$$

and, with respect to definition 1.435b, the interphase flux is given by

$$j = c^{*II} \sqrt{\frac{D^{II}}{\pi t}} \tag{1.516}$$

Comparing this with eqn. 1.452 ($\Delta c = c^{*II} - 0 = c^{*II}$) it follows that the mass transfer coefficient is

$$K = \sqrt{\frac{D^{II}}{\pi t}} \tag{1.517}$$

Such an expression is suitable for the mathematical approximation of many non-steady transfer problems (D.A. Frank-Kamenetskii, 1967). Considering eqns. 1.471 and 1.478 for dimensionless criteria, the flux, (eqn. 1.516) is determined as

$$Sh = \sqrt{\frac{Fo_D}{\pi}} \tag{1.518}$$

The result is that the non-steady flux is, in general, inversely proportional to the square root of time.

Further problems arise, however, with a more or less precise specification of the interphase boundary:

(i) the concentration at the boundary should obey the distribution law, eqn. 1.301, and therefore the concentration profiles in phases I and II are related through distribution coefficients; because of this relation, the two equations 1.512 should be solved simultaneously for both phases,

(ii) for a rigid interphase (gas-solid, liquid-solid) the hydrodynamics are relatively simple; however, **free interphase turbulence** occurs on the contact of two fluid phases, i.e. gas-liquid (*C.J. King, 1964*) or liquid- liquid — Figure 36.

In the latter case the interphase boundary increases and the general eqn. 1.482 should include a **coefficient of dynamic change** f,

$$Sh = C\, Re^m\, Sc^n (1 + f) \qquad (1.519)$$

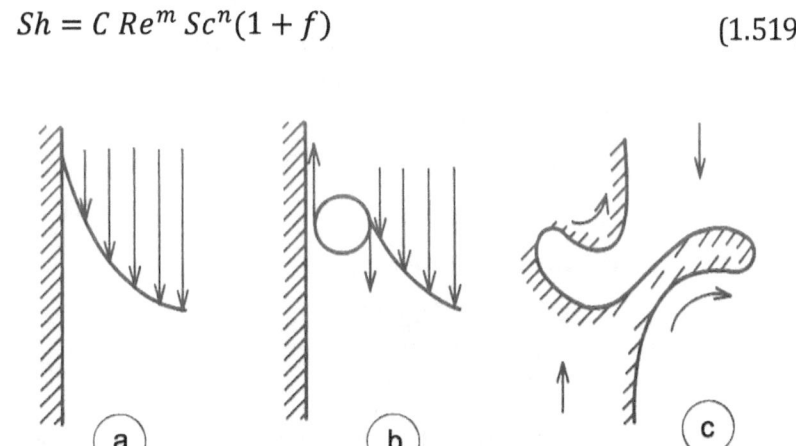

Figure 36. Velocity vectors at interface: a — laminar flow at fixed boundary, b — eddy (turbulent) flow, c — eddies at free boundary.

The coefficient depends on the phase flux ratio, phase densities and the dynamic viscosities of the phases

$$f = \text{const} \left(\frac{w_I}{w_{II}}\right)^p \left(\frac{\rho_I}{\rho_{II}}\right)^q \left(\frac{\eta_I}{\eta_{II}}\right)^r \qquad (1.520)$$

In this instance, function f has the properties of a **similitude criterion for biphasic systems** (*V.V. Kafarov, 1960, 1962*). For example, for dispersion of phase II in phase I, the radius of drops of phase II being the fundamental dimension, there occurs at $Re < 200$ (a laminar flow)

$$Sh_{II} = 0.65\, Re_{II}^{0.5}\, Sc_{II}^{0.5} \left(1 + \frac{\eta_I}{\eta_{II}}\right)^{-0.5} \qquad (1.521a)$$

and at $Re > 200$ (turbulences behind the drops)

$$Sh_{II} = 0.32\, Re_{II}^{0.6}\, Sc_{II}^{0.5} \left(1 + \frac{\eta_I}{\eta_{II}}\right)^{-0.5} \qquad (1.521b)$$

(*A.M. Rozen, 1965*). Another example is flow traveling down the separation membrane; the transfer flux changes at a distance z from the feed in the laminar regime to

$$Sh = C \left(\frac{d_0}{z}\right)^{0.33} Re^{0.35}\, Sc^{0.3} \qquad (1.522a)$$

or in the eddy diffusion regime,

$$Sh = C \left(\frac{d_0}{z}\right)^{0.33} Re^{0.65}\, Sc^{0.3} \qquad (1.522b)$$

where d_0 is the fundamental dimension, i.e. the size of the planar membrane or the diameter of the tube membrane (*S. T. Hwang, K. Kammermeyer, 1981*).

In these equations, the Reynolds number Re is the dimension of the **effective interphase surface,** the Schmidt number Sc remains as reverse value of the **effective thickness of the**

diffusion layer and the Sherwood criterion Sh serves as the **effective transfer coefficient**, according to eqn. 1.453.

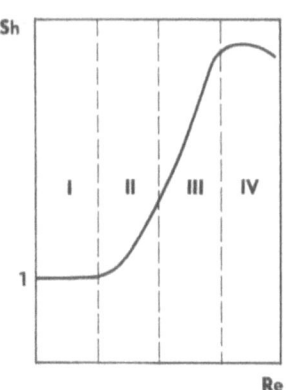

Figure 37. Transfer rate (Sherwood number) as a function of flow velocity (Reynolds number): I — region of laminar flow, II —region of intermediate regime (diffusion and eddy dispersion transfer), III — region of eddy dispersion, IV—strong eddy dispersion.

Generally, transfer is proportional to agitation of the fluid phase. In the region of laminar flow, Sh has the lowest and constant value ($Sh = 1$), until it increases because of the appearance of turbulences. However, in the region of a strong eddy dispersion ($D^E \gg D^M$), transfer does not depend on molecular viscosity and its efficiency may even decrease (Figure 37).

In the derivation of eqn. 1.516, it is implied that the concentration at the boundary, c^{*II}, is known. This is true, however, only if, for example, the interphase flux does not decrease the interphase concentration due to fulfilment of the equality of fluxes in both phases, eqn. 1.513, which occurs either when

—the distribution coefficient $K_D = c^{*II}/c^{*I}$ differs greatly from unity, or

—phase II becomes saturated with diffusant at the boundary, i.e. the interphase flux is low.

In other cases, as mentioned above, the two eqns. 1.512 with common boundary conditions should be solved. The simplest way to accomplish this is with the theory of steady-state linear concentration gradients in two diffusion layers detached to an interphase, the so-called **two-film Lewis-Whitman diffusion**

model (*W.G. Whitman, 1923; W.K. Lewis, W.G. Whitman, 1924*). The main postulates of the two-film theory are:

(i) a diffusion layer (with a linear concentration gradient) is formed in each phase in accordance with the model in Figure 31 — see Figure 38;

(ii) a distribution equilibrium exists at the interphase boundary (according to the distribution law, eqn. 1.300);

(iii) each diffusion layer has its own resistance towards mass transfer and the resistances are additive.

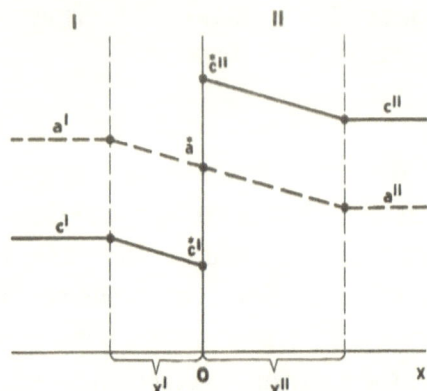

Figure 38. Two-film diffusion model.

With respect to condition (i) and eqn. 1.513, the equality of fluxes is projected onto the absolute activity scale — see eqn. 1.89a and Fig.38 — as follows

$$\kappa^{I}(a^{I} - a^{*I}) = \kappa^{II}(a^{*II} - a^{II}) = \kappa(a^{I} - a^{II}) \quad (1.523a)$$

If the products of the transfer coefficient and dimensionless activities should have the dimensions of molar flux, κ also has the dimensions mol m^{-2}s^{-1}. When the concentration gradients are expressed in the molarity scale, we have

$$\kappa^{I} = 10^{3} K^{I}/\gamma^{I} \quad (1.523b)$$

$$\kappa^{II} = 10^{3} K^{II}/\gamma^{II} \quad (1.523c)$$

Because at interphase equilibrium — condition (ii) — the interface activities are linked in accordance with eqn. 1.299:

$$a^{*I} = a^{*II} = a^* \quad (1.524)$$

and from the four equations 1.523a and 1.524, the three activities are eliminated and

$$\frac{1}{\kappa} = \frac{1}{\kappa^I} + \frac{1}{\kappa^{II}} \quad (1.525)$$

The reciprocal values of the transfer coefficients have the meaning of transfer resistances, i.e. postulate (iii) of the two-film theory is proven.

The continuity equation in the concentration scale is

$$K^I(c^I - c^{*I}) = K^{II}(c^{*II} - c^{II}) = K(c^I - c^{II}) \quad (1.526)$$

and

$$c^{*II} = Dc^{*I} \quad (1.527)$$

Considering equations 1.301, 1.309 and 1.523b, c, we find

$$c^{*I} = \frac{K}{K^I}(c^{II} - c^I) + c^I \quad (1.528a)$$

$$c^{*II} = \frac{K}{K^{II}}(c^I - c^{II}) + c^{II} \quad (1.528b)$$

and

$$\frac{1}{K} = \frac{1}{K^I} + \frac{1}{D}\frac{1}{K^{II}} \quad (1.529)$$

The total transfer coefficient

$$K = \frac{K^I K^{II} D}{K^I + K^{II} D} \quad (1.530)$$

increases with increasing D (decreasing activity coefficient in phase II). The ratio $\frac{K^I}{K^{II}}$ plays the role of similitude criterion and is usually considered as a **Biot diffusion number**

$$Bi_D = \frac{K^I d^{II}}{D^{II}} \tag{1.531a}$$

which expresses the ratio of the transfer resistance in phase I to that in phase II, where the fundamental dimension (film thickness) is d^{II} and the diffusion coefficient D^{II} (see eqn. 1.453). The Biot diffusion number is obviously the two-phase analogue of the Sherwood criterion, eqn. 1.478. At high distribution ratios (strong interactions in phase II), when $K^{II} \gg K^I (Bi_D \ll D)$, the result is

$$K \approx K^I \tag{1.531b}$$

i.e. the resistance of phase II becomes negligible and the diffusion in phase I the single controlling factor of mass transfer. At a very low distribution ratio (low moving force), when $Bi_D \gg D$,

$$K \approx DK^{II} \tag{1.531c}$$

i.e. the total transfer coefficient should also be very small and transfer from phase I to phase II is inefficient — cf. eqn. 1.39.

Both eqns. 1.525 and 1.529 represent the expressions for the general **rule of diffusion resistance additivity** which is the basis for the two-film theory. In practice, however, the presence of a third film (surface-active solute and mechanical barriers, i.e. third-phase solid) combined with spontaneous turbulences are responsible for the deviation from the simple additivity law. The structure of the interphase can be characterized as at least three layers (A.F.H. Ward, L.H. Brooks, 1952; G.A. Yagodin, V.V. Tarasov, 1974; K. Stamberg, J. Cabicar, L. Havlicek, 1980) — Figure 39. The adsorption layer — see section 1.3.2. — is of molecular size ($\delta_1 \approx 0.1 - 1$ nm). The interfacial layer, where the phase properties change continuity, can be much greater (δ_2 is up to 10 µm), but its diffusional

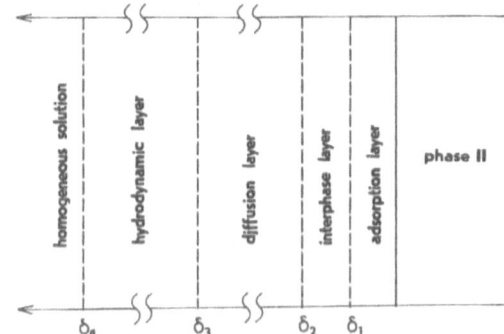

Figure 39. Structure of boundary layer in flowing phase.

resistance is still small compared with the unmoving diffusional layer ($\delta_3 \approx 10 - 100$ µm) and the non-homogeneous hydrodynamic layer (δ_4) where eddy dispersion occurs. Each diffusant has it "own" characteristic diffusional layer. With respect to definition 1.453 and the empirical eqns. 1.521 or 1.525, δ_3 can be projected to be proportional to $(D^M)^{1/2}$ or $(D^M)^{1/3}$.

Some other approaches to mass transfer have attempted to avoid the undefined diffusion layer thickness since other speculative considerations were needed to make the calculations agree with experimental data which often indicate a dependence of the transfer coefficient $K \propto (D^M)^{1/2}$. In addition to the film models of diffusion, the **theory of surface recovery** (*R. Higbie, 1935*) has often been applied to mass transport calculations. Higbie's theory results from the model of perpetual exchange of the volume elements at the interface, proceeding with certain frequency; such elements can be chosen large enough to include diffusion from it as diffusion from a semi-infinite medium —eqn. 1.517. Various frequency functions,

$$f(\tau) = \frac{1}{A_0} \frac{dA(\tau)}{d\tau} \qquad (1.532)$$

are used for the lifetime of the surface elements, i.e. $A(\tau)$, a distribution related to the total interfacial area A_0. For various functions $A(\tau)$ and integration constants, the transport coefficient is typically expressed as the time-averaged value. For

$$f = 1/t_r = \text{const} \qquad (1.533)$$

it is

$$\langle K \rangle = 2\sqrt{\frac{D^M}{\pi t_r}} \qquad (1.534a)$$

(*R. Higbie, 1935*) where t_r is the characteristic residence time of the volume element at the surface. For the function

$$f \propto e^{-s\tau} \qquad (1.534b)$$

we have

$$\langle K \rangle = \sqrt{D^M s} \qquad (1.534c)$$

(*P. V. Danckwerts, 1950*), where s is the characteristic exchange frequency of the volume elements. The parameters t_r and s can be formally connected with the previously discussed parameters of unsteady diffusion (compare eqn. 1.532 with eqn. 1.517) or film theory (eqns. 1.533 and 1.453) so that the models do not avoid hydrodynamic factors and their advantage consists mainly in the convenience of mathematical treatment of the diffusion equations.

Chemical reactions proceeding at the interface or in bulk phases considerably change the concentration profiles of diffusants. A new term connected with the actual reaction rate $r(c_1 \ldots c_i, x, y, z, t)$ in a given phase element appears in the diffusion equation. For two-dimensional convective diffusion of the i-th component, the eqn. 1.512 is extended to

$$\frac{dc_i}{dt} = -\vec{w}_z \frac{\partial c_i}{\partial z} + D_i \frac{\partial^2 c_i}{\partial x^2} + r_i(c_i, t) \tag{1.535}$$

When other diffusants take part in chemical reactions, the system of the equations should be solved. However, an analytical solution of the equations can be obtained only for particular boundary and initial conditions (*W. Nernst, 1904; S. Hatta, 1928; R. Higbie, 1935; Ya.B. Zeldovich, 1939; D.W. Van Krevelen, P.J. Hoftijzer, 1948; P.V. Danckwerts, 1950; G. Astarita, 1962; F. Helfferich, 1965; L.L. Tavlarides, G.O. Benjamin, 1969; R.L. Pigford, G. Sliger, 1973; V. Hlavacek, V. Vaclavek, M. Kubicek, 1979; L.L. Tavlarides, M. Stamatoudis, 1981; S. Gordon, B.J. McBride, 1994*). Numerical methods of solution have become efficient and universal with the introduction of computers; programs are available as standard software for most computers.

As was pointed out at the beginning of this section, there is no need to calculate the entire concentration profiles for the rate of interfacial transport but only their interfacial slopes, eqn. 1.514.

Let us consider the influence of a **chemical reaction in receiving phase II** on the transport of the i-th component, with the conditions indicated in eqn. 1.466, i.e. the equation of steady state diffusion

$$D_i \frac{\partial^2 c_i}{\partial x^2} - r_i = 0 \tag{1.536}$$

(the superscripts II at D, t and x will be omitted for simplicity). Substituting $dc/dx = p$ and differentiating, $d^2c/dx^2 = dp/dx = dp/(dc/p) = (1/2)d(p^2)/dc$, we arrive at

$$\frac{\partial(p^2)}{\partial c_i} = \frac{2}{D_1} r_i \tag{1.537}$$

or, in integrated form

$$p = \frac{\partial c_i}{\partial x} = \sqrt{\frac{2}{D_i} \int r_i \, dc_i} + \text{const} \tag{1.538}$$

Many chemical reactions may be simulated by the rate, r, given by

$$r_i = k_v c_i^{v_i} \tag{1.539}$$

The steady-state flux of diffusant in the semi-infinite phase, once it enters chemical reaction (e.g. adsorption of sulphur dioxide in aqueous alkali) results in the **Zeldovich equation** (*Ya.B. Zeldovich, 1939*)

$$j_r = D_i \left(\frac{\partial c_i}{\partial x}\right)_{x=0} = \sqrt{2 D_i k_v (c_i^*)^{v_i+1}/(v_i + 1)} \tag{1.540}$$

where c_i^* is the concentration at the interphase of the receiving phase II. For the particular case of first-order reactions ($v = 1$) we have

$$j_r = c_i^* \sqrt{D_i k_v} \tag{1.541}$$

In analogous conditions without chemical reaction it is impossible to obtain steady-state diffusion; here the diffusant, according to eqn. 1.437b, penetrates into the depth of the receiving phase proportionally to $\sqrt{t}$ — see also eqn. 1.516, Hence, for steady diffusion with chemical reaction, the ratio of the fluxes with and without reaction,

$$E = \frac{j_r}{j} \tag{1.542}$$

characterizes the acceleration of diffusion due to exhaustion of the diffusant on its diffusion path by chemical reaction, which is called the **acceleration coefficient**. From eqns. 1.539 and 1.516 it follows that

$$E = \sqrt{2\pi k_v (c_i^*)^{v_i-1}/(v_i + 1)} \sqrt{t} \tag{1.543}$$

i.e. the acceleration coefficient is much more important for lengthier phase contacts; it depends on the diffusant concentration except in the case of a first-order reaction ($v_i = 1$).

In respect of transfer resistance, the chemical reaction in the receiving phase decreases the resistance of phase II and can be modelled as an addition of a parallel resistance to mass transport in phase II (Figure 40).

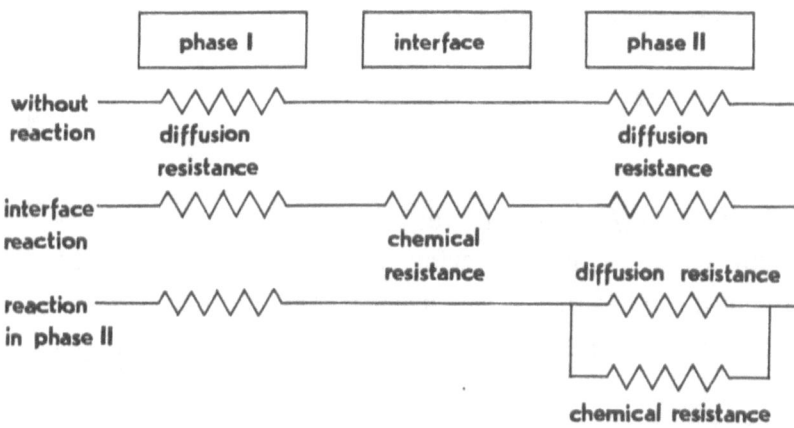

Figure 40. "Electric" model of resistances for diffusion with chemical reaction at interface or in bulk of phase II.

In spite of the first-order reaction, **chemical resistance** is a complicated function of time, depending on the topical concentration of all reacting and diffusing species. For many separation systems the reaction between two diffusants A ($i = 1$) and B ($i = 2$) usually may be considered as a fast reversible reaction

$$v_1 A_1 + v_2 A_2 \rightleftarrows v_3 A_3 + v_4 A_4 \tag{1.544}$$

with a reaction rate at local equilibrium

$$r = k_{12} c_1^{v_1} c_2^{v_3} = k_{34} c_3^{v_3} c_4^{v_4} \tag{1.545}$$

The acceleration coefficients for the reaction were discussed by R.M. Secor and J.A. Beutler (*1967*) for various combinations of stoichiometric coefficients, and rate and equilibrium constants. An analytical solution was obtained, e.g. for the particular case of irreversible bimolecular reaction (*D. W. Van Krevelen, P.J. Hoftijzer, 1948*) or local bimolecular equilibrium (*D.R. Olander, 1960*). In the latter case ($v_1 = v_2 = v_3 = 1, v_4 = 0$) when all diffusion coefficients are equal and there is no A_1 in the bulk, the **maximal acceleration coefficient** becomes

$$E = 1 + \sqrt{\frac{K}{4}} \left(\sqrt{\frac{K}{4} + \frac{c_2}{c_1^*}} - \sqrt{\frac{K}{4}} \right) \tag{1.546}$$

where $K = k_{12}/k_{34}$ is the equilibrium constant, c_1^* is the concentration of A_1 at the interface, and c_2 is the bulk concentration of A_2. For large K (a practically irreversible reaction) the maximal value of the acceleration coefficient results (*S. Hatta, 1928*)

$$E = 1 + \frac{c_2}{c_1^*} \tag{1.547}$$

In this case, component A_1 diffuses from the interface (where it occurs at concentration c_1^* into the bulk phase containing reactant A_2 and at some depth it disappears due to the progressing of chemical reaction ($c_1 = 0$) – Figure 41.

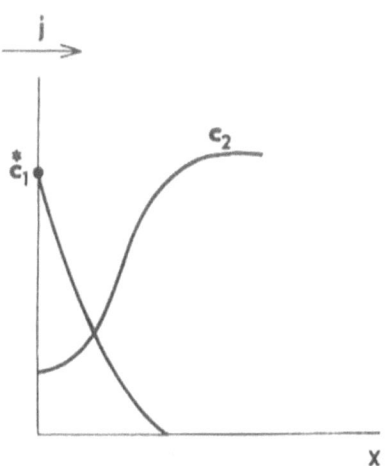

Figure 41. Concentration profiles of solute 1 interacting with solute 2: solute 1 diffuses from boundary to bulk of solute 2 and is completely exhausted during the diffusion path.

Efficient stoichiometric coefficients of the reaction participating in mass transfer can be obtained experimentally by investigating the transfer rate at various phase compositions. If the chemical reaction in the bulk of phase II is rate-determining, there follows a steady state condition for the i-th diffusant

$$r_i = j_i a \tag{1.548a}$$

where a is the specific interphase area ($m^2/m^3 = m^{-1}$), or

$$\frac{dc_i^{II}}{dt} = j_i a \tag{1.548b}$$

In this way the interfacial area and mass transfer coefficients in two-liquid systems can be measured by chemical methods (M.M. Sharma, P. V. Danckwerts, 1970). If a is held constant (under constant hydrodynamic conditions), then according to 1.534, deriving the logarithmic relations

$$\frac{\partial \log j_i}{\partial \log c_i} = \frac{\partial \log r_i}{\partial \log c_i} = v_i \tag{1.549}$$

i.e. the effective stoichiometric coefficient v_i can be obtained from the bilogarithmic function $\log j_i$ vs. $\log c_i$ (R.D. Noble,

1983). Though the derivatives, eqn. 1.549, do not depend on the interphase, under various conditions various steps of chemical interaction may prevail. For example, during solvent extraction of metal M^{2+} with the chelating agent HL, under slow mixing, the flux of metal corresponds to the coefficients $v_1 = 1$ and $v_2 = 2$, i.e. the heterogeneous reaction

$$M^{2+} + 2(HL)_{II} \rightleftarrows (ML_2)_{II} + 2H^+ \tag{1.550}$$

seems reasonable. However, under stronger emulsifying conditions it was found that $v_1 = 1$ and $v_2 = 1$, i.e. the rate determining step may be better considered as a reaction in the bulk phase

$$M^{2+} + L^- \rightleftarrows ML^+ \tag{1.551}$$

It is impossible to find a simple additivity rule for the calculation of the resistance. G. *Damkhöler* (*1939*) proposed applying similitude theory in transport systems with chemical reactions using dimensionless criteria, the **Damköhler numbers**:

$$Da = d \left(\frac{v_i r_i}{c_i D_i}\right)^{1/2} \tag{1.552}$$

or

$$Da_I = d \frac{v_i r_i}{c_i w} \tag{1.553}$$

and

$$Da_{II} = d^2 \frac{v_i r_i}{c_i D_i} \tag{1.554}$$

where r_i is the rate of the reaction (mol dm^{-3} s^{-1}) which the i-th component enters with reaction order v_i at concentration c_i (mol dm^{-3}), d is the characteristic length (m), and D_i is the diffusion coefficient (m^2s^{-1}). The first Damköhler number, Da_I, gives the relation between the time when the diffusant is exhausted by chemical reaction and the actual contact time. The second Damkhőler number, Da_{II} characterizes the ratio between the reaction rate and the diffusion transport.

Although it can be seen that bulk reactions increase mass transfer, **chemical reactions at the interface (ss-reactions)** work in the opposite direction because the step considered normally as an immediate "jump" of diffusant in this case needs a truly positive time (*M.E. Wadsworth, 1975*). The reaction rate r_s (mol $m^{-2}s^{-1}$) of a pseudo- monomolecular reaction

$$A^I \rightleftarrows A^{II} \tag{1.555}$$

is given by the difference between the activities at the interface — cf. eqn.1.523 —i.e.

$$r_s = k_s(c^{*I} - c^{*II}/D) \tag{1.556}$$

where k_s (10^{-3} m s^{-1}) is the surface rate constant. The continuity eqns. 1.526 are completed by the equality

$$j = r_s \tag{1.557}$$

i.e. the lowest value from the diffusion flux or the surface reaction rate is rate-determining. In connection with eqns. 1.526–1.529, we have

$$\frac{1}{K} = \frac{1}{K^I} + \frac{1}{D}\left(\frac{1}{K^{II}} + \frac{1}{k_s}\right) \tag{1.558}$$

This means that the *ss*-reaction increases the resistance to mass transfer and is equivalent to the introduction of a further diffusion layer — Figure 40 (*P.R. Danesi at al., 1978*). Although the *K*'s vary with hydrodynamic conditions (viscosity and agitation) and only slightly with temperature, ks. can increase dramatically with temperature.

Eqns. 1.555 and 1.557 can serve as the basis for analysis of a heterogeneous separation process such as **crystallization** (*A.A. Noyes, W.R. Whitney, 1897; R.S. Tipeon, 1950; N.S. Tavare, 1995; A.S. Myerson, 2001; A.G. Jones, 2002; W. Beckmann, 2013*). When single crystal growth is controlled by diffusion from solution (bulk concentration c) to the crystal surface (equilibrium concentration at saturation c^*), the diffusion flux is

$$j = K(c - c^*) \tag{1.559}$$

where $c - c^*$ represents the **supersaturation** of solution (as the driving force for the crystallization process). Hence, the mass of crystal increase of a crystal area A and molar mass M (kg mol^{-1}) is

$$\frac{dm}{dt} = AMK(c - c^*) \tag{1.560}$$

(*I.I. Andreev, 1908; A.J. Berhoud, 1912; I.I. Valeton, 1923*). For a cubic crystal of side l and density p,

$$m = \rho l^3 \tag{1.561a}$$

$$A = 6l^2 \tag{1.561b}$$

and the **linear growth** ($c > c^*$) or **dissolution rate** (when $c < c^*$) becomes

$$\frac{dl}{dt} = \frac{2MK}{p}(c - c^*) \tag{1.562}$$

When crystals are small, however, the Gibbs energy of a solution with "nuclei" may be still higher than that of a supersaturated solution. The kinetic equation in this case should include a nucleation rate, proportional to $(c - c^*)^n$ where n is an empirical parameter. The crystallization can be related to a surface reaction and then, according to a one-film model ($K^{II} = \infty$) and eqn. 1.559, we have

$$\frac{dl}{dt} = \frac{2M}{p(1/K^{I}+1/k_s)}(c - c^*) \qquad (1.563)$$

Increased stirring rate and temperature have positive effects on diffusion and the kinetic mechanism of crystal growth (*J.D. Jenkins, 1925; V.I. Danilov, E.E. Pluzhnik, B.M. Teverkovskii, 1939; J.W. Mullin, 1972; A.S. Myerson, 2001*).

To illustrate the simplest physico-chemical separation **by fractional crystallization, precipitation, zone melting** or **cementation** in the diffusion regime (*I. V. Melikhov, L.B. Berliner, 1981*), eqns. 1.559 or 1.561 may be used; to solve them the mass balance of the solid phase should be involved. If we neglect the amount of diffusant in the diffusion layer, at an initial concentration c_0 (mol dm^{-3}) and total volume (m³), the result is

$$m = 10^{-3} MV(c_o - c) \qquad (1.564)$$

When introducing

$$\vartheta = c - c^* = c_0 - \frac{10^3 m}{MV} - c^* \qquad (1.565)$$

as the driving force of crystallization (identical with supersaturation) — see section 2.1. — integration of the following equation, obtained from eqn. 1.560,

$$\frac{d\vartheta}{dt} = -\frac{AK}{V}\vartheta \qquad (1.566)$$

gives

$$\ln(\vartheta/\vartheta^0) = -aKt \qquad (1.567)$$

where ϑ^0 is the extent of supersaturation at $t = 0$ and $a = A/V$ is the specific interface of the liquid phase. When thermodynamic equilibrium is achieved, at $t = \infty$, supersaturation must disappear and $\vartheta = 0$. If mutual crystallization, i.e. **co-crystallization** of two components, A and B occurs in the diffusion regime with local equilibrium at the interface, an expression for the rate of crystallization is obtained by dividing two equations of type 1.567 for each component as the **kinetic separation equation**

$$\frac{\ln(\vartheta_A/\vartheta_A^0)}{\ln(\vartheta_B/\vartheta_B^0)} = \frac{K_A}{K_B} \tag{1.568}$$

or

$$\ln(\vartheta_A/\vartheta_A^0) = \lambda \ln(\vartheta_B/\vartheta_B^0) \tag{1.569}$$

where

$$\lambda = \frac{K_A}{K_B} \tag{1.570}$$

is the **crystallization coefficient**. In the diffusion regime, which is far from the total equilibrium (*I. V. Melikhov, 1979*), i.e. when $c^* \ll c$, and therefore $\theta \approx -c$, we find

$$\ln(c_A/c_A^0) = \lambda \ln(c_B/c_B^0) \tag{1.571a}$$

or

$$\ln \frac{m_A^\infty - m_A}{m_A^\infty} = \lambda \ln \frac{m_B^\infty - m_B}{m_B^\infty} \tag{1.571b}$$

where m and m^∞ are the actual and maximal amounts of components in the crystals. Eqn. 1.571b is known as the **Doerner-Hoskins equation** (*N.A. Doerner, W.M. Hoskins, 1925*). Under practical conditions, co-crystallization is controlled by surface interaction and rarely by the kinetics of chemical reactions in the bulk solution (*I. V. Melikhov, S.S. Berdonosav, 1974; I. V. Melikhov, M.S. Merkulova, 1975; L.A. Zhukova, 1981*).

Obviously, the kinetic equation 1.568 contains a universal meaning and can be applied to other separation processes controlled by mass transfer (non-equilibrium separations). For instance, the kinetics of **adsorption** (*Ya.L. Zabezhinskii, 1940*) can be described by an equation equivalent to eqn. 1.558,

$$\frac{d\Gamma}{dt} = K(c - c^*) \tag{1.572}$$

where Γ is the amount adsorbed (mol m^{-2}) and the relation between and Γ can be given by one of the adsorption isotherms, eqns. 1.321–1.326, if a local adsorption equilibrium is predicted.

On the **electrolytic deposition (cementation)** of ions with charge z, the electric migration flux of the ions ($A\ m^{-2}$) at the concentration polarized electrode ($c^* \neq 0$) is

$$i = zFK(c - c^*) \tag{1.573}$$

and at the fully depolarized electrode ($c^* = 0$), the limiting ion current density is in accordance with the Faraday law

$$i_d = zFKc \tag{1.574}$$

($F = 96\ 500$ C mol^{-1}). Electrolysis proceeds at voltage φ_p which is higher than the equilibrium value 1.122, the difference

$$\eta = \varphi_p - \varphi \tag{1.575}$$

being the overvoltage. Taking into account eqns. 1.122, 1.573 and 1.574, the value of the overvoltage is

$$\eta = \frac{RT}{zF} \ln\left(\frac{c^*}{c}\right) = \frac{RT}{zF} \ln\left(1 - \frac{i}{i_d}\right) \tag{1.576}$$

which, in accordance with eqn. 1.494, leads to

$$i = i_d[1 - \exp(\eta u/D)] \tag{1.577}$$

(see, e.g. *E. Raub, K. Müller, 1968*). On the electrodeposition of two ions (*O.A. Esin, 1935*), the ratio of their migration fluxes is

$$\frac{i_A}{i_B} = \frac{c_A K_A}{c_B K_B} \tag{1.578}$$

where

$$K_i = \exp\left(\frac{-\alpha_i z_i F \eta_i}{RT}\right) = \exp\left(\frac{-\alpha_i \eta u_i}{D_i}\right) \tag{1.579}$$

and where $i \equiv A$ or B, k_i are constants and α's are kinetic coefficients ($\alpha \leq 1$).

A phenomenological transfer coefficient for the deposition of an ion on an electrode washed with electrolyte flowing parallel to the electrode with velocity w can be derived as

$$K = \frac{1}{f} \frac{w^m}{v^n} D_i^{2/3} \tag{1.580}$$

where f is the geometrical factor, and v is the kinematic viscosity of the electrolyte solution. For laminar flow, empirical values are $m = 0.50$ and $n \approx 0.16$, while for turbulent flow $m = 0.90$ and $n = 0.57$ (*K.J. Vetter, 1967*).

1.4.5 Combined convective and interphase transport. Chromatography

Space-oriented convection and interphase transport, which we discussed in the previous section, is the basis of a wide class of powerful **chromatographic separation methods** invented by M.S.Tsvet in 1906 during his research of plant pigments (see the reviews: *G. Damköhler, H. Theile, 1943; M.S. Tsvet, 1946; E. Cremer, R. Müller, 1951; J. Janak, 1953; A.I.M. Keulemans, A. Kwantes, P.Zaal, 1957; J.H.Knox, 1962; I.M. Hais, K. Macek, 1963; D. Abbott,*

R.S. Andrews, 1969; A.B. Littlewood, 1970; P.R. Brown, 1973; G.R. Lowe, P.D.G. Dean, 1974; R.P.W. Scott, 1976; V.G. Berezkin, 1976; H.F. Walton, 1976; C. Horvath, W. Melander, 1977; L.R. Snyder, J.J. Kirkland, 1979; O. Mikes, 1979; A.S. Said, 1981; M.S.W. Hearn, 1984; W.D. Conway, 1990; M. Lederer, 1992; G. Ganetsos, P.E. Barker, 1992; W.J. Lough, I.W. Wainer, 1995; J. Sherma, B. Fried, 1999; C. Poole, 2003; S. Fanali et al., 2013; M. Naushad, M.R. Khan, 2014). The techniques are based on the unequal distribution of the solutes between two phases which move relative to one another. Usually one of the phases is fixed or anchored (solid phase, or liquid phase on a solid carrier), called the **stationary phase**, and the second, or the **mobile phase**, flows over its surface. Under these conditions, different solutes move with the mobile phase matrix (phase I in Figure 35) at different speeds because they can be partially bonded to the surface of the stationary phase (phase II at Figure 35) or, to some extent, penetrate its depth. In **displacement chromatography**, the mobile phase represents, or contains a principal component, the **eluent** with a considerably higher distribution than any of the other components of the separated mixture and pushes them in front of its own zone. When an eluant is weak in action, the solutes are eluted gradually by **elution chromatography**.

According to the mode of interaction of the solutes with the stationary phase, chromatographic separation methods can be classified as follows:

1) **size exclusion chromatography, SEC (gel permeation chromatography, GPC)** — separation according to accessibility of the pores to molecules of different size: solute molecules enter the gel pores by diffusion, the small molecules penetrate the interior of gel network while large molecules are unable to enter the pores and are unretained (*G.H. Lathe, C.R. Ruthven, 1956; J. Porath, P. Flodin, 1959; J.C. Giddings et al., 1968; H. Determan, 1969; D.Berek et al., 1983; D. Hunkeler et al., 1996*);

2) **adsorption chromatography** — the adsorption of solutes from a mobile gas or liquid phase on the surface of the stationary phase differs (*M.S. Tsvet, 1903, 1906*); according to the combination of the mobile and stationary phases, adsorption chromatography is divided to subgroups:
    - **Gas-solid chromatography, GSC** (*A. Eucken, H. Knick, 1936; J. Janak, 1953; J.H. Knox, 1962; J.H. Purnell, 1962; J.C. Giddings et al., 1968; H.G. Struppe, 1968; A.V. Kiselev, Ya.I. Yashin, 1969; R.E. Clement, 1991*).
    - **Gas-liquid chromatography** (*A.T. James, A.J.P. Martin, 1952; G.C. Blytas, D.L. Peterson, 1967*) and **liquid-solid chromatography, LSC** (*M.S. Tsvet, 1910; R. Kuhn, E. Lederer, 1931*); the stationary phase is usually maintained in columns or capillaries (*J.H. Purnell, 1959; L. Sojak et al., 1981*) through which the mobile phase is passed (*E. Leibnitz, H.G. Struppe, 1984*). A horizontal **coiled tube arrangement**, likewise a countercurrent system, can be used (*Y.Ito, W.D. Conway, 1984*).
    - **Gas-liquid-solid chromatography** (*V.G. Berezkin et al., 1968*).
    - **Supercitical fluid chromatography** (*R.M. Smith, 1993; G.K. Webster, 2014*).
    - Paper sheets and adsorbent thin layers (*K. Macek, I.M. Hais, 1965*) with the mobile phase moving by capillarity are used for flat, open bed **paper chromatography, PC** (*A.H. Gordon, A.J.P. Martin, R.L.M. Synge, 1943; R. Consden, A.H. Gordon, A.J.P. Martin, 1944; I.M. Hais, K. Macek, 1963*) and **thin-layer chromatography, TLC** (*N.A. Izmailov, M.S. Shraiber, 1938, E. Stahl et al., 1956*); the **two-dimensional** variants of these techniques are especially promising (*J.C. Giddings, 1984*).
3) **partition chromatography** or **liquid-liquid chromatography, LLC**,- the solutes distribute between two immiscible liquids and are retarded by the phase, firmly fixed (anchored), proportionally to the distribution

constants; originally the aqueous phase was anchored to a solid hydrophilic carrier while a mobile organic solvent was used (*A.J.P. Martin, R.L.M. Synge, 1941; L.C. Craig, 1950*). As a result, the arrangement with a carrier impregnated with organic solvent is called **reversed phase chromatography, RPC**, or **extraction chromatography** (*G.C. Howard, A.J.P. Martin, 1950; S. Siekierski, B. Kotlinskaya, 1959; S. Siekierski, I. Fidelis, 1960; C. Horvath, W. Melander, I. Molnar, 1977; P.R. Brown, A.M. Krstulovic, 1979; P.R. Brown, E. Grushka, 1980; M.T.W. Hearn, 1984*).

4) **ion exchange chromatography (IEC, IX)** — ionogenic solutes enter electrostatic interactions with ionogenic functional groups of the polymeric stationary phase, the ion exchanger, during their diffusion inside the ion exchanger gel (*O. Samuelson, 1939; F.H. Spedding et al., 1947; E.R. Tompkins, J.H. Khym, W.E. Cohn, 1947; S. Moor, W. Stein, 1948; G.E. Moore, K.A. Kraus, 1949; W.Rieman III, H.F. Walton, 1970; J.H. Khym, 1974; K. Dorfner, 1991*); its variants are **ligand-exchange chromatography, LEC** (*V.A. Davankov, S.V. Rogozhin, 1971; H.F. Walton, 1973; J.D. Navratil, E. Murgia, H.T. Walton, 1975; V.A Davankov, J.D. Navratil, H.F. Walton, 1988*) and ion exchange in liquid-liquid systems (*J. Mikulski, I. Stronski, 1965; M. Kyrs, L. Kadlecova, 1968*).

5) **precipitation chromatography** — solutes separate due to the difference in solubility products of the compounds which arise on chemisorption or formation of a solid phase (*E.N. Gapon, T.B. Gapon, 1948; K.M. Olshanova et al. 1963*); that which is based on specific covalent interactions of biopolymeric molecules is called **affinity (bioaffinity) chromatography** (*R. Axen, J. Porath, S. Ernback, 1967; P. Cuatrecases, M. Wilchek; C.B. Anfinsen, 1968; P.C. Wankat, 1974; C.R. Lowe, P.D.G. Dean, 1974; J. Turkova, 1978, 1993; V.I. Lodzinski, S.V. Rogozhin, 1980; G. Johansson, 1984; R.R. Walters, 1985*).

The latter methods can be considered as the specific ones of a broader class of the chromatography of the solutes chemically interacting with chromatographic bed, or **reaction chromatography** (E.M. Magee, 1963; *V.G Berezkin, V.S.Kruglikova, N.A Belikova, V.E. Shirayeva, 1966; D.E. Martire, P. Riedl, 1968; M. Van Swaay, 1969; D.F. Cadogan, J.H. Purnell, 1969;; V.G.Berezkin, 1980*). Specific forms are those based on metal chelate affinity in **immobilized metal affinity chromatography, IMACS** (*J. Porath, J. Garlsson, I. Olsson, G. Belfrage, 1975; J. Porath, 1988*).

Inter- and intramolecular conversion of separated **chemical species** during transportation and transfer is important in solutions of metal-organic and bio-inorganic complexes, and also for conformations of biopolymers (*A.A. Stepanov, A.D. Gedeonov, 1976; H.P. Van Leewen 1989; T. Milton, M.T.W. Hearn, B. Auspach, 1990; M. Hutta, M. Moskalova, M. Zemberyova, M. Foltin, 1996*).

Electrokinetic chromatography (*S. Terabe et al., 1984, 1989*) represents an interface between electrophoresis and chromatography: the solutes are distributed between a solution and charged micelles and transferred by their flow.

Hydrodynamic chromatography, HDC (*H. Small, N.A. Langhorst, 1982*) is a specific chromatographic technique because it is performed in one liquid phase, followed by the separation of microscopic (colloidal) particles due to their different hydrodynamic resistance in laminar flow (non-zero gradient $\partial w/\partial x$).

Abstracting from the chemical mechanism of distribution of solutes between mobile and stationary phases, a general description of all chromatographic processes can be derived on the basis of eqn. 1.420 for convective diffusion and the interfacial flux balance, eqn. 1.526 (*J.C. Giddings, 1965*). In spite of its importance, however, chromatography is only one particular transport process which will be discussed in full detail in section 2.4.1. To avoid duplication and to give the key to a generalized system of separation methods (*J.C. Giddings, 1978*), see section 1.8.; here it will be useful to give an even more universal characterization of the separation as a combination of convection and interphase transfer.

For the sake of convenience, the **transport of a solute with one phase in contact with a second phase** is appropriately modelled by the one-dimensional convective diffusion equation of type 1.429, instead of the two-dimensional equation 1.512, reflecting the situation illustrated by Figure 35. This can be done relating the transport of the solutes to the bulk phase I which moves with velocity w (m s^{-1}) and passes in time Δt a distance L,

$$L = w\Delta t \tag{1.581}$$

During this time, the i-th solute is distributed between phase I and phase II so that it spends an average time $\langle \Delta t_i^I \rangle$ in phase I and $\langle \Delta t_i^{II} \rangle$ in phase II:

$$\Delta t = \langle \Delta t_i^I \rangle + \langle \Delta t_i^{II} \rangle \tag{1.582}$$

If the average velocity of solute transport is $\langle v_i \rangle$, the distance passed by the solute is

$$L_i = \langle v_i \rangle \Delta t = w \langle \Delta t_i^I \rangle \tag{1.583}$$

The ratio, the **retardation factor** R_i

$$\frac{L_i}{L} = \frac{\langle v_i \rangle}{w} = R_i \tag{1.584}$$

or

$$R_i = \frac{\langle v_i \rangle}{\Delta t} = \frac{1}{1+\langle \Delta t_i^{II} \rangle/\langle \Delta t_i^{II} \rangle} \tag{1.585}$$

can be considered to be constant if

$$\langle \Delta t_i^{II} \rangle/\langle \Delta t_i^{II} \rangle = \text{const} \tag{1.586}$$

i.e. if the probability of occurrence of the i-th solute molecules in certain phases is constant. The value of 1.584 is also called the **retention factor** (R_F) and

$$t_{R_i} = \frac{L}{\langle v_i \rangle} \tag{1.587}$$

is referred to in separation as the **retention time (elution time)** of the i-th component in a chromatographic column with length L.

According to its physical meaning, the residence time ratio, eqn. 1.586, is proportional to the mean mass distribution ratio, eqn. 2.15, of the separated solute

$$\langle \Delta t_i^{II} \rangle / \langle \Delta t_i^{II} \rangle = \langle D_i^m \rangle \equiv k_i \tag{1.588}$$

and in chromatography this is called the **capacity factor**. According to eqns. 1.585 and 1.588,

$$R_i = \frac{1}{1+k_i} \tag{1.589}$$

and because we also have

$$R_i = \frac{\langle v_i \rangle}{w} = \frac{L}{t_{R_i} w} \tag{1.590}$$

the fundamental relation for chromatographic retention is

$$t_{R_i} = t_M (1 + k) \tag{1.591a}$$

where the ratio

$$t_M = \frac{L}{w} \tag{1.591b}$$

is called the **dead retenion time** (the time necessary for the mobile phase to pass a column with length L).

Consequently, eqn. 1.419 is modified to a one-dimensional equation

$$\frac{dc_i}{dt} = -R_i w \frac{\partial c_i^I}{\partial z} + D_i \frac{\partial^2 c_i^I}{\partial z^2} \qquad (1.592)$$

for transport in the direction of movement of the mobile phase (the z-axis).

In the model for **ideal linear chromatography** (*J.N. Wilson, 1940*), it is assumed that Henry's law is valid for the equilibrium distribution of the solute and that no longitudinal diffusion occurs in the mobile phase. Thus,

$$\frac{dc_i}{dt} = -v_i \frac{\partial c_i}{\partial z} \qquad (1.593a)$$

where

$$v_i = \frac{w}{1+K_i^D r} \qquad (1.593b)$$

This indicates that the initial concentration profile moves with a velocity v_i which is constant, and is lower the higher the value of the distribution constant K_i^D between the stationary phase II and the mobile phase I (and also increases with their volume ratio $\frac{V^{II}}{V^I}$) — compare with discussion of eqn. 1.429 at high Péclet numbers. The discussion can be generalized for multi-phase equilibria, e.g. solid-liquid-gas chromatography (*V.G.Berezkin, et al., 1968*).

In the previous model it was assumed that the solute band in chromatographic column moves without any change in shape. However, longitudinal molecular diffusion in the mobile phase leads to **spreading of the concentration profiles** (Figure 42) in accordance with eqn. 1.437b, so that

$$\langle \Delta z_i \rangle_M = \sqrt{2D^I L/w} \qquad (1.594)$$

The non-uniformity of flow of the mobile phase in the interpartide space due to turbulences and micropore structure, and on eddy dispersion given by eqn. 1.451, introduces a broadening

$$\langle \Delta z \rangle_E = \sqrt{\lambda L d^{II}} \qquad (1.595)$$

where λ is the **obstruction (tortuosity) factor** ($\lambda = 2 - 10$) — see discussion of Darcy's eqn. 1.463, and d^{II} is the diameter of the stationary phase particles (*R. Aris, N.R. Amundson, 1957*).

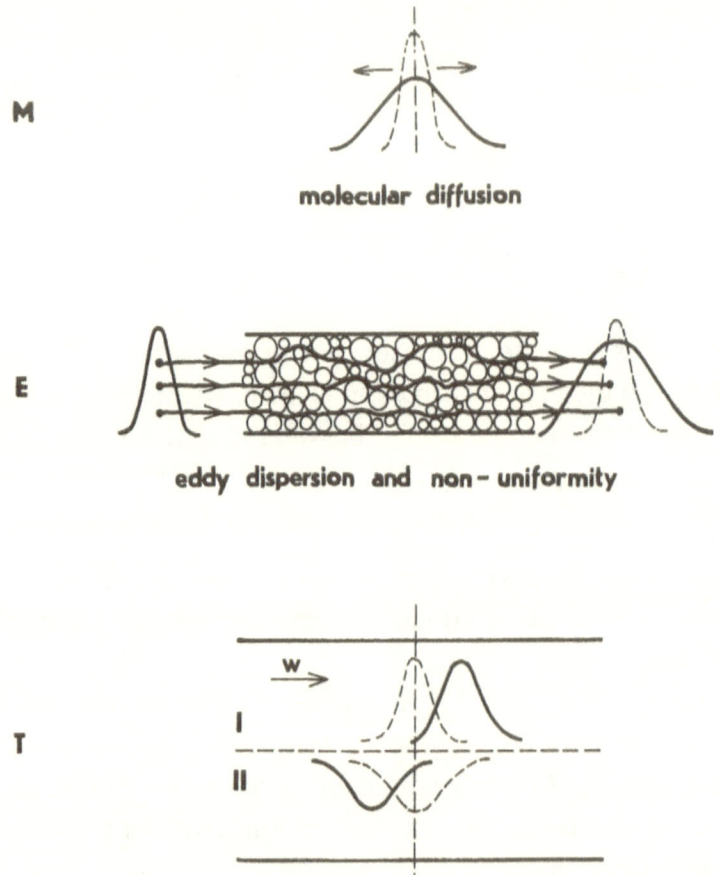

Figure 42. Principal factors influencing the longitudinal spreading of concentration band: M —molecular diffusion, E — eddy dispersion

and bed heterogeneity, T — non-equilibrium distribution between phases.

Departure from equilibrium due to a finite rate of transport in phase II contributes to longitudinal variation as follows

$$\langle \Delta z \rangle_T^2 = \sqrt{2R_i(1-R_i)Lw/K^{II}} \tag{1.596}$$

The squares of the individual spreading factors, eqns. 1.594–1.596, are additive (*J.J. Van Deemter et al., 1956; Ya.V.Shevelov, 1957*)

$$\langle \Delta z \rangle^2 = \langle \Delta z \rangle_E^2 + \langle \Delta z \rangle_M^2 + \langle \Delta z \rangle_T^2 \tag{1.597}$$

i.e.

$$\langle \Delta z \rangle^2 = C_E + \frac{C_M}{w} + C_T w \tag{1.598}$$

which is known as the Van Deemter equation. On the basis of the Péclet and Biot similitude criteria, eqns. 1.430 and 1.531, this can be presented as

$$\langle \Delta z \rangle^2 = C_E' + \frac{C_M'}{Pe_D} + C_T' Bi_D \tag{1.599}$$

The Van Deemter equation is applicable when the mobile phase is gaseous (GC). In liquids the velocity profile is important in respect of the flow in capillary channels and diffusion between layers moving with different velocities (*J.C. Giddings, R.A. Robinson, 1962*). This is included in the term

$$\langle \Delta z \rangle_H = d^{II} \sqrt{Kw/D^I} \tag{1.600}$$

and the resulting **Giddings equation** is

$$\langle \Delta z \rangle^2 = C_T w + \frac{C_M}{w} + \frac{C_E C_H w}{C_E + C_H w} \tag{1.601}$$

(Fig. 43). It can be seen that if $C_H w \gg C_E$, then the eqns. 1.598 and 1.601 lead to identical results. Optimum is about at $Pe \approx 3$.

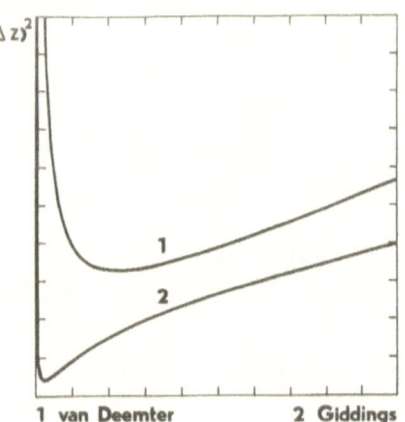

Figure 43. Spreading of chromatographic band $\langle \Delta z \rangle$ as a function of flow velocity (w) at different diameter of sorbent particles: 1 — van Deemter equation 1.598, 2 — Giddings equation 1.601.

Additional spreading is introduced by longitudinal diffusion in the mobile phase, and non-equilibrium in the mobile phase and in the space within the pores of the stationary phase (J.C. Giddings, 1961; J.C. Giddings, R.A. Robinson, 1962; J.C. Giddings, 1965).

A general problem of **non-equilibrium chromatography** (J.J. Van Deemter, F.J.Zwiderweg, A Klinkenberg, 1956; E. Kucera, 1965; A.S. Said, 1981; L.N. Snyder, 1983; T. Gu, 1995) consists of the simultaneous solution of eqn. 1.419 for transport in the direction of the z-axis and eqn. 1.514 for transfer in the direction perpendicular to the interface, i.e. in the direction of the x-axis. In the absence of longitudinal diffusion, this means that the following equation should be solved:

$$\frac{\partial c^{II}}{\partial t} + r \frac{\partial c^{II}}{\partial t} = -w \frac{\partial c^{I}}{\partial z} \tag{1.602}$$

where $r = V^{II}/V^{I}$ is the phase ratio, and the equation is as follows

$$\frac{\partial c^{II}}{\partial t} = aD^{II} \left(\frac{\partial c^{II}}{\partial x} \right)_{x=0} \tag{1.603}$$

When the concentration gradient in the stationary phase is linear, the equation simply transforms to:

$$\frac{\partial c^{II}}{\partial t} = aK^{II}(c^{II} - c^{*II}) \tag{1.604}$$

The mutual solution of the differential equations is rather difficult. The expressions which result from their Laplace transformations are frequently too complicated to invert to the time domain. However, by the mathematical technique of analysis of moments the inversion becomes unnecessary to obtain sufficient criteria for separation, e.g. on fluid-solid adsorption with porous particles, GLC and GC (B.J. McCoy, 1986). Many practically important cases have been solved by numerous investigations (H. Thomas, 1944; J.B. Rosen, 1952; F. Helfferich, 1965; E. Kucera, 1965; R.D. Fleck, Jr. et al., 1973) not only in respect of separation problems but generally for heterogeneous kinetics of adsorption and catalysis.

For instance, the deviation from equilibrium concentration in the stationary phase which is gradually saturated by a solute can be characterized by the coefficient ∈ (0,1):

$$\gamma = \frac{c^{II}}{c_{eq}^{II}} = 1 - \sum_{n=1}^{3} B_n \exp(-\mu_n^2 \tau) \tag{1.605}$$

where τ is the dimensionless time for a spherical particle with diameter d and B_n, μ_n are coefficients ($0 < B_n < 1, 0 < \mu_n < 10$) obtained from transcendental equations containing Biot's diffusion number. If the outer resistance in phase I is negligible ($Bi_D = \infty$) γ is dose to unity for a longer diffusion time τ ($\tau > 0.3$). If $Bi_D \leq 0.1$, we have $\gamma \approx 1 - \exp(\mu_1^2/\pi^2)$, where $\mu_1 < 0.54$ (D.P. Timofeev, 1962). For brief penetration into the stationary phase, the diffusion flux can be simulated by a model with a constant outer concentration — eqn. 1.515.

For the separation of macrocomponents, when the concentration of solute in the stationary phase is close to the adsorption capacity, dynamic factors become less important and the **non-linearity of distribution isotherms** plays a major role (*A.I. Kalinitchev, 1996*), causing asymmetrical deformation of chromatographic zones. When the distribution ratio depends on the concentration of solute (Figure 23), in accordance with eqn. 1.539, the relative velocity of transport will be

$$R_i = \frac{1}{1 + r\frac{dc_i^{II}}{dc_i^{I}}} \tag{1.606}$$

From the differential

$$dc_i^{II} = c_i^I dD_i + D_i dc_i^I \tag{1.607a}$$

we get

$$\frac{dc^{II}}{dc^{I}} = D_i + \frac{dD_i}{dc_i^I} c_i^I = D_i \left(1 + \frac{d \ln D_i}{d \ln c_i^I}\right) \tag{1.607b}$$

and

$$R_i = \frac{1}{1 + rD_i}\left(1 + \frac{d \ln D_i}{d \ln c_i^I}\right) \tag{1.608}$$

where R_i is a function of the distribuend concentration c_i^I. If the distribution isotherm (as a rule) is concave, i.e. $d^2 D_i/(dc_i^I)^2 < 0$, the retardation factor is smaller at higher concentrations. As a favorable result (focusing effect), the concentration isotherm peak moves faster than the front and the spreading is smaller in the tail of the zone, and vice versa (Figure 44).

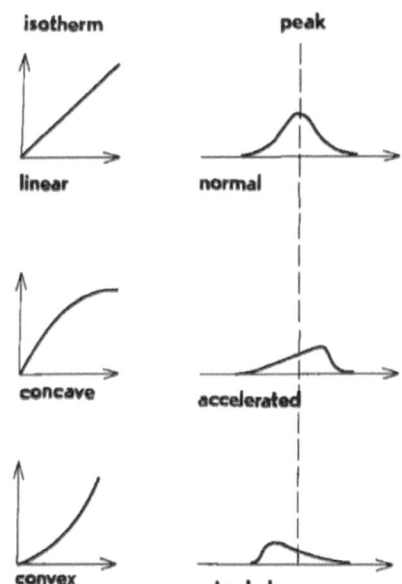

Figure 44. Typical deformation of elution chromatographic bands for various shapes of absorption isotherms.

The relative forces in chromatography are the interactive forces normal to the interface of the stationary and mobile phases, i.e. the flow and force are perpendicular. In other separation methods (e.g. filtration and reverse osmosis), the flow carries the solute across the interface and their mutual action is parallel.

For a two-dimensional arrangement, as in **continuous electrophoresis** (*K. Hannig, 1961*) and **annular chromatography** (*M. Taramasso, 1970; J.D. Navratil, A. Murphy, D. Sun, 1990*), eqn. 1.592 contains two further terms of perpendicular convection and diffusion.

The importance of the relative orientation of the flow/force vectors in separation science will be discussed in general — see section 1.8. At the moment it is important to obtain a single equation that describes both separation arrangements.

The transport rate of the solute $\vec{v}$ is a vector sum of matrix (bulk) displacement $\vec{w}$ and relative solute displacement in the chemical potential gradient, $\vec{u}$ — eqn. 1.41:

$$\vec{v} = \vec{w} + \vec{u} \tag{1.609}$$

For the perpendicular force-flow arrangement (Figure 35),

$$\vec{v} = \vec{w} + \vec{u}, \tag{1.610}$$

and, as discussed in connection with the approximation of the v_z component of vector $\vec{v}$ along the flow axis, it can be assumed that

$$v_z = Rw_z \tag{1.611}$$

This is formally equivalent (isomorphous) to a parallel force/flow arrangement:

$$\vec{v} = \vec{w}_z + \vec{u}_z \tag{1.612}$$

or stated simply

$$v_z = w_z + u_z = u_z(1 + w_z/u_z) \tag{1.613}$$

Therefore, for general purposes — see section 1.8 — the convective transport can again be modelled by movement in a single direction (z-axis), eqn. 1.420 applied in this case being

$$\frac{dc}{dt} = -\frac{v_z}{L}\frac{\partial c}{\partial \zeta} + \frac{D}{L^2}\frac{\partial^z c}{\partial \zeta^2} \tag{1.614}$$

where zeta is the dimensionless coordinate, the dimensionless distance along separation path L

$$\zeta = z/L \tag{1.615a}$$

If the total increment of the chemical potential used to implement separation at the length is $\Delta\mu$, the **dimensionless chemical potential** (μ_0 at the length $L=0$)

$$\varepsilon = \frac{\mu - \mu_0}{\Delta\mu} = \frac{\mu - \mu_0}{\mu_0 - \mu_L} \qquad (1.615b)$$

can be introduced. If we take into account that the velocity w_z is controlled by the chemical potential gradient, eqn. 1.41, and consider eqns. 1.613, 1.615 and also the dimensionless time $\tau = Dt/L^2$ (cf. eqn. 1.428), eqn. 1.616 becomes

$$\frac{dc}{d\tau} = -\frac{w_z L}{D}\left(1 + \frac{\Delta\mu}{w_z fL}\frac{d\varepsilon}{d\zeta}\right)\frac{dc}{d\zeta} + \frac{\partial^2 c}{\partial\zeta^2} \qquad (1.616)$$

Because the frictional coefficient f is given by Einstein's equation 1.405, by using the Péclet diffusion criterion, eqn. 1.430, with L as a fundamental measure, we find

$$\frac{dc}{dr} = -Pe\left(1 + \frac{\Delta\mu}{RT}\frac{1}{Pe}\frac{\partial\varepsilon}{\partial\zeta}\right)\frac{\partial c}{\partial\zeta} + \frac{\partial^2 c}{\partial\zeta^2} \qquad (1.617)$$

The term $Pe\ \Delta\mu/RT$ gives the relative strength of the chemical potential increment $\Delta\mu$ structuring the separation compared to diffusion (*J.C. Giddings, 1978*) and plays role of the Stanton diffusion criterion, eqn. 1.480. It should be stressed, however, that the Péclet criterion in this case is related to the entire separation path L which is usually much longer than the width d of the moving zone of substance. When the Péclet number is related to the width of the zone, d, as the fundamental dimension, we have

$$\frac{dc}{d\tau} = -Pe\left(\frac{d}{L} + \frac{\Delta\mu}{RT}\frac{1}{Pe}\frac{\partial\varepsilon}{\partial\zeta}\right)\frac{\partial c}{\partial\zeta} + \frac{\partial^2 c}{\partial\zeta^2} \qquad (1.618)$$

For ideal solutions and in the absence of external fields, according to eqn. 1.37, $\Delta\mu = RT\Delta\ln c$, i.e. $\Delta\mu/RT = \Delta\ln c \approx \Delta c$, cf. eqns. 1.399–1.403.

The chemical potential gradient in the dimensionless coordinates, $d\varepsilon/d\zeta$, can be considered a constant in many separation processes — see section 1.8. For the i-th component, the $\Delta\mu$ value is a specific characteristic as is the Péclet number, due to a specific diffusion coefficient. The dimensionless coefficient before the derivative $\partial c/\partial \zeta$ is the dimensionless velocity υ (upsilon),

$$\upsilon = Pe \left(\frac{d}{L} + \frac{\Delta\mu}{RT} \frac{1}{Pe} \frac{\partial \varepsilon}{\partial \zeta} \right) \quad (1.619)$$

— cf. eqn. 1.430. Now eqn. 1.614, after the following transformations

$$\frac{dc_i}{d\tau} = -\upsilon \frac{\partial c_i}{\partial \zeta} + \frac{\partial^2 c_i}{\partial \zeta^2} \quad (1.620)$$

and

$$\zeta' = \zeta - \upsilon\tau \quad (1.621)$$

gives

$$\frac{dc_i}{d\tau} = \frac{\partial^2 c_i}{\partial \zeta^2} \quad (1.622)$$

Eqn. 1.622 is identical to eqn. 1.431 and is solved in the same way. For instance, if the separated zone is extremely narrow at the beginning, it takes the form of a Gaussian curve, eqn. 1.435c. With respect to eqn. 1.621, the **diffusion spread of the moving zone** is given by the equation

$$\chi = \frac{1}{\sqrt{4\pi\tau}} \exp\left[-\frac{(\zeta-\upsilon\tau)}{4\tau}\right] \quad (1.623a)$$

where

$$\chi = c/c_0 \quad (1.623b)$$

is the dimensionless concentration, related to the initial concentration of the component in the separated zone. The mean spread (eqn. 1.437a) of the concentration zone is

$$\langle \Delta\zeta \rangle = \sqrt{2\tau} \tag{1.624}$$

As usual, the separation is performed in a time when $\tau \ll 1$, because whole separation path, $\zeta = 1$, is achieved at $\tau = 1/v$. E.g. the effect in $\langle \Delta\zeta \rangle^2$ is smaller than 0.01% when the zone is moving in a capillary of radius $r > 700\ L/Pe$ and it is utilized in diffusion coefficient measurements *(P.W.M. Rutten, 1992)*

At zero chemical potential gradient, $\Delta\mu = 0$, the eqn. 1.618 is identical to eqn. 1.429 for the diffusional spreading of a concentration band. However, it is more universal because it was obtained from 1.614 under the condition that the bulk phase is moving, $w_z \neq 0$. Vice versa, for **zero-flow separation**, like centrifugation, electrophoresis etc. ($w_z = 0, Pe = 0, v_z = u_z$)

$$v = \frac{\Delta\mu}{RT}\frac{d\varepsilon}{d\zeta} \tag{1.625}$$

For instance, the separation of a singly-charged ion ($z = 1$) is performed in an electric field strength $500\ Vm^{-1}$ at a fundamental length $L = 0.05\ m$ and at a temperature = $300K$ ($RT = 2.5$ kJ mol^{-1}). The electro-chemical potential difference is, in ideal solution ($\Delta\mu_i^0 = 0$), according to eqn. 1.109, $\Delta\mu_i = zF\varphi = 1 \times 96500 \times 25$ C mol^{-1} V $= 2413$ kJ mol^{-1}. Because ε varies linearly from 0 to 1 for unit change of the separation coordinate, the value of $d\varepsilon/d\zeta$ is also unity and the dimensionless velocity $v = (\Delta\mu/RT)(d\varepsilon/d\zeta) = (2413/2.5)\ 1 = 9.7 \times 10^3$. The ratio d/L is even for a broad zone of separated ion (e.g. 1/5 of the separation path), small enough to be neglected in eqn. 1.619, if the Péclet number remains sufficiently low (say, < 1000. Diffusion broadening of the zone is then significant only at its sharp edges, where $\partial \ln c/\partial \zeta \geq \Delta\mu/(Pe\ RT)$.

Generally, to be distinguished from other moving zones of components having analogous concentration profiles, N zones with a half width Δz can be situated along the separation path L,

$$2N \langle \Delta z \rangle = L \qquad (1.626a)$$

or

$$2N \langle \Delta \zeta \rangle = 1 \qquad (1.626b)$$

at the moment when the fastest moving zone reaches the end of the separation path ($z = L$ or $\zeta = 1$). Because at this moment

$$\tau = 1/v \qquad (1.626c)$$

according to eqns. 1.624 and 1.625

$$N = \sqrt{\frac{v}{8}} \qquad (1.627)$$

The **number of zones separable** (resolvable), i.e. the **peak capacity**, is for the v given for zero-flow separation by eqn. 1.625 ($d\varepsilon/d\zeta = 1$), hence

$$N = \sqrt{\frac{\Delta \mu}{RT}} \qquad (1.628)$$

(J.C. Giddings, K. Dahlgren, 1971). For instance, to resolve the concentration peaks of two separated solutes ($N = 2$) the minimal difference $\Delta \mu$ at $T = 300K$ is $= (2)^2 \times 8.3 \times 300$ J mol^{-1} $\cong$ 10 kJ mol^{-1}. This number is bigger at GC and LC methods, but lower at SEC chromatography (J.C. Giddings, 1967).

1.5 MEMBRANE PROCESSES

A **membrane** in separation science is defined as a thin, usually polymeric, film which exhibits **permselectivity**, i.e. a much higher permeability for one species over another. From such a definition, it is the <u>function</u> of the membrane which is more important than its form or substance. From the point of view of formal thermodynamics, the membrane is an intermediate zone between two homogeneous subsystems, e.g. phase I and phase II. Permselectivity, derived from different transport rates from one phase to another, was first used in separation by *T. Graham (1854, 1863)*.

1.5.1 Passive transport

In the simplest case, when the transport proceeds as the Knudsen diffusion of gases A and B, the permselectivity of a membrane results from the ratio of fluxes given by eqn. 1.454, i.e. it is the product of their relative molecular masses M_A and M_B:

$$\frac{j_A}{j_B} = \sqrt{\frac{M_B}{M_A}} \tag{1.629}$$

(T. Graham, 1863). The membrane in this case plays the passive role of an intermediate zone with small holes through which the gases cross from the feeding to the receiving compartment *(S. Weller, W.A. Steiner, 1950; A.E. Applegate, 1984; T.Matsuura, 1994)*.

Diffusion concentration profiles for passive membrane transport *(S.N. Kim, K. Kammermeyer, 1970)* have been described by eqns. 1.442 and shown in Figure 30. In the past, the major use of the membrane separation of gases was to separate uranium isotopes *(K. Cohen, 1951)*; it is now used to recover nitrogen from air, helium and carbon dioxide from gas wells and hydrogen from ammonia plants *(K. Kammermeyer, 1976; G. Schulz, N. Werner, 1984; W.J. Schell, 1985; M. Mulder, 1996)*. For example, for N_2/O_2

separation the ratio of fluxes is $\sqrt{32/28} = 1.07$, i.e. the nitrogen molecule transfer is 7% faster than that of oxygen. Though the selectivity can be increased by suitable membrane materials, a cascade of membranes is necessary to achieve sufficient recovery — see section 2.3.

The selective diffusion of liquid and vapors across membranes increases the efficiency of distillation (e.g. of azeotropic mixtures). The separation process involving the contact of a liquid mixture on one side of a membrane is called **pervaporation** (*R. Rautenbach, R. Albrecht, 1980*). During pervaporation, the liquid diffuses through a membrane and evaporates at the permeate side of the membrane, the separation being determined by the rate of liquid diffusion. In this respect, pervaporation is an analogue of molecular distillation — see eqns. 1.456 and 1.457. **Membrane distillation** is a process in which the membrane is not wetted by liquids (*K. Smolders, A.C.M. Franken, 1989*). Even if there is no gas-gap and the membrane is in direct contact with the liquid, the vapor is transported through the membrane, the vapor-liquid equilibrium is not altered and the driving force is the partial-pressure gradient. The efficiency depends on the counter-current mode of evaporation (*K. Kammermeyer, D.H. Hagenbauer, 1955; R.C. Binning, F.E. James, 1958; P. Schissel, R.A. Orth, 1984; M. Mulder, 1991*).

At equilibrium, the fugacities of the separated gases in the compartments divided by membrane become the same and no separation results when the gases are ideal (i.e. when fugacities are equal to partial pressures). The equilibrium concentration from the different sides of the membrane would differ only when there are unequal activity coefficients of the diffusing species in the resulting solutions (e.g. in solutions of strong electrolytes) or if the membrane is not permeable towards one of the species, which is rejected to the feed solution due to its charge or large size (e.g. colloid particles or cells), or both. The measure of rejection is determined by the ratio of the passing flux to the flux coming to the membrane surface, i.e. the **reflection coefficient** ($0 \leq \sigma \leq 1$).

For example, if a solution of a macromolecular substance A which is larger than the pore size of the membrane flows through the membrane, the macromolecular solute can be completely rejected by the membrane ($\sigma_A = 0$) and its diffusion flux is zero. This is called **ultrafiltration** (H.Bechhold, 1907). Industrial ultrafiltration for pressure drops of 50–500 *kPa* has become a major separation process in the last two decades for the concentration of proteins and separation of colloids (oil from water, aqueous latex, electrophoretic paints, haze in fruit juices) which have sizes of 1–100 nm (A.S. Michaels, 1965, 1968; W.C. McGregor, 1986; J. Asenjo, 1990; J.A. Howell, V. Sanchez, R.W. Field, 1993).

The flux j_S (m s^{-1}) of solvent S would normally be proportional, according to Darcy's law, eqn. 1.463, to the outer pressure difference at both sides of the membrane,

$$j_S = L_{S1}(p^I - p^{II}) = L_{S1}\Delta p \tag{1.630}$$

However, because the concentration of solute A is different in the feeding and receiving compartments ($c_A^I = c_A^I$, $c_A^{II} = 0$), the osmotic pressure difference will work in the opposite direction. At equilibrium, the chemical potential of the solvent must be equal at different inner pressures p_{in}:

$$\mu_S^o + p^I \overline{V}_S + RT \ln x_S = \mu_S^o + p_{in}^{II} \overline{V}_S \tag{1.631}$$

where $\overline{V}_S$ is the molar volume (m³mol^{-1}) and x the molar fraction of solvent. The difference

$$p_{in}^{II} - p_{in}^I = \Pi = -\frac{RT}{\overline{V}_S} \ln x_S \tag{1.632}$$

is the **osmotic pressure**. In dilute solutions $x_S \cong 1$. As a result, $\ln x_S \cong 1 - x_S = x_A$. According to eqn. 1.51, $x_A/\overline{V}_S = 10^3 c_A$ (c_A in mol dm^{-3}) and therefore

$$\Pi \cong 10^3 RT c_A \qquad (1.633)$$

known as **van't Hoff's equation**.

For example, at 300 K, 0.15 M NaCl should have an osmotic pressure $\pi \cong 10^3 \times 8.31 \times 300 \times 2 \times 0.15 = 7.48 \times 10^5$ Pa (7.48 bar), because in 0.15 M NaCl the total concentration of ions is 2 × 0.15 = 0.3 M. The actual osmotic pressure is slightly lower due to the negative non-ideality of the solution.

The pressure difference in two-phase aqueous systems, such as water-protein-polysaccharide systems, can be applied for the transfer of water and preconcentration of the components: this is called **membraneless osmosis** (N.A. Zhuravskaya et al., 1986).

Given the identical phenomenological coefficient L_{S1} as in eq. 1.630, this results in

$$j_S = -L_{S1}(\Pi^I - \Pi^{II}) = -L_{S1}\Delta\Pi \qquad (1.634)$$

Summarizing eqns. 1.630 and 1.634, the resulting flux of solvent becomes

$$j_S = L_{S1}(\Delta p - \Delta\Pi) \qquad (1.635)$$

The presence of the pressure-driven volume flow j_S (mol m^{-2} s^{-1}) and concentration gradient Δc_A (mol dm^{-3}) is the reason why the low-molecular solute permeates through the membrane both as a result of the permeability (transfer) coefficient L_A(m s^{-1}), and the solvent flow j_S with an efficiency given by the dimensionless reflection coefficient σ_A. At a concentration $c_A *$ adjacent to the membrane, the total solute flow is given by the **Kedem-Katchalsky equation**

$$j_A = (1 - \sigma_A)10^3 c_A^* j_S + L_A 10^3 \Delta c_A \qquad (1.636a)$$

(O. Kedem, A. Katchalsky, 1958) of solute-solvent co-diffusion. The first term characterizes the solute transfer at hydrostatic (osmotic) pressure difference and for specific reflective properties of the membrane. The second term describes the diffusion of solute in the absence of a static pressure gradient, but at a solute concentration gradient; in this case, the diffusion of the solute disturbs the overall osmotic equilibrium. As a result of the osmotic gradient $\Delta\Pi$, an osmotic volume flow of solvent occurs in the same direction, according to eqn. 1.634,

$$j_S = L_S(\Delta p - \sigma \Delta \Pi) \qquad (1.636b)$$

For a reflection coefficient $\sigma_A = 0$ the solute is identical to the solvent and passes freely, and consequently no osmotic gradient is created. For $\sigma_A = 1$ the membrane is permeable only to the solvent and ultrafiltration occurs. The coefficients L_A and L_S can be estimated from intermolecular interactions (*A. Katchalsky, P.F. Curran, 1967; G.S. Manning, 1972*). Ultrafiltration ceases ($j_S = 0$) when the pressure applied and the corresponding flow j_v is balanced by the osmotic pressure. Therefore this process is often called **reverse osmosis** (*S. Sourirajan, 1970*). Typical pressure drops applied are 2-10 MPa.

For example, the reverse osmosis of sea water ($c_A = 30$ g dm^{-3} = 0.51 M NaCl = c_A^*) at a filtration rate $j = 0.8$ cm^{3}m^{-2} s^{-1} = 8×10^{-7} m s^{-1} and a reflection coefficient of the semipermeable (ion-exchange) membrane $\sigma_A = 0.99995$ results in the co-transport of salt $(1-0.99995) \times (10^3 \times 0.51) \times (8 \times 10^{-7}) = 2.0 \times 10^{-8}$ mol m^{-2}s^{-1} For a diffusion coefficient $D_A = 2 \times 10^{-10}$ m^2 s^{-1} and membrane thickness $\Delta x = 8 \times 10^{-4}$ m, $L_A = 2.5 \times 10^{-8}$ m s^{-1} (see eqn. 1.453) and the maximal partial flux into the pure solvent zone becomes $(2.5 \times 10^{-8})(10^3 \times 0.51) \cong 1.27 \times 10^{-5}$ mol m^{-2}s^{-1}. The salt concentration in the ultrafiltrate is therefore $(2.0 \times 10^{-8} + 1.27$

$\times 10^{-5}$ mol m^{-2}s^{-1}):(8 × 10^{-7} m s^{-1}) ≅ 16 mol m^{-3} = 0.016 M. This concentration, about 1 g dm^{-3} is, incidentally, acceptable for drinking water. The osmotic pressure difference at T = 300 K is, according to eqn. 1.633, $10^3(8.31 \times 300) \times (0.51 - 0.016)$ = 2.46 × 10^6 Pa ≅ 2.5 MPa and a hydrostatic pressure several times higher (up to 10-13 MPa) must be applied to produce a positive filtration flux of solvent.

During ultrafiltration, the solute accumulates at the membrane surface and the actual concentration and osmotic pressure at the interface may be much higher than that which corresponds to the bulk concentration. This process, the **concentration polarization of the membrane**, is typical for all processes with non-zero reflection coefficients and decreases the efficiency of ultrafiltration. Concentration polarization can be minimized by a flow parallel to the membrane interface which sweeps the accumulated solute, leaving only a thin layer near the surface (*T.K. Sherwood et al., 1965*) —see also eqn. 1.522. If the thickness of the remaining diffusion layer is d, the steady concentration of solute at the surface, c_A^{*I}, is obtained from the steady-state condition

$$j_A = c_A j_S = -D_A \frac{\partial c_A}{\partial z} \tag{1.637}$$

(cf. eqn. 1. 445b) and the boundary conditions $c_A = c_A^{*I}$ at $z = 0$ and $c_A = c_A^{I}$ at $z = d$, so that

$$c_A^{*I} = c_A^{I} \exp\left(\frac{j_S d}{D_A}\right) \tag{1.638}$$

where $j_S d/D_A$ is the Péclet diffusion number (Pe) according to eqn. 1.430. For example, under the experimental conditions $j_S = 8 \times 10^{-7}$ mol s^{-1}, $d = 8 \times 10^{-4}$ m and $D_A = 2 \times 10^{-10}$ m^{2}s^{-1}, Pe is 3.2 and the polarization is not negligible, $c_A^{*I} = 24.5\, c_A^{I}$.

Polarization due to an increase of concentration at the surface is often serious and may cause additional transfer resistance due to **membrane fouling** by the deposition of solids and gel-like layer formation. Ferric hydroxide, bacteria and proteins are the most frequent foulants but these can be removed by acids, hypochlorite or caustic detergents, respectively.

Further cases of selective permeation, often met in practice, consist of the diffusion of **polyelectrolytes** having one of their counter-ions macro-molecular and non-permeable. Such behavior was demonstrated by F.G. Donnan (1911) and is known as the **Donnan equilibrium**.

Let us consider two compartments of solutions of electrolyte $K_{\nu_K}A_{\nu_A}$ consisting of cation K^{z_K} and anion A^{z_A} with only one of the compartments containing the macromolecular ion (L). This might be a bulk solution divided by a semipermeable membrane, or an ion-exchange resin contacting a solution. The presence of the macromolecular ion in one compartment results in an unequal distribution of the electrolyte K A. This distribution can be derived from the equations of electrochemical potential equality,

$$a^I_{KA} = a^{II}_{KA} \tag{1.639}$$

and the electroneutrality of each compartment:

$$z_K c^I_K + z_A c^I_A + z_L c^I_L = 0 \tag{1.640a}$$

$$z_K c^{II}_K + z_A c^{II}_A = 0 \tag{1.640b}$$

which is valid for strong electrolytes KA and KL. Taking into account the expression 1.106 for the ionic activity for electrolyte $K_{\nu_K}A_{\nu_A}$, this system of non-linear equations (*when* $\nu_K, \nu_A \neq 1$) can be solved as done for the general form of chemical equilibria in section 1.2.3. From eqn. 1.639 results

$$\frac{(c_K^{II})^{\nu_K} (c_A^{II})^{\nu_A}}{(c_K^{I})^{\nu_K} (c_A^{I})^{\nu_A}} Y = 1 \tag{1.641}$$

where, corresponding to eqns. 1.162 and 1.120,

$$Y = \frac{(\gamma_K^{II})^{\nu_K} (\gamma_A^{II})^{\nu_A}}{(\gamma_K^{I})^{\nu_K} (\gamma_A^{I})^{\nu_A}} = \left(\frac{\gamma_\pm^{II}}{\gamma_\pm^{I}}\right)^{\nu} \tag{1.642}$$

From the equilibrium eqn. 1.641 and the charge balances 1.640, an implicit equation is obtained

$$\left(\frac{c_A^I}{c_A^{II}}\right)^{\nu_A} = Y \left(\frac{z_A c_A^{II}}{z_A c_A^I + z_L c_L^I}\right)^{\nu_K} \tag{1.643}$$

For dilute solutions and a large concentration of macro ion (e.g. in an ion-exchange resin) when $c_L^I \gg c_A^{II}$, the concentration c_A^I becomes negligible. For example, distribution ratio of co-anion A ($z_A=-1$) between cation exchanger L ($z_L=-1$) and own solution of salt KA ($z_K=+1$, $\nu_A=\nu_K=1$) follows from this equation

$$D_A = \frac{c_A^I}{c_A^{II}} \approx Y \frac{c_A^{II}}{c_L^I}$$

i.e. it is proportional to equilibrium concentration of the anion in solution and inversely proportional to the cation exchanger capacity.

The Donnan effect is decreased by decreasing the activity coefficients because of ion association between the counter-ions K and A, or K and L. If $z_A = -z_K = -z_L$, then $\nu_K = \nu_A = 1$ and eqn. 1.643 becomes

$$\frac{c_A^I}{c_A^{II}} = \sqrt{Y\left(1 - \frac{c_L^I}{c_K^I}\right)} \tag{1.644}$$

For quasi-ideal solutions, $Y = 1$. This is known as the **Donnan equilibrium equation** (*F. G. Donnan, 1911*), which indicates that the concentration of permeating ion (A) of the same sign as the polyelectrolyte ion (L) is lower on the membrane side containing the high-molecular ion, $c_A^I < c_A^{II}$, i.e. it is ejected by that co-ion (Fig. 45).

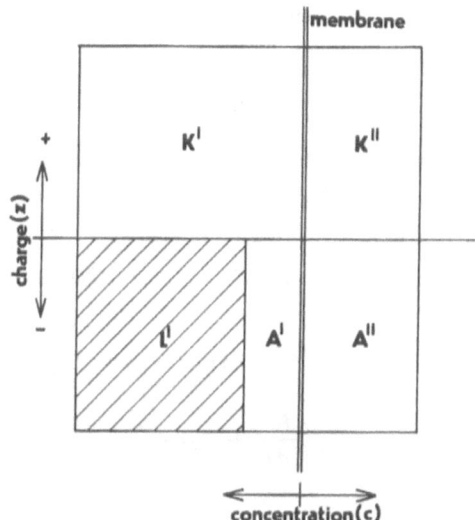

Figure 45. Schematic mass and charge balance at Donnan equilibrium (K — cation, A —anion, L — macromolecular anion).

The equality of activity of the electrolyte KA, at the same time, indicates that in general the concentration of small ions is unequal between the compartments separated by the membrane. The Donnan effect can be used for the mutual separation of low- and high-molecular components and to separate salts and microsolutes from mixtures. Thus, the salts used for salting out proteins (section 1.2.4.) can be removed by **dialysis**, i.e. contact of the source (feed) solution with the water compartment disengaged by the membrane. In absence of solid membrane, **membrane-less dialysis** can proceed in two immiscible aqueous phases (*F. Macasek, P. Gerhart, A. Malovikova, 1994*).

With respect to eqn. 1.643, the higher the concentration of protein (i.e c_L^I), the smaller is the fraction of salting-out anion A remaining in compartment I. The receiving phase, called the **dialysate**, usually flows on the opposite side of the membrane to take advantage of the counter-current arrangement (section

2.4.). A major use of dialysis is the artificial kidney for blood purification, **hemodialysis**, from low-molecular (mainly urea, kreatinin and ureic acid) and medium-molecular metabolites (*W.J. Kolff, H.T.J. Berk, 1946; E. Klein, 1978*).

If the **transport of ions** occurs within the membrane, both the equilibria and dynamic parameters are influenced by electrostatic interactions. Because of the different diffusion rates of ions, an electric potential gradient, i.e. the **diffusion potential gradient** ($d\phi/dx$), appears within the membrane. In limiting cases, the membrane is non-permeable at least towards one of the ions present, i.e. an Donnan-type equilibrium occurs; such a potential is called the **Donnan potential**.

The mass flux of a particular ion, j_A and j_K is given by the **Nernst-Planck ion transport equation** (see eqn. 1.425),

$$j_K = -D_K \left(\frac{dc_K}{dx} + z_K c_K \frac{F}{RT} \frac{d\varphi}{dx} \right) \tag{1.645a}$$

$$j_A = -D_A \left(\frac{dc_A}{dx} + z_A c_A \frac{F}{RT} \frac{d\varphi}{dx} \right) \tag{1.645b}$$

where x is the coordinate perpendicular to the membrane surface. The equation can be solved numerically (*V.M. Aguillela et al., 1986*). From the electroneutrality condition,

$$z_A c_A + z_K c_K + z_L c_L = 0 \tag{1.646}$$

and for a non-diffusing polyelectrolyte (c_L = const), the ion gradients become

$$z_A \frac{dc_A}{dx} = -z_K \frac{dc_K}{dx} \tag{1.647}$$

If no electric current flows through the membrane, according to Faraday's law, eqn. 1.500

$$i = Fz_K j_K + Fz_A j_A = 0 \tag{1.648}$$

After substitution of eqns. 1.647 and 1.648 into eqn. 1.645, the summation gives for the **diffusion potential gradient**

$$\frac{d\varphi}{dx} = -\frac{RT}{F} \frac{(D_K-D_A)z_K}{D_K z_K^2 c_K + D_A z_A^2 c_A} \tag{1.649}$$

The steady-state Donnan potential is obtained from eqn. 1.645 at zero flux for a particular ion, when the concentration diffusion is inhibited by the potential formed,

$$j_K = 0 \tag{1.650}$$

Then, from 1.645b we can assume

$$\frac{dc_K}{c_K} = -\frac{z_K F}{RT} d\varphi \tag{1.651}$$

and integration gives

$$\varphi^{II} - \varphi^{I} = \varphi_{Don} = \frac{RT}{z_K F} \ln \frac{c_K^I}{c_K^{II}} \tag{1.652}$$

When the solutions are not ideal, the logarithm refers to the ratio of activities, a_K^I/a_K^{II}. A tenfold difference between concentrations, i.e. when $c^I/c^{II} = 10$, corresponds, for univalent ions ($z = 1$) at $T = 300K$, to a potential difference of 58 mV. In the steady state of ion flux, e.g. for anion A, we have

$$j_A = -\frac{D_A D_K (z_A^2 c_A + z_K^2 c_K)}{D_K z_K^2 c_K + D_A z_A^2 c_A} \frac{dc_A}{dx} \tag{1.653}$$

while the same expression containing the gradient dc_K/dx is obtainable for j_K. When the membrane is completely non-permeable to co-ion K, from eqns. 1.651, 1.647 and 1.645a we find

$$j_A = -D_A \left(1 - \frac{z_A}{z_K}\right)\frac{dc_A}{dx} \qquad (1.654)$$

This indicates that the term before the concentration gradient is an effective diffusion coefficient of the electrolyte in the membrane matrix — see eqn. 1.408.

The inter-ionic effect which drags highly-mobile ions and accelerates the transport of slow ions is obviously the direct effect of the electroneutrality of ionic mixtures. If a mixture of various (*i*) electrolytes is to be considered, regardless of the presence of a polyelectrolyte, a potential difference $\Delta\phi_L$ occurs, as is demonstrated by solving the generalized eqn. 1.648 for zero electric current flow:

$$\sum F z_i j_i = 0 \qquad (1.655)$$

and for the electroneutrality equation,

$$\sum z_i c_i = 0 \qquad (1.656)$$

which can be substituted into eqn. 1.425, and the **liquid junction potential** difference for multi-ion systems results:

$$\frac{d\varphi_L}{dx} = -\sum \frac{t_i}{z_i} \frac{d \ln c_i}{dx} \qquad (1.657)$$

where t_i is the **transference, (or transport) number** of the i-th component,

$$t_i = \frac{z_i^2 u_i c_i}{\sum z_j^2 u_j c_j} \qquad (1.658)$$

The integration of eqn. 1.657 is complicated and needs classification of the ions according to the charges, z_i and z_j (M. Planck, 1890; R. Schlögl, 1954; F. Helfferich, 1958, 1959).

When the membrane contains a large concentration of polyelectrolyte ions, i.e. in an **ion exchange membrane** (J. Vacik,

J. Kopecek, 1975; D.S. Flett, 1983; G. Scibona, C. Fabian, B. Scuppa, 1983), the concentration of permeating co-ion is negligible. For example, when diffusion occurs in a cation-exchange membrane, we have $c_A \ll c_K$ and the flux of the cation,

$$j_K \cong -D_A \frac{dc_K}{dx} \tag{1.659}$$

depends on the diffusion coefficient of the anion A, as does its own flux.

The electrical potential can be used to effect membrane separation in a variety of **electromembrane processes** (R.E. Lacey, 1972). Ion transport under an electric field across the membrane is used in **electrodialysis** (K.H. Meyer, W. Strauss, 1940; V.D. Grebenyuk, 1976; H. Strathmann, 1985) to remove selectively ionic species. The Coulombic (current) efficiency η_c of electrodialysis is given by the ratio of the current j_A transferred by the separated solute A to the total electric current passing through the membrane:

$$\eta_c = \frac{F z_A j_A}{i} \tag{1.660}$$

In absence of the flux of co-ion, the limiting ion current density — see eqn. 1.574 — is obtained from eqn. 1.654 as

$$i_d = \frac{F D_A}{d}\left(1 - \frac{z_A}{z_K}\right) c_A^* \tag{1.661}$$

or, in general

$$i_d = \frac{F D_A z_A}{d(t_A - t_K)} c_A^* \tag{1.662}$$

where t_A and t_K are the transference numbers of ions A and K respectively — eqn. 1.504. This indicates that the current efficiency of electrodialysis is given by the transference number in the membrane and decreases with concentration of solute. The use of a stack of ion-exchange membranes reduces gas evaluation on the electrodes and increases the efficiency of the process. Electrodialysis is usually applied for 0.01–0.5M solutions, highly dilute solutions having too high an electrical resistance while in concentrated solutions diffusion transport dominates over electromigration. For example, for $D_A = 10^{-9}$ m^{2}s^{-1}, $c_A = 0.01$ mol dm^{-3} and d $= 10^{-4}$ m, in a uni-uni valent electrolyte ($z_A = -z_K = -1$), the limiting current density is (96500 C mol^{-1}) $(10^{-9}$ m^2 s$^{-1})(10$ mol m$^{-3})(1 + 1)/(10^{-4}$ m$) \cong 20$ A m^{-2}. The limiting current is increased by the membrane resistance which for ion-exchange membranes is usually from 0.7 to 3 Ω m.

The major use of electrodialysis is the desalination of sea and brackish water (*E.D. Howe, 1974; I.S. Al-Mutaz, 1986; K.S. Scott, R. Hughes, 1993*).

1.5.2 Carrier-facilitated transport

The strong electrostatic interactions at the ion-exchange membrane/solution interface change the concentration profiles when compared to an indifferent membrane (Fig. 46a). There is a sudden change of concentration at the interface (Fig. 46b) corresponding to the distribution ratio D^I of the ion between the source solution and membrane, and the jump in the opposite direction, corresponding to the distribution ratio D^{II} on the receiver side (*J.W. Ward III, 1970; J.M. Bloch et al., 1988; M.J. Bedzyk et al., 1990*). The concentration gradient in a membrane (for a given concentration outside the membrane) becomes higher as D^I is increased and D^{II} is decreased. Consequently, a positive concentration gradient in the membrane can be achieved even if the external gradient is negative (Fig. 46c) and the membrane "pumps" a solute up-hill (*E.L. Cussler, 1971*). On the activity scale, however, each model is identical with that discussed in the two-film model — section 1.4.4 and Fig. 38 — because the shifts are caused by different activity coefficients and a local equilibrium exists at the interface. The flux through the membrane is always given by the gradient of activities (cf. eqn. 1.523):

$$j = \kappa(a_I^{*M} - a_{II}^{*M}) \qquad (1.663)$$

where a_I^{*M} and a_{II}^{*M} are the activities at the membrane interfaces with phases I and II, respectively. In the ideal (or pseudo-ideal) membrane phase

$$j = K(c_I^{*M} - c_{II}^{*M}) \qquad (1.664)$$

The boundary concentrations in the local equilibria

$$a^{*I} = a^{*M} \qquad (1.665a)$$

$$a^{*II} = a^{*M} \qquad (1.665b)$$

can be expressed through the respective concentrations in the bulk phase and the distribution ratios D^I and D^{II} (eqn. 1.309):

$$j = K(D^I c^{*I} - D^{II} c^{*II}) \qquad (1.666)$$

The flux is positive when

$$c^{*I} > \frac{D^{II}}{D^I} c^{*II} \qquad (1.667)$$

i.e. it is maintained at small D^{II} values even when $c^I < c^{II}$ (Fig. 46c).

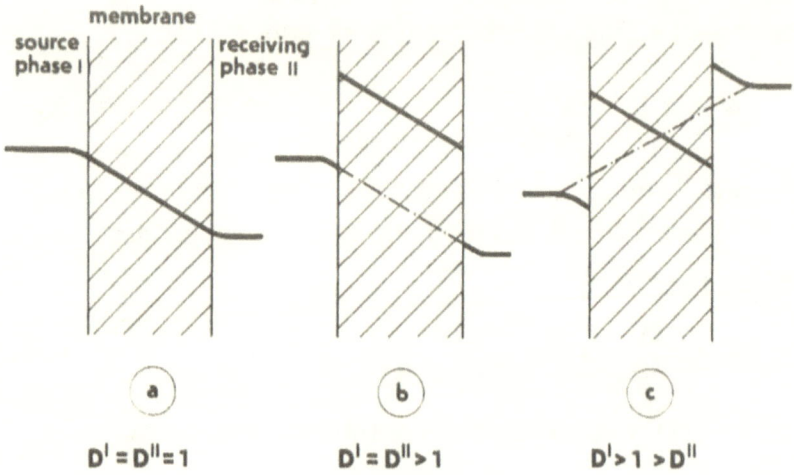

Figure 46. Concentration profiles at membrane interface; a) ideal phases b) negative ideality in membrane, c) strong negative ideality in receiving phase (facilitated transport).

According to eqns. 1.309 and 1.305, the distribution ratio is proportional to the ratio of the concentration activity coefficients ($D^I \propto \dot{\gamma}^{II}/\dot{\gamma}^M$) and strong negative non-ideality of the receiving phase ($\dot{\gamma}^{II} \ll 1$) is necessary to facilitate membrane transport. At the same time, negative non-ideality of the membrane compared to the source phase ($\dot{\gamma}^M \ll \dot{\gamma}^I, D^I \gg 1$) diminishes the contribution of the membrane to transport resistance, cf. eqn. 1.530.

Hence, **facilitated transport** makes use of reversible reactions between the solute and a carrier which is mobile in the membrane. Classical examples are hemoglobin as the carrier for oxygen transport (*P.F. Scholander, 1960*) or valinomycin as an ionophorous agent for alkali metal ions (*G.J. Pedersen, 1967; M.M. Shemyakin et al., 1969*) in non-polar membranes. Such a reaction normally occurs at the membrane interface since the solute (A) and carrier (C) are usually soluble only in the source (or receiving) and membrane phase, respectively, or throughout the membrane phase. There are three different mechanisms of transport (*J.S. Schultz, J.D. Goddard, S.R. Suckdeo, 1974*):

(i) **carrier (uniport) transport** of solute A due to formation of a permeable complex with carrier C:

$$A^I + C^M \rightleftarrows (AC)^M \rightleftarrows A^{II} + C^M \qquad (1.668a)$$

      feed   *membrane*   *receiver*

One example is oxygen-hemoglobin transport (*P.F. Scholander, 1960*):

$$O_2^I + Hb^M \rightleftarrows (O_2Hb)^M \rightleftarrows O_2^{II} + Hb^M \qquad (1.668b)$$

(ii) **coupled (symport) transport** or **co-transport** (*W.C. Babcock et al., 1980; R.M. Izatt et al., 1983*) uses the common diffusion of a solute (A) with another solute (B), e.g. cation and counter anion, and carrier (C), with the reactions

$$A^I + B^I + C^M \rightleftarrows (ABC)^M \rightleftarrows A^{II} + B^{II} + C^M \qquad (1.669)$$

      feed   *membrane*   *receiver*

The flux of A is linked (coupled) to the flux of B, and proceeds from a common source to the receiving phase; for example, the transport of carbon dioxide with a tertiary amine

$$(CO_2)^I + (H_2O)^I + (R_3N)^M \rightleftarrows (H_2CO_3R_3N)^M \rightleftarrows (CO_2)^{II} + (H_2)^{II} + (R_3N)^M \qquad (1.669a)$$

(iii) **counter (antiport) transport, exchange diffusion** in which the solute A is transported by carrier C and displaced by solute B,

$$A^I + (BC)^M \rightleftarrows (AC)^M + B^I \qquad (1.670a)$$

$$(AC)^M + B^{II} \rightleftarrows (BC)^M + A^{II} \tag{1.670b}$$

and are both transported in opposite directions because, in the beginning, A and B prevails in different phases; for example, the counter-transport of the uranyl chelate with 8-hydroxyquinoline (oxine, HOx):

$$(UO_2^{2+})^I + 2(HOx)^M \rightleftarrows (UO_2Ox_2)^M + 2(H^+)^I \tag{1.670c}$$

$$(UO_2Ox_2)^M + 2(H^+)^{II} \rightleftarrows (UO_2^{2+})^{II} + 2(HOx)^M \tag{1.670d}$$

Because the equilibrium distribution of the solute in a closed three-phase system (phase I — membrane — phase II) is given by the universal equality of chemical potentials:

$$\mu^I = \mu^M = \mu^{II} \tag{1.671}$$

from the equilibrium condition, eqn. 1.155, the result for
(i) equilibrium, eqn. 1.668, of carrier transport is

$$a_A^I = a_A^{II} \tag{1.672}$$

(ii) equilibrium, eqn. 1.669, of coupled transport

$$a_A^I a_B^I = a_A^{II} a_B^{II} \tag{1.673}$$

(iii) equilibrium, eqn. 1.670, of counter transport

$$a_A^I / a_B^I = a_A^{II} / a_B^{II} \tag{1.674}$$

This indicates that, as in Donnan equilibria, the properties of the carrier and membrane are irrelevant from the point of view of equilibrium, unless the mass balance of the diffusant is considered. The major role of the membrane is to provide specific features to the kinetics of transport, according to eqn. 1.662, i.e. to be a "**separation catalyst**". The final success of facilitated transport, according to eqns. 1.672–1.674, is always given by a strong negative non-ideality of solute A in phase II and, in coupled transport, by high activity of solute B in source phase I or, in counter transport, by the latter in the receiving phase II. In the examples given, eqn. 1.669a can be achieved, for example, by making alkaline the receiving phase or adding carbonic anhydrase to it. The transport of uranium according to eqn. 1.670, can be accomplished both by complexation masking of the uranyl ion in phase II or by increasing the acidity of the receiving phase.

As was mentioned in passive transport, membrane separation is not accomplished by influence of the membrane on the equilibrium but on a driving force. Hence, the relation between the steady fluxes of two separated solutes, A and B, j_A and j_B, is of principal interest and can be obtained from equations of type 1.666. However, even for binary mixtures the equations may be mathematically very complex, because the distribution ratios depend on the composition of the feed and receiving phases (see section 2.2.1). According to the two-film theory — see eqn. 1.530 — **chemical selectivity** can disappear at high distribution ratios when resistance in the feed solution and diffusion becomes the controlling factors (*P.R. Danesi, R. Chiarizia, M. Muhammed, 1978; P.R. Danesi et al., 1981*). When the distribution ratio at the membrane-strip interface is much lower than at the membrane-feed interface, and the concentration in the receiving phase does not greatly exceed that in the feed (e.g. when the interface is washed by fresh stripping solution), then $D^I \gg D^{II} c^{*II}/c^{*I}$ and instead of eqn. 1.666, the result is

$$j = K D^I c^I \qquad (1.675)$$

while the **permeation ratio** is given by

$$\frac{j_A}{j_B} = \frac{K_A D_A^I}{K_B D_B^{II}} \frac{c_A^I}{c_B^I} \qquad (1.676)$$

where K_A and K_B are the transport coefficients and D_A^I, D_B^I are the distribution ratios at the membrane-feed interface (*P.R. Danesi, L. Reichley-Yinger, 1986*). When the resistance of the feed aqueous phase cannot be neglected (*K.H. Lee, D.F. Evans, E.L. Cussler, 1978; L. Boyadzhiev, T. Sapundzhiev, E. Bezenshek, 1977*) the two-film theory, eqn. 1.530, for the permeation coefficient results in the following

$$j = \frac{1}{1/K^{II}+D^I/K^I} D^I c^I \qquad (1.677)$$

where the parameters K^I and K^{II} are measurable under varying hydrodynamic conditions (stirring) of the feed phase (*P.R. Danesi, 1984*). Assuming that K^I and K^{II} are approximately the same and that for the main solute A we have $D_A^I \gg K_A^I K_B^{II}$ and, conversely, for the minor solute $D_A^I \gg K_A^I K_B^{II}$, it follows that the ratio of their fluxes is equal to

$$\frac{j_A}{j_B} = \frac{K_A^I}{K_B^{II}} \frac{1}{D_B^I} \frac{c_A^I}{c_B^I} \qquad (1.678)$$

(*R. Chiarizia, E.P. Horwitz, 1990*), i.e. the separation efficiency is regulated by the permeation ability of A in the source phase and that of B in the membrane, which corresponds to the different roles of the boundary layers according to eqn. 1.530.

Generally speaking, facilitated membrane transport becomes successful if chemical resistance controls the transfer, and vice versa. On integration of eqn. 1.675 under the conditions $c^I = c_0^I$ at $t = 0$, the **kinetics of pertraction** results in:

$$\ln \frac{c^I}{c_0^I} = -\frac{A}{V_I} D^I K t \qquad (1.679)$$

where A is the membrane area and V_I is the volume of feed solution. However, when the distribution ratio at the strip interface, D^{II} is not negligible and the volume V_{II} is not very large as compared to the volume of membrane, V_M, a **pertraction factor**

$$p = 1 + \frac{1}{D^{II} r_{II}} \tag{1.680}$$

which includes the volume ratio of the stripping to the membrane phase, $r_{II} = \frac{V_M}{V_{II}}$, enters eqn. 1.679, resulting in (*F. Macasek et al., 1984*).

$$\ln \frac{c^I - c^I_\infty}{c^I_0 - c^I_\infty} = -\frac{A}{V_I} K \frac{2(1+pD^I r_I)}{(2p-1)r_I} t \tag{1.681}$$

where c^I_∞ is the equilibrium concentration in feed at $t = \infty$. It can be seen that at $p \gg 1$ and $c^I_\infty = 0$, eqns. 1.679 and 1.681 give the same results. Further, when permeation proceeds by saturation of the carrier with solute A, the distribution ratio D^I is not constant; if the distribution at the feed interface can be described **by a Langmuir-type isotherm**, eqn. 1.323, and the strip is full, the approximation (*F. Macasek, 1989*) is valid

$$\frac{1}{(D^I)_0} \ln \frac{c^I}{c^I_0} - q_0 r_I (1 - \frac{c^I}{c^I_0}) = -\frac{A}{V_I} Kt \tag{1.682}$$

where q_0 is the initial molar ratio of solute and carrier and D^I_0 is the distribution ratio at a large molar excess of carrier, when $q_0 \cong 0$.

Carrier-mediated transport can be performed by using
(i) supported liquid membranes and flowing film,
(ii) double emulsion liquid membranes,
(iii) creeping films,
(iv) bulk liquid membranes (on the laboratory scale).

Solid-**supported liquid membranes** (SLM) consist of a solvent, immiscible with the source, and a receiving solution, anchored on a porous sheet or hollow fiber material. A high liquid membrane/water interfacial tension is important to keep the solvent steady in the capillary pores once the membrane is soaked; small pore radii are necessary to maximize the capillary forces and prevent a leaky membrane (*Zh. Fan, X.F. Zhang, X.S. Su, 1988; P.R. Danesi, 1988*). Solvent can also be removed from the pores by emulsion formation caused by lateral shear forces at a high flow rate of the outer phases (*A.M. Neplenbroek, D. Bargeman, C.A. Smolders, 1988*). Supports for SLM are made from hydrophobic materials — PTFE, polypropylene, polyethylene, etc., of thickness usually 10 − 100 µm; thicker membranes have a greater resistance while thinner ones are useless because the resistances of the outer solutions limit the transport rate. The pore diameter is 0.02 − 1 µm and the free area up to 75%.

In **flowing film membranes** (FFM), the membrane solvent flows in a thin channel between two support membranes which can be arranged as a spiral-type module. When the flow rate of solvent is high, the mass transport is several times higher than that of SLM and the stability is also better (*M. Teramoto et al., 1988*).

Creeping film membranes (CFM) are formed by keeping the 3–5 mm organic film between two hydrophilic vertical porous (fibrous) plates, 0.5–5 mm thick, to support both the feed and receiving aqueous phase which flow down (*L. Boyadzhiev, Z. Lazarova, 1987*). CFM are notable for the greater stability of mass transport under continuous operation, their ability to achieve highly-concentrated strip solutions (up to 5000-fold preconcentration) and greater throughput (3 − 10 m^3 per 24 hours per m^2 of pertractor).

Electrostatic pseudo liquid membranes (ESPLIM) consist of a continuous oil phase divided by two vertical perforated plates, or one baffle plate, into three compartments (extraction, membrane and stripping cells). The feed and strip aqueous solutions are introduced to the top of the two outside compartments across which an electrostatic field of several *kV* is applied, the baffle plate being the earthed electrode. Due to this field, the aqueous phase is dispersed into 0.1–0.3 mm droplets, i.e. it functions under "electrostatic agitation". A complex formed in the extraction cell diffuses through the baffle plate into the stripping cell. The electrostatic field also assists in the separation of phases in the stripping section (*Q.J. Zhou, Z.M. Gu, 1988; Z.M. Gu, 1990; Z.M. Gu, J.D. Navratil, H. Cheng, 1990*).

The double **emulsion liquid membrane** (ELM) was introduced by *N.N. Li (1971)* for the selective separation of hydrocarbons by an aqueous membrane. *L. Cussler (1971)* used a complexing carrier to transport metals through an organic membrane. ELM can also be used in biochemical separations (*H.P. Thieu, T.A. Hatton, D.I.C. Wang, 1986*). The organic membranes arise when a water-in-oil (W/O) emulsion is dispersed in a continuous feed aqueous phase (or, in case of aqueous membranes, O/W emulsion in the organic phase): between the outer aqueous solution and the inner aqueous solution a semi-mobile organic (oil) layer is formed — Fig. 47. Such double emulsions are tailor-made by the addition of an emulgator whose hydrophilic-lipophilic balance — eqn. 1.250 — is between 3 and 6 (*E.L. Cussler, D.F. Evans, 1974*).

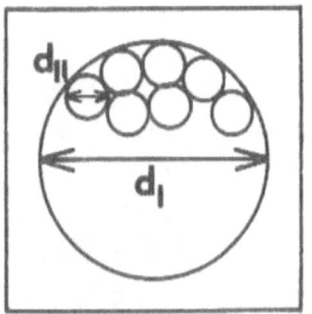

Figure 47. Liquid emulsion globule (d_I — outer diameter, d_{II} — diameter of inner dloplets).

The prediction of the mass transport rate in the double emulsion systems is rather complicated because the geometries of the outer and the inner interphase are determined by the emulsion globules and the inner droplets' size and arrangement. If V_I, V_M and V_{II} are the volumes of the external, membrane and internal phases, then the interfacial area between the external aqueous phase and the globules of W/O emulsion or the interfacial area between the membrane and the internal droplets is

$$a = \frac{6}{\langle d \rangle (1-v)} \tag{1.683}$$

where $\langle d \rangle$ is the Sauter mean diameter of the globules or droplets respectively, obtained as the ratio of the mean volume and surface of the particles,

$$\langle d \rangle = \frac{\langle V \rangle}{\langle A \rangle} = \frac{\sum n_i V_i}{\sum n_i A_i} \tag{1.684}$$

(*M. Teramoto et al., 1983*) and v is the volume fraction of the particles, i.e. v_I of the globules for the outer interface:

$$v_I = \frac{V_{II} + V_M}{V_I + V_M + V_{II}} \tag{1.685a}$$

or v_{II} for the inner interface:

$$v_{II} = \frac{V_{II}}{V_M + V_{II}} \tag{1.685b}$$

The ELM separation process is, in many respects, a three-phase analogue of double solvent extraction (*P.C. Wankat, 1980; S. Schlosser, E. Kossaczky, 1980; F. Macasek et al., 1984*) and is often called **(emulsion liquid) membrane extraction** or **pertraction**. Its kinetics can be roughly approximated by eqns. 1.675–1.682, using the value $a = A/V_I$ from eqn. 1.683. However, an **advancing front model, multilayer shell model** or **cluster of cubic elements model** may be more realistic (*R. Marr, A. Kopp, 1980; M. Teramoto, 1983, 1986; A.L. Bunge, R.D. Noble, 1984; T. Takaoka et al., 1989*). In the multilayer shell model, the globule of W/O emulsion is considered to consist of layers of spherical shells, between which mass transfer proceeds due both to diffusion through the membrane (permeation coefficient

K_i) and the adjacent jth and $(j + 1)$ th shells (K_{ij}). The latter can be derived from the thickness of the jth shell and the effective diffusivity (D_i) of the i-th species as a function of division of the globule to n shells of equal volume (*M. Teramoto, 1985*):

$$K_{ij} = \frac{D_i}{d^I}\left[\left(\frac{j+1}{n}\right)^{1/3} - \left(\frac{j-1}{n}\right)^{1/3}\right] \tag{1.686}$$

This approach considerably reduces the time for numerical integration of a set of equations of the type 1.665.

The stripping phase composition exerts only a small influence on the pertraction kinetics, because when the pertraction factor, eqn. 1.680, varies from 1 to ∞, the permeation rate increases at most twice under reasonable yields, as can be seen from the rate-determining term in eqn. 1.681. However, when compared with solvent extraction, ELM has advantages when a high preconcentration for a limited amount of organic phase or a low capacity of the latter is required, e.g. at a low concentration of carrier and low extractant-to-feed volume ratio. This is often met both for economic or ecological reasons. (*F. Macasek et al., 1984; M. Teramoto et al., 1986; P. Rajec, F. Macasek, J. Belan, 1986; F. Macasek, 1989*). The maximum obtainable yield for batch pertraction is

$$R_{ELM} = \frac{pD^I r_I}{1+pD^I r_I} \tag{1.687a}$$

where r_I is the volume ratio

$$r_I = \frac{V_M}{V_I} \tag{1.687b}$$

D^I is the distribution ratio at the outer interphase and p is the pertraction factor, eqn. 1.680. A high pertraction factor leads to yields that are equivalent to an N-fold repeated batch solvent extraction — see section 2.4. — and N can be found from

$$N = \frac{\log(1+pD^I r_I)}{\log(1+D^I r_I)} \tag{1.688}$$

This makes pertraction a superior preconcentration technique (*F. Macasek, 1989*). A comparison of staged and continuous arrangements — section 2.4. —has also been made (*S. Schlosser, E. Kossaczky, 1980; P.C. Wankat, 1980; P. C. Wankat, R.D. Noble, 1984*).

A special type of three-phase system with the organic phase working as a separation membrane is represented by **microemulsions** (*C. Tondre, A. Xenakis, 1984; L.M. Gan et al., 1987*). Microemulsions are transparent, thermodynamically stable dispersions in water, oil, surfactant and cosurfactant (usually a higher alcohol) in small droplets with a diameter of 5-80 nm, i.e. much smaller than the globules in ELM. They can simultaneously solubilize relatively large amounts of both hydrophobic and water molecules. The dissolution process involves the diffusion-controlled transport between feed and microemulsion droplets. Microemulsions can also occur in solvent extraction systems with surface-active extractants.

Organized **surfactant assemblies** are promising tools of future separation techniques (*W.J. Hinze, D.W. Armstrong, 1987*). Non-living biological **cell walls** of plants (polysaccharides, cellulose-lignin complexes) exhibit ion-exchange properties (*J. Stary, K. Kratzer, 1982; J. Stary, K. Kratzer, J. Prasilova, 1986*). Liposomes, lipid **microvesicles** suspended in an aqueous phase, resemble a primitive cell (*H.T. Tien, 1974; D. Lichtenberg, Y. Barenholz, 1988*)

Cell membranes are essentially liquid-crystalline **bilayer lipid membranes** (*J.F. Danielli, H. Davson, 1934*); cellophane membranes impregnated with organic solvents (*A. Ilani, 1963*) were used as model systems, but the explanation of ion selectivity and gating in hydrophic channels deserves more thorough investigation (*W.D. Stein, 1967; M.M. Shemyakin et al., 1969; E.N. Lightfoot, 1974; A.A. Lev, 1976; B. Hille, 1975,*

1977). Though the lipid bilayer is the key structural element of biological membranes, proteins, carbohydrates and other minor constituents are responsible for a host of functions: ion-selectivity, metabolite transport, photosynthesis, etc. The active transport of ions (H^+, Na^+, K^+, etc.) plays a vital role in living cells and is directly related to the fundamental process of ATP synthesis from ADP and inorganic phosphate (*S.J.D. Karlish, D. W. Yates, I.M. Glynn, 1978; Y. Kagawa, 1978; H. Schindler, L. Nelson, 1982*). The electrochemical H^+-gradient results from an oriented redox reaction across the mitochondrial membrane,

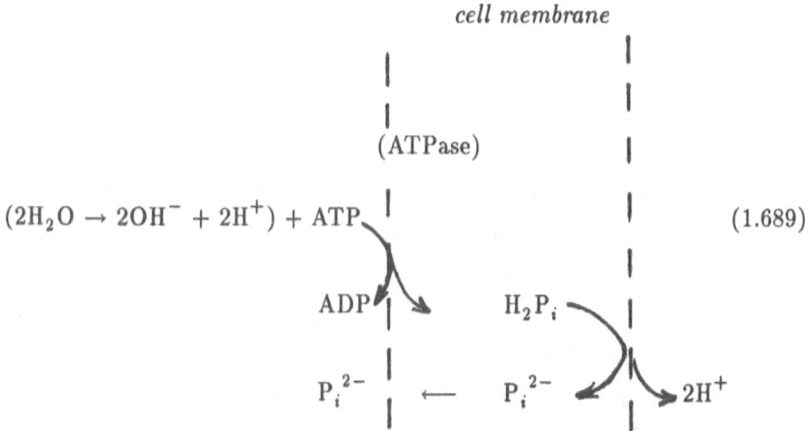

(1.689)

(P_i is inorganic phosphate) and H^+-pumps occupy a central role in membrane biology (*P. Mitchell, 1961, 1965*).

In addition to polysaccharides and polynucleic acids, about 5000 different proteins are known to play a specific function in cellular and subcellular membranes: these can stimulate the search for molecular design of modern separation units. In this respect, **biomimetic membranes** represent a bridge between the areas of separation science and cell physiology (*A. Kotyk, K. Janacek, 1977; J.H. Fendler, 1982, 1987; H. Kuhn, 1983; K. Heckmann, C. Strobl, S. Bauer, 1983; G. Menestrina, F. Pasquali, 1985; E. Sada, M.Terashima, 1985, 1989; V.S. Gevod, O.S. Ksenzhek, L.L. Reshetnyak, 1988*).

1.6 PHOTOCHEMICAL SEPARATION

In chapters 1.2.3. and 1.3.2., we stressed the importance of reversible chemical interactions for separation processes. Selectively-induced activation of irreversible chemical reactions can be used for the separation of such components concerning which their chemical state is irrelevant. Practically speaking, there are two classes of components which can be separated in a new chemical state:
(i) isotopes (preparation of a particular chemical state is only a small fraction of the cost involved in isotope separation),
(ii) impurities (from the standpoint of pure matrix preparation, the chemical state in which the impurities are removed is irrelevant).

The first successful attempt at the selective separation of isotopes was achieved by photochemically induced oxidation of mercury isotopes (*K. Zuber, 1935*). Photodissociative processes which combine selective IR and UV photon absorption became efficient following the introduction of laser systems (*R. V. Ambartsumian, V.S. Letokhov, 1977; R.J. Jensen, J.A. Sullivan, F.I. Finch, 1980; K. Thyagarajan, Ajoy Ghatak, 2010*).

Of the great variety of photochemically-induced, irreversible chemical reactions (*R.B. Berstein, 1974; J.I. Steinfeld, 1981; V.S. Letokhov, 1983*), the following may be considered for separation purposes:
1) ionization or spontaneous dissociation of an excited molecule:

$$A^* \rightarrow A_1 + A_2 \tag{1.690}$$

(A_1, A_2 are atoms, molecules, radicals or ions).
2) quenching, with formation of a new molecule:

$$A^* + M \rightarrow AM_1 + M_2 \tag{1.691}$$

3) quenching, with association on surfaces:

$$A^* + M \to AM \tag{1.692}$$

Separation systems using irreversible reactions are characterized by the rate of formation of easily separable products P_A, e.g. solids or ions from a gas, $P_A \equiv A_1(A_2)$, AM_1 or AM.

The selectivity and efficiency of photochemical separation depends on (i) the excitation yield of one chemically-transformed component (P_A) and (ii) the yield of the desired reaction with a minimum number of competing processes. The basic **law of photochemical equivalence** (*J. Stark, 1908; A. Einstein, 1912*) postulates that in the primary photochemical act, each molecule is activated by the absorption of one quantum of light (photon). In this case, the **quantum yield** Φ(dimensionless) given by the ratio of the chemically-activated molecules $N(A^*)$ and the number of photons $N(p)$ absorbed

$$\Phi = \frac{N(A^*)}{N(p)} \tag{1.693}$$

is unity. In cases where the quantum yield is expressed through the number of moles of excited molecules, $n(A^*)$ per quantity of energy of one mole of photons, known as the **Einstein (E)**, the result is a yield (mol J^{-1}):

$$\Phi = \frac{n(A^*)}{E} \tag{1.694a}$$

One Einstein is linked to one photon with frequency $v(s^{-1})$ or wavelength $\lambda(m)$ by the relations

$$1E = N_A h v = 3.987 \times 10^{-10}\, v\, \text{J s} = \frac{1.195}{\lambda} \text{J mol}^{-1} \tag{1.694b}$$

where N_A is Avogadro's number (mol^{-1}) and h is Planck's constant (J s). In the typical range of applied wavelength from 10 μm (infrared) to 0.1 μm (ultraviolet), the value of E varies from 120 to 12000 kJ mol⁻¹.

If the activation of the separation of substrate A, followed by a reaction with the acceptor product P, proceeds directly as

$$A + h\nu \xrightarrow{k_0} A^* + M \xrightarrow{k_1} P \tag{1.695}$$

the accumulation of P proceeds with a rate

$$\frac{d[P]}{dt} = k_1[A^*][M] \tag{1.696}$$

If the reacting species A and M are present in excess and their concentration is constant, then the concentration of activated species can be obtained from the steady-state condition

$$\frac{d[A^*]}{dt} = k_0 - k_1[A^*][M] = 0 \tag{1.697}$$

where

$$k_0 = \Phi I \tag{1.698}$$

is a constant volume rate of photon absorption at a photon flux I (E m⁻³ s⁻¹ = J m⁻³ s⁻¹) and Φ is the quantum yield of the first step of reaction 1.695. In this case, from eqns. 1.696 – 1.698, the accumulation of the product simply follows zero-order kinetics (P.J. Robinson, K.A. Holbrook, 1975)

$$\frac{d[P]}{dt} = \Phi I \tag{1.699}$$

The separation of the two components, A and B, giving products P_A and P_B will be

$$\frac{[P_A]}{[P_B]} = \frac{k_0^A}{k_0^B} = \frac{\Phi^A}{\Phi^B} \tag{1.700}$$

However, part of the excited molecules can undergo deactivation processes by a radiative transition (fluorescence, phosphorescence) or the non-radiative dissipation of energy. For an average life-time of the excited species A of about 10^{-8} s, they undergo in a gas at normal pressure about 100 collisions with surrounding molecules and the fluorescence is quenched. Competing reactions become more prevalent when some quenching agent (Q) is present. Then, instead of eqn. 1.697, there exists the steady-state condition

$$k_0 - k_1[A^*][M] - k_2[A^*][Q] = 0 \qquad (1.701)$$

and product accumulation rate is

$$\frac{d[P]}{dt} = \Phi I \frac{k_1[M]}{k_1[M]+k_2[Q]} \qquad (1.702)$$

i.e. the apparent yield will always be lower than that for primary excitation. The same result happens when the reaction is partially reversible (e.g. the geminate recombination of radicals). Conversely, activated molecules can induce a chain reaction, thus giving vastly increased (up to a factor of 10^5) yields.

Typical lasers can produce light pulses with an energy about 1 E in pulses with a frequency of about 1 kHz; even at a high quantum yield $\Phi = 1$, the production can reach, according to eqn. 1.696, not more than 10^{-3} mol s^{-1}. Hence, a truly valuable product must be obtained to make the process economically feasible. The first major production application of lasers promises to be in isotope separation (*I.N. Knyazev at al., 1978; R.J. Jensen, J.A. Sullivan, F.T. Finch, 1980*). Laser light induced drift (LID) of atoms in basic and excited state causes diffusion of the atoms in opposite direction what was used for separation of sodium isotopes (*Yu.P. Gangrsky et al., 1992*).

UV-photochemical excitation of isotopically-substituted molecules is insufficiently selective, the isotope effects in electronic spectra being rather small. Selective excitation is possible through the differences in the vibrational and rotational states of isotopic molecules, although at higher temperatures the vibrational levels are anharmonic and often degenerate. This, together with collisional deactivation effects, leads to broadening of the levels and loss of selectivity. Multiple-photon and two-frequency processes were developed to cope with this problem (Fig. 48).

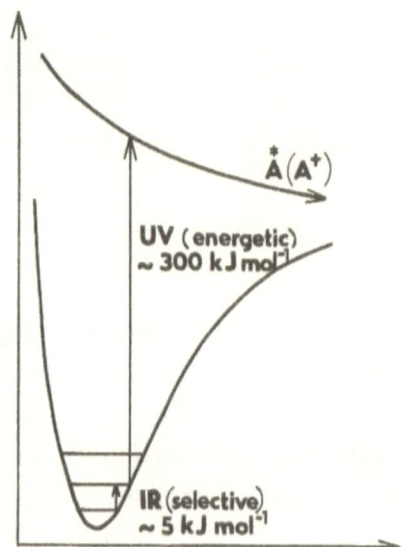

Figure 48. Energetic diagram of molecule (potential energy vs. interatomic distance) for multiple (two-photon) IR and UV excitation.

In the first step of the process, highly-selective vibrational excitation is achieved by infrared frequencies. Excited molecules have dramatically increased absorption both for IR or UV photons, which should ensure their ionization or dissociation. Such processes were successfully applied to the separation of $^{12}C/^{13}C$ isotopes via the photolysis of CF_3I (*I.N. Knyazev et al., 1977*) and the selective ionization or dissociation of $^{235}UF_6$ looks promising for commercial uranium isotope enrichment, e.g. the AVLIS (Atomic Vapor Laser Isotope Separation) project (*R.J. Jensen et al., 1980*).

With achievements in laser technology, especially the rapid development of rare-gas-halide lasers, the selective removal of impurities also appears quite competitive: laser add-on purification of silanes (*J.H. Clark, R.G. Anderson, 1978*) holds great promise for the production of semiconducting silicon. Also the photoreduction and oxidation processes are perspective (*T. Donohue, 1977; Lu-Fu Qiu, Xi-Hui Kang, Tong-Sheng Wang, 1991*).

1.7 SPECIFIC NUCLEAR PHENOMENA IN SEPARATION

The differences in the nuclear (isotope) composition of mixtures leads to **distribution processes without chemical changes** in which entropic factors strongly prevail. The instability of radioactive nuclides is connected with phenomena which do not originate in the interactions of electrons, either weak, electrostatic or chemical. Nuclear composition and nuclear transformations are the sources of effects which are specific for the separation of both stable and radioactive radionuclides. Conversely, the isotope labelling technique, e.g. radiotracer work, is a powerful tool in studying many transport phenomena and separation processes (*B.Z. Iofa, An.N. Nesmeyanov, G.I. Kireev, 1970; M.A. Hughes, V. Rod, 1984; G.J. Hanna, R.D. Noble, 1985; J. Tőlgyessy, M. Kyrs, 1989*).

1.7.2 Effects of nuclear composition (heterogeneous isotope exchange)

If two compartments (phases) W and P (or I and II) contain a constant amount (q_T) of element (compound) in two isotopic modifications A and A*, respectively, (Fig. 49) the following material balances (in moles) for the individual isotope are valid:

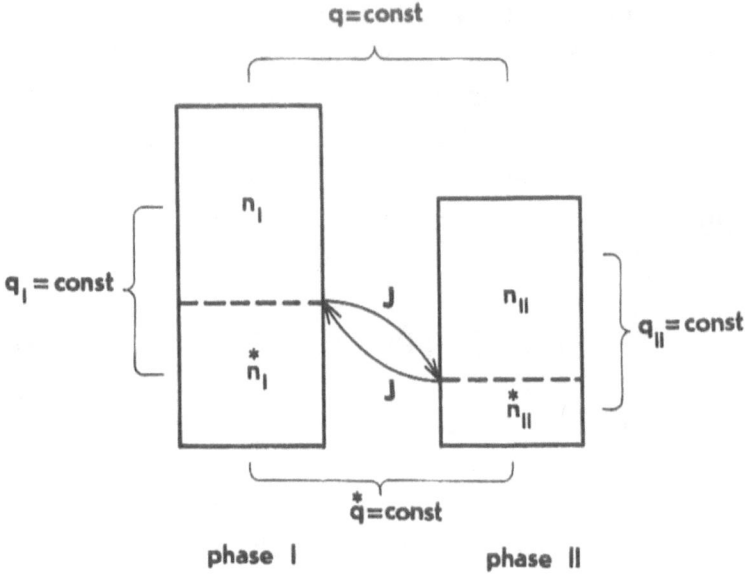

Figure 49. Schematic material balance for interphase isotope exchange.

$$n_\mathrm{I} + n_\mathrm{II} = q \qquad (1.703a)$$

$$n_\mathrm{I}^* + n_\mathrm{II}^* = q^* \qquad (1.703b)$$

and for the individual phases,

$$n_\mathrm{I} + n_\mathrm{I}^* = q_\mathrm{I} \qquad (1.704a)$$

$$n_\mathrm{II} + n_\mathrm{II}^* = q_\mathrm{II} \qquad (1.704b)$$

as well as for the entire system

$$q_\mathrm{I} + q_\mathrm{II} = q + q^* = q_\mathrm{T} \qquad (1.705)$$

The entire system is in chemical equilibrium and the mass fluxes J of the isotope-containing substances are equal in both directions. However, at a different isotopic phase composition, the system is not thermodynamically stable and entropic factors lead to equal partitioning of the isotopes between the two compartments (phases). This process is called **isotope exchange**.

The **kinetics of heterogeneous isotope exchange** can be derived from different models (*N. Ye. Brezhneva, S.Z. Roginskii, A.I. Shilinskii, 1937; S.Z. Roginskii, 1940; H.A.C. McKay, 1938, 1943*).

For a **heterogeneous chemical reaction**

$$A_I^* + A_{II} \rightleftarrows A_{II}^* + A_I \qquad (1.706)$$

there is a kinetic equation for the extent of reaction (eqn. 1.153)

$$\frac{dn_{II}^*}{dt} = k_1 n_I^* q_{II} - k_{-1} n_{II}^* q_I \qquad (1.707a)$$

or

$$\frac{dn_{II}^*}{dt} = k_1' n_I^* - k_2' n_{II}^* \qquad (1.707b)$$

with biphasic rate constants k_1 and k_{-1} (mol^{-1} s^{-1}):

$$k_1 = k_{-1} = k \qquad (1.707c)$$

$$k_1' = k_1 q_{II} \qquad (1.707d)$$

$$k_{-1}' = k_{-1} q_I \qquad (1.707e)$$

Another model is **derived from the driving force**, which is proportional to the difference between the instaneous (n) and equilibrium (n^{eq}) amounts,

$$\vartheta = n_{II}^* - n_{II}^{eq} \qquad (1.708)$$

and gives

$$\frac{dn_{II}^*}{dt} = k' q_I q_{II} \vartheta \qquad (1.709)$$

where the product of the constant k' (mol^{-2} s^{-1}) and $q_I q_{II}$ characterizes a "gross rate", i.e. the relative extent (s^{-1}) of the heterogeneous reaction 1.706. This process resembles other chemical processes such as crystal growth from supersaturated solutions — eqn. 1.567 — and in this respect the phase where $n^* > n^{eq}$ is "isotopically supersaturated"

The third possible model of isotope exchange is derived from **diffusion transfer with chemical reaction** — eqn. 1.526 — which gives the velocity

$$\frac{dn_{II}^*}{dt} = J = aK \left(n_I^* - \frac{n_{II}^*}{D_m(A^*)} \right) \tag{1.710}$$

where J is the interphase flux (mol s^{-1}), a is the specific surface area (m^{-1}), K is the transfer coefficient (m s^{-1}) and

$$D_m(A^*) = \left(\frac{n_{II}^{*eq}}{n_I^{*eq}} \right) \tag{1.711}$$

is the equilibrium mass distribution ratio — eqn. 2.15 — of the isotope between phase II and phase I.

All the models can be shown to lead to the same exponential law, which is commonly referred to as the **McKay equation**

$$\frac{n_{II}^*}{q_{II}} = \frac{n^*}{q} (1 - e^{-kqt}) \tag{1.712}$$

or

$$-\ln(1 - F) = kqt \tag{1.713a}$$

where

$$F = \frac{n_{II}^*}{q_{II}} \tag{1.713b}$$

is the **fraction of isotope exchange** (fractional attainment of equilibrium) and the ratios n^*/q represent the molar fractions of radionuclides (or specific activity — see eqn. 1.58) in the system and its compartments. The constant $k(\text{mol}^{-1}\text{s}^{-1})$ corresponds to

$$k \equiv k_1 \equiv k_{-1} \tag{1.713c}$$

in the first model,

$$k \equiv k' \, q_I \tag{1.713d}$$

$$k = k'_1 + k'_2 \tag{1.713e}$$

for the model with eqn. 1.709 or

$$k \equiv aK/n_{II} \propto a(D_M + D_T)/(\Delta x n_{II}) \tag{1.713f}$$

according to eqns. 1.710 and 1.453.

For **ideal isotopic mixtures**, the **equilibrium isotope composition** of each phase is equal,

$$\frac{n_I^*}{n_I} = \frac{n_{II}^*}{n_{II}} \tag{1.714}$$

and therefore the equilibrium distribution of the isotopic form A^* corresponds to

$$R_{A^*} = \frac{n_{II}^*}{n_I^* + n_{II}^*} = \frac{n_{II}}{n_I + n_{II}} = R_A \tag{1.715}$$

This indicates that isotope exchange can be used for the separation of a radionuclide from phase I in the system, where the inactive form A predominates in the product phase II (*F. Paneth, 1922; A.P. Ratner, 1936; K. H. Lieser, P. Gütlich, I. Rosenbaum, 1965*), e.g. separation with crystalline precipitates contacted with radioactive solution is the isotopic analogue of the adsorption process, **isotope sorption**.

The equality, eqn. 1.715, is valid only for ideal isotope exchange. In real isotopic mixtures, there exists a minute difference between the isotopic modifications of the molecules and the ratio

$$\frac{n_I^* n_{II}^*}{n_I n_{II}} = \alpha \tag{1.716}$$

differs from unity ($\alpha = 1.001 - 1.03$). By multiple isotope exchange, efficient **isotope separation** can be achieved (H. Urey et al., 1937; H.G. Thode, H.C. Urey, 1939; G.M. Panchenkov, V.D. Moiseev, A.V. Makarov, 1961; J.P. Butler, 1980; M.C. Chalabreyse, R. Neige, J. Aubert, 1980; J. Cabicar et al., 1983).

When solid-phase diffusion is the controlling factor of isotope exchange, instead of eqn. 1.712 which is valid for the steady diffusion model, the equation is

$$-\ln(1 - F) = k_0 + k\tau \tag{1.717a}$$

where τ is the dimensionless time — see equation 1.428 — and k_0 is a constant connected with geometry and the time of establishment of the steady diffusion layer (K. Zimens, 1945; G. Berthier, 1955). Under non-steady diffusion (eqn. 1.518), good agreement is achieved by the equation

$$-\ln(1 - F) = k\tau^{1/2} \tag{1.717b}$$

(F. Helfferich, 1959).

1.7.3 Effects of nuclear transformations

Nuclear transformations may introduce certain perturbations in the separation process. Separation systems with radioactive isotopes, which are closed in respect of their chemical composition and isolated in respect of energetic relations, become **pseudo-closed and pseudo-isolated** due to the spontaneous transformation of the separated solutes by radioactive decay. Even the simplest decay of parent radionuclide A to daughter B

$$A \to B + \text{energy} \tag{1.718}$$

is followed by

(1) perturbation of the material and energy balances if the time of separation is of the same order as the half-life of the radionuclides, i.e. their concentration in the separation system is a function of time.
(2) stabilization of the daughter nuclide B in unusual chemical states, if particular oxidation states and chemical bonds can be formed. A classic example is the liberation of radioactive iodine (^{128}I) atoms from (^{127}I)ethyliodide irradiated by neutrons which became known, after its discoverers, as the **Szilard-Chalmers effect** (*L. Szilard, T.A. Chalmers, 1934; W. Herr, 1952; T. Matsuura, T. Sasaki, 1967; B.S. Tomar et al., 2008*).
(3) establishment of a **special form of steady-state**, "radioactive equilibrium" instead of a true equilibrium when both A and B are radioactive (*F. Soddy, 1910*).
(4) **dissipation of the energy** released in radioactive decay in the matrix which causes a variety of radiation, or autoradiolysis effects.

If the parent radionuclide A decays with a decay constant $\lambda_A (s^{-1})$ and the daughter B has a decay constant λ_B, the total amounts of the radionuclides in a system is determined by the kinetic equations (see e.g. *O. Navratil et al., 1990*):

$$\frac{dN_A}{dt} = -\lambda_A N_A \tag{1.719}$$

$$\frac{dN_B}{dt} = \lambda_A N_A - \lambda_B N_B \tag{1.720}$$

where the products

$$\lambda_A N_A = A_A \tag{1.721a}$$

$$\lambda_B N_B = A_B \tag{1.721b}$$

represent the decay rates, activities (s^{-1}) as a measure both of the amount of radionuclide (see section 1.2.1.) and of the energy produced in the system.

The general solution of eqns. 1.719 and 1.708, if at time $t = 0$ no daughter is present, $A_B^0 = 0$, results in

$$A_A = A_A^0 e^{-\lambda_A t} \qquad (1.722a)$$

$$A_B = A_A^0 \frac{\lambda_B}{\lambda_B - \lambda_A} \left(e^{-\lambda_A t} - e^{-\lambda_B t} \right) \qquad (1.722b)$$

where A_A^0 is the activity of A at time $t = 0$. It means that the ratio A : B changes even if no physico-chemical separation takes place. Initially pure A always becomes contaminated by accumulation of the daughter product B. Conversely, if the parent nuclide A is short-lived in comparison with daughter, $\lambda_A > \lambda_B$, pure B is obtainable after a time which sufficiently exceeds the average life of the parent nuclide, $t \gg 1/\lambda_A$.

The situation when the parent nuclide is relatively long-lived compared with the daughter, $\lambda_A < \lambda_B$, is of greater practical importance, because after some time, when $t \gg 1/\lambda_B$, a steady state occurs. From eqn. 1.710b the steady state is characterized by a constant ratio of activities of A and B.

$$\frac{A_A}{A_B} = \frac{\lambda_B}{\lambda_B - \lambda_A} \qquad (1.723)$$

This is called **radioactive equilibrium**. It occurs widely in radiochemical practice and is often used, e.g. for medical purposes, for the production of short-lived daughter radionuclides by separation from a long-lived parent. The production of ^{99m}Tc ($\lambda_B = 0.115$ h^{-1}) as a radiopharmaceutical from parent ^{99}Mo ($\lambda_A = 0.0108$ h^{-1}) is a typical example: in an alumina column containing (^{99}Mo) molybdate the accumulation of 75-94% equilibrium activity of ^{99m}Tc occurs after 12-24 hrs. and pertechnetate can be eluted by NaCl solution. The separation devices enabling the separation of a short-lived radionuclide from its parent are called **radionuclide generators**. Usually these devices consist of an adsorbent with an immobilized parent nuclide, as in the example above (*W.D. Tucker, M.W. Greene, A.P. Murrenhof, 1962; N.B. Mikheev, M. El-garhy, Z. Moustafa, 1964; J.F. Allen, 1965; M. El-Garhy, S. El-Bayomy, S. El-Alfy, 1967; F. Sebesta, J. Stary, 1974; B.S. Tomar et al., 2008; A.G. Denkova et al., 2013*) or an organic solution, immiscible with the aqueous phase, of the parent radionuclide (*V. Mikulaj, F. Macasek, M. Steinerova, 1977*).

In two-phase systems used for separation, however, the process of daughter accumulation is accomplished both by the unusual chemical state of the daughter radionuclide (*V. Mikulaj, P. Rajec, V. Faberova, 1974; P. Rajec, V. Mikulaj, 1974*) and by its parallel rapid and slow redistribution processes (*F. Macasek et al., 1976*). If the parent radionuclide is in the source phase I and the separation of the daughter is expected in the receiving phase II, the separation is complete only if all generated B is immediately and irreversibly transported to this phase; this is however more the exception than the rule.

In biphasic systems of parent and daughter nuclides, rapid chemical **stabilization of the nascent species** B (within 10^{-7} s) can lead to a "prompt" distribution of daughter between the phases, and a fraction α of the nascent species is retained in source phase I. This is followed by transport of the chemically-stabilized nascent forms of B to receiving phase II at a heterogeneous rate constant k_2 (s^{-1}). An opposite transfer of chemically stabilized forms of B (the fraction $1 - \alpha$, promptly appears in the receiving phase II) from the phase II back to phase I has the rate constant k_1 (s^{-1}). Then the rate of accumulation of B in the receiving phase II is given by the kinetic equation

$$\frac{dN_B^{II}}{dt} = (1 - \alpha) N_A^I \lambda_A + N_B^I k_2 - N_B^{II}(\lambda_B + k_1) \qquad (1.724)$$

Thus, by analogy with eqn. 1.710b the fraction of the **daughter radionuclide retained in the source phase**,

$$R_B = \frac{R_A^{eq}}{N_B^I + N_B^{II}} \tag{1.725}$$

is obtained as

$$R_B = R_A^{eq} \frac{\alpha(\lambda_B - \lambda_A) + k_1}{\lambda} \left(1 - e^{-\lambda t}\right) \tag{1.726a}$$

where R_A^{eq} is the fraction of the parent nuclide in the source phase,

$$R_A^{eq} = \left(\frac{N_A^I}{N_A^I + N_A^{II}}\right)_{eq} \tag{1.726b}$$

and

$$\lambda = k_1 + k_2 + \lambda_B - \lambda_A \tag{1.726c}$$

The aim of radionuclide generators is to work at low R_B's to reach high yields, and at high R_A's to avoid the contamination of daughter B with a long-lived parent A.

For practical purposes, the relation between the steady state (R_B^∞) and equilibrium (R_B^{eq}) values, i.e. the distribution of the daughter radionuclide which gives the relative yield of daughter radionuclide, is also important. Because in the biphasic system as a whole, the steady-state with respect to radioactive decay (radioactive equilibrium) is still preserved during separation operations,

$$\frac{d(N_B^I + N_B^{II})}{dt} = 0 \tag{1.727}$$

and the distribution of the daughter radionuclide at equilibrium is

$$\left(\frac{N_B^I}{N_B^I+N_B^{II}}\right)_{eq} = R_B^{eq} = \frac{k_1}{k_1+k_2} \tag{1.728}$$

Now we have

$$\frac{R_B^\infty}{R_A^{eq}} = \frac{\alpha\sigma}{z+\sigma} + \frac{z}{z+\sigma}\frac{R_B^\infty}{R_A^{eq}} \tag{1.729a}$$

where

$$\sigma = \frac{\lambda_B-\lambda_A}{\lambda_B} \tag{1.729b}$$

and

$$z = \frac{k_1+k_2}{\lambda_B} \tag{1.729c}$$

(F. Macasek et al., 1976). Usually $\lambda_B \gg \lambda_A$ and σ is close to unity (0.90– 0.99 in most practical cases). The sum of the constants k_1 and k_2 can be obtained from the distribution kinetics, particularly from the rate of heterogeneous isotope exchange — cf. eqns. 1.710 and 1.713. Eqn. 1.727 gives a comparison of the values of the enrichment factors (see section 2.1.1) in the steady state, $S^{st} = R_B^\infty/R_A^{eq}$, and at ideal equilibrium, $S^{eq} = R_B^{eq}/R_A^{eq}$ (Fig. 50). The lower the enrichment in phase I, the more efficient is the generator. Conversely, the separation efficiency is lower for the daughter atoms with a long mean life-time. At $z \gg 1$, the daughter atoms B will be distributed as if by a true chemical equilibrium but at $z < 1$ the distribution will strongly depend on the fraction of prompt retention in phase I (α) which reflects the peculiarities of nascent (hot) atom chemistry.

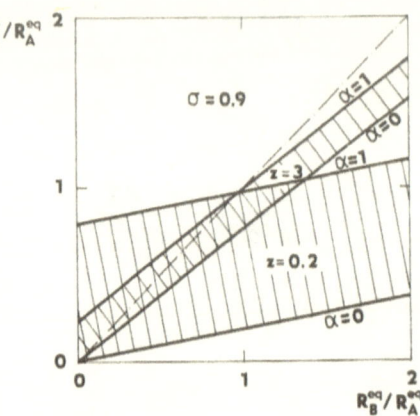

Figure 50. Enrichment factor R_B/R_A on separation of parent A and daughter B radionuclides, as compared with equilibrium enrichment. at various branching ratios of specific reactions (α) and rate of interphase exchange (z).

Dissipation of the energy released by radioactive decay in the matrices of separation systems is obvious during the separation of highly-active (HLA) solutions in reprocessing spent nuclear fuel, and also during the production of radionuclides and labelled compounds of high specific activity. The consequences of **radiolysis of biphasic separation systems** are often deduced and well known from autoradiolysis and radiolysis of homogeneous gas-, liquid- or solid-phase systems (*H.A.C. McKay, 1956, 1967; I.L. Jenkins, 1984*). The behavior of biphasic systems, however, may differ because of transfer of both substrates and the products of radiolysis between the two phases (*B.F. Warner et al., 1974*). Particularly, in the liquid-liquid systems widely used for solvent-extraction reprocessing of nuclear fuel, such specific phenomena are unavoidable and the decomposition of the basic components and accumulation of interfering impurities play a vital role (*F. Macasek, R. Cech, 1984*).

The rate of indirect destruction of the separation agent A due to attack by free radicals R arising from radiolysis is given in the bimolecular reaction

$$A + R \rightarrow \textbf{decomposition products} \tag{1.730}$$

where R's originate from radiolysis reactions of a precursor, such as water or an organic molecule:

$$\text{precursor} \rightsquigarrow R \tag{1.731}$$

The radiation yields of radicals are characterized by their radiation yield in molecules $N(R)$ per 100 eV of absorbed energy of radiation,

$$G(R) = \frac{N(R)}{100 \text{ eV}} \tag{1.732}$$

(cf. photochemical yield, eqn. 1.693). The radiation yield of destruction of the separation agent which is often a highly important characteristic of radiation damage in the separation system, is expressed in the same way. When the matrix (e.g. ion exchanger, solvent, etc.) acts as the principal factor of separation, its destruction proceeds via eqn. 1.731, where the matrix functions as the precursor. Thus **matrix radiolysis** proceeds according to the pseudo-zero order reaction:

$$-\frac{d[A]}{dt} = fG(-A)\dot{D} \tag{1.733}$$

where $\dot{D}$ is the radiation dose rate (J kg^{-1} s^{-1}) and f is the proportionality factor, $f = 1.04 \times 10^{-7}$ mol eV J^{-1}. On integration of eqn. 1.733, if at $t = 0$ concentration $[A] = c_A$, we find

$$[A] = c_A - fG(-A)\dot{D}t \tag{1.734}$$

When agent A is a minor component, the **radiolysis of solutes** takes place by the indirect action of radicals formed from major components, being decomposed in a pseudo-second-order reaction

$$-\frac{d[A]}{dt} = k_A[A][R] \tag{1.735}$$

If the radiolysis proceeds at a steady state, and the radical concentration is constant, a pseudo-first order reaction occurs,

$$-\frac{d[A]}{dt} = k'_A[A] \tag{1.736}$$

where k'_A is proportional to the product of $k_A \times G(R)$ and inversely proportional to the concentration of other radical scavengers; the concentration of the agent decays exponentially:

$$[A] = c_A \exp(-f\dot{D}k'_A t) \tag{1.737}$$

One should note that one radioactive particle carries about $10^5 - 10^6$ eV of energy. However, its action is neither selective nor efficient: about 90-95% is dissipaled by thermal processes and the normal rates of radical generation in separation systems of highly radioactive substances are generally within the range $10^{-8} - 10^{-6}$ mol dm^{-3} s^{-1}. Direct radiolysis is usually not risky with respect to the linear decrease of the amount of the principal component (matrix) of separation systems, but results in a broad scale of reactive products of radiolysis which have specific properties (surface active, complexing, redox, etc.) and become undesirable impurities (*S. Cechova, F. Macasek, 1991*).

Radiolysis of a biphasic system has two further features (*F. Macasek, R. Cech, 1984*):

(i) during the distribution of the separation agent A between the two phases, the decomposition reaction 1.730 occurs in both phases and proceeds, due to various reactions 1.731, with various radicals and at different total velocities,

(ii) because the distribution of the agent is usually non-symmetrical, its concentration in one of the phases is usually small; hence, its destruction becomes indirect, and it can be diffusionally-controlled by the mass flux of the agent from the second phase, acting as a reservoir.

The **gross radiation yield** of substrate decomposition, $G_T(-A)$ in two-liquid systems (*F. Macasek, 1984*) is

$$G_T(-A) = v_I G_I(-A) + z v_{II} G_{II}(-A) \tag{1.738a}$$

where v's are volume fractions which can be replaced by the phase ratio $r = V_{II}/V_I$ as

$$v_I = 1/(1+r) \text{ and } v_{II} = r/(1+r) \tag{1.738b}$$

and $G(-A)$'s are the partial radiation yields of decomposition in the particular phase. The factor z

$$z = \frac{\dot{D} \rho_{II}}{\dot{D} \rho_I} \tag{1.738c}$$

gives the ratio of radiation dose rates $\dot{D}$ (J kg^{-1} s^{-1}) in the different phases, the densities of which are ρ (kg m^{-3}).

In agitated systems, particularly in emulsions, the rate of decomposition of the separation agent is a slow reaction compared with diffusion (the Damköhler criterion, eqn. 1.554 is small). Radiolysis proceeds in a chemical reaction regime and the additivity rule, eqn. 1.738, should be followed. Moreover, during good phase contact, even with low concentrations of the reagent in one phase, the second phase acts as a pool for the reagent and the kinetics follow the law given in eqn. 1.722 (c_A = const). In motionless, layered liquid-liquid systems, or in all liquid- solid systems, including an ion-exchanger, radiolysis in the phase with a low content of separation component is diffusionally controlled; the relation between the radiation yields depends on the specific interfacial area, dose rate, limiting separation agent concentration, and the apparent reaction order with respect to the component decomposition (*F. Macasek, R. Cech, 1983*). The radiolysis of separation systems is accompanied not only by destruction of the matrix, which becomes significant (roughly above 5%) according to eqn. 1.722, after radiation doses

(J kg^{-1} = 1 Gray) of about $D > 5 \times 10^5 c_A/G(-A)$ (*R. Chiarizia, E.P. Horwitz, 1986*), but also by the interference of surface-active substances at doses of about 10^5 J kg^{-1} and at higher doses (10^6-10^8 J kg^{-1}) by the effective production of new agents, influencing in particular the separation of impurities. Separation systems working under such conditions need periodic purification from interfering radiolysis by-products, e.g. tributyl phosphate used in the PUREX solvent-extraction process (*H.A.C. McKay, 1956, 1967; T.H. Siddal, 1959; I.L. Jenkins, 1984*) and alternative extraction reagents, such as bidentate organophosphorus compounds (*T.H. Siddal III, 1963*) and in particular carbamoylalkyl phosphonates in the TRUEX process (*A.C. Muscatello, J.D. Navratil, M.E. Killion, 1983; W.W. Schulz, J.D. Navratil, 1984; E.P. Horwitz et al., 1985*) which have been applied to improve the process.

1.8 GENERALIZATION AND CLASSIFICATION OF SEPARATION PROCESSES

> *By and large the history of separation science has been a history of diverging pathways. The consequence has been much redundancy in the independent design and optimization of systems that are based on common theory, and lost opportunities in technological spinoff from the more advanced methods to those at a less sophisticated stage of development.*
>
> J. C. Giddings, 1978

Mathematical taxonomy, on which a classification should be based, looks for the grouping of objects, taxones, by n common characteristics; the classification using $n = 1$ characteristics is called monothetic; at > 1, the systems are polythetic.

In separation science, the classification systems commonly in use are based on the differences in physico-chemical properties of the separated components and the forces causing their separation in various phases, state conditions, flows and geometries. Within a class of separation methods, classification is mostly monothetic, stressing the technical aspect and construction geometry of the separation device, the use of external fields and forces, the feed capacity and throughput, the time of separation, and the mode of monitoring of separation.

Table 2 - Classification of separation methods by physico-chemical properties of separated components, driving forces and mode of transport in various one- or two-phase systems. G — gas, L — liquid, S — solid.

System	Property	Principal driving force	Displacement mode (D-diffusion, F-flow)	Separation technique
G-G	diffusivity	conc. gradient	D	gas diffusion, membrane separation
		temp. gradient	F(G)	thermodiffusion
	charge/mass	elmagn. field	F	elmagn. ion separation
G-L	volatility	thermal energy	D	distillation
	diffusivity	thermal energy	D	rectification
	velocity	thermal energy	D	molecular distillation
	solubility	weak interaction	F(G)	gas-liquid chromatography
G-S	volatility	thermal energy	expansion	sublimation
	adsorption	surface interaction	D	gas adsorption
	adsorption	weak interaction	F(G)	gas-solid chromatography
	charge	electric field	D	electrodeposition
	chem. bond	photochemical	D	photochemical deposition
	diffusivity	conc. gradient	F(L)	gel permeation chromatography

System	Property	Principal driving force	Displacement mode (D-diffusion, F-flow)	Separation technique
(G-S)	diffusivity	centrifuge. field	D	ultracentrifugation
L-L	diffusivity	conc. gradient	D	dialysis, osmosis
	charges	electr. field	migration	electrodialysis, electrophoresis
	solubility	weak interaction	D	solvent extraction
		weak interaction	F(L)	liquid-liquid chromatography, reversed phase chromatography
	size	weak interaction	F(L) +centrifug. field	hydrodynamic chromatography, field-flow fractionation, flotation
	size	weak interaction	F(L)	ultrafiltration, reverse osmosis
L-S	size	geom. Barrier	F(L)	filtration
	solubility	thermal energy	D	crystallization
	solubility	chemical inter.	D	precipitation
	solubility	chemical inter.	F(L)	precipitation chromatog.
	solubility	elstatic inter.	D	salting-out precipitation
	solubility	weak interaction	D	gel extraction
	charge	lattice inter.	F(L)	ion exchange
	adsorption	weak inter.	F(L)	liquid-solid chromatography, affinity chromatography
	nuclear composition	entropy	D	isotope sorption

J.C. Giddings (*1978, 1991*) made an admirable attempt to classify the separation systems constituted on such fundamental transport phenomena leading to separation as the profiles of driving forces and flow.

The common approach is to classify the separation methods as shown in Table 2. Obviously, many of the characteristics could be generalized (e.g. van der Waals interactions instead of volatility, solubility etc.) and none exists in "pure" form (e.g. interaction with a surface or carrier gas). On the other hand, diffusion processes include a wide spectrum of forms of convective transport characterized in eqns. 1.404 and 1.405, e.g. molecular and eddy diffusion, effusion, electromigration, gravitational and centrifugal settling, etc.

The **Giddings classification** (*J.C. Giddings, 1978*) results from the basic transport equations 1.419 and 1.614 as well as two main taxonomic characteristics:

(1) the spatial characteristics of the profile of driving forces (the profile of the dimensionless chemical potential ε, eqn. 1.613) across the separation path,
(2) the spatial characteristics of transport (orientation of flow $\vec{w}$ and force $\vec{u}$ vectors, Figure 35. and eqn. 1.607).

Two basic categories of underlying forces driving the component through the matrix (bulk) were distinguished:

c — continuous forces, which exert a steady "pull" on the solute component in the separative region (external fields and gradients),

d — discontinuous forces, which originate at the interfacial discontinuity between two phases.

The discontinuity results from ignoring the entropic term of chemical potential — see eqn. 1.37 — and stressing the different interactions of component molecules in different phases, i.e. the enthalpic term. Such discontinuity exists only before actual contact of the phases; thereafter, due to local equilibrium and compensation of the enthalpic term by the entropic (concentration) term, the chemical potential is continuous: the equality of the chemical potential at the interface is a general concept of phase equilibria — see discussion concerning eqns. 1.38, 1.268, 1.298 and 1.302. Hence, the concept of discontinuous forces actually characterizes the situation before separation, whether the concentration discontinuity belongs to the phase boundary, a membrane or a rectangular solute pulse, or, in the macrokinetic sense, as a diffusion in the depth of the bulk phases. An example of chemical potential continuity in a heterogeneous system is given by the curves 1 and 3 in Fig. 51; they correspond to the diffusion model characterized by eqn. 1.438 and illustrated in Fig. 29b (for non-zero contact time, $= 6.25 \times 10^{-2}$).

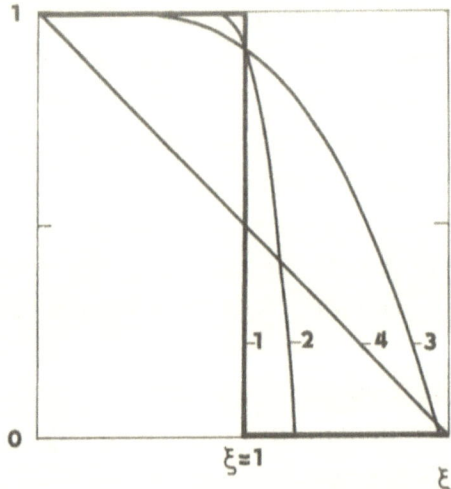

Figure 51. Gradients of concentration and dimensionless thermodynamic potential at interface (dimensionless coordinate $\xi = 1$):

1 — chemical potential before contact,
2 — concentration after dimensionless contact time $\tau = 6.25 \times 10^{-2}$,
3 — chemical potential at the same τ,
4 — chemical potential for linear gradient.

The discontinuity of forces stresses the case where the diffusion layer d is negligibly small compared with the total separation path L — see eqns. 1.615 and 1.616. In some cases (electrodialysis, electrostatic precipitation) neither the continuous (electromigration) nor discontinuous (transfer, discharge) component of the driving forces can be neglected and an intermediate category results: cd — continuous-discontinuous force.

The spatial arrangement of flow and force has been demonstrated in section 1.4.5.; to keep the same symbols as in Fig. 35, flow displacement will be considered further along the z-axis (velocity w_z).

The thermodynamic force driving the component through its surrounding medium can be parallel (in the direction of the z-axis) or perpendicular (in the direction of the x-axis). The stirring and turbulent motion are irrelevant, because they result in homogenization of the bulk and affect only the driving force gradient in the diffusion layer, without any particular displacement of the separated components, i.e. their involvement does not change the static (nonflow) character of the systems.

Most separation methods are categorized into six basic groups (Table 3).

Table 3—Classification of separation methods: $\vec{w}$ — matrix (bulk) displacement, $\vec{u}$ – relative solute displacement by driving forces

Flow and force orientation	Driving forces	
	Continuous	Discontinuous
Static (random)	<ul><li>sedimentation</li><li>isopycnic centrifugation</li><li>electrophoresis</li><li>isoelectric focusing</li><li>electrodialysis</li><li>elstatic precipitation</li></ul>	<ul><li>dialysis</li><li>single stage distribution -extraction, -adsorption,</li></ul>
Parallel $(\vec{w}\|\|\vec{u})$	<ul><li>counter-current electrophoresis</li><li>elutriation</li><li>hydrodynamic chromatography</li></ul>	<ul><li>filtration</li><li>ultrafiltration</li><li>reverse osmosis</li><li>zone melting</li></ul>

Perpendicular $(\vec{w} \perp \vec{u})$	■ thermodiffusion-convection ■ electrophoresis-convection ■ field-flow fractionation ■ electromagnetic ion separation	■ chromatography ■ counter-current distribution - rectification - extraction - adsorption - crystallization ■ flotation

The classes of separation are characterized by the same convective transport equations. The transport in static (random) and parallel type groups is given by eqn. 1.620, i.e. they are isomorphous with purely diffusive one-dimensional *l* form, Fick's second law, eqn. 1.417. The perpendicular type of separation can be described by the two-dimensional convective diffusion eqn. 1.512 but also with the pseudo-one-dimensional eqn. 1.592. Their real difference lie in the detailed chemical, physical and mechanical performances that depend on diverse composition and amounts, chemical stability and properties of the mixtures, economic demands and engineering potentialities.

2 Operation and optimization of separation

A **separation technique** is the practical realization of a separation principle under actual conditions of contact and disengagement of phases, amounts and flows of chemical components, and bringing the influence of external fields on systems.

The subject of pure chemistry is to find the optimal conditions and chemical composition of matrices applied to the isolation of pure components by considering the initial matrix composition and avoiding the irreversible destruction of the solutes, if the aim of separation is preparative, or to enable the estimation of amounts of the individual components of mixtures, i.e. if the goal is analytical.

The **task of chemical engineering** is to perform the separation for various quantitative parameters such as the amounts of mixtures and products, size and geometry of the apparatus and packing, mass and heat flows, and time scale, but also to take into account the demands of automation, waste recycling, conditions of safety and sterility, consumption of energy and reagents and total costs, including amortization of investments (*see e.g. W.H. Walker, W.K. Lewis, W.H. McAdams, 1924; H.W. Cremer, T. Davies, 1956; M. Benedict, T.H. Pigford, 1957; E.D. Oliver, 1966; A.S. Michaels, 1968; A.G. Kasatkin 1971; C.J. King, 1971; A.I. Boyarinov, V.V. Kafarov, 1975; C.J.O.R. Morris, P. Morris, 1976; H.R. Null, 1980; P.A. Belter, E.L. Cussler, W.S. Hu, 1988; A.S. Grandison, M.J. Lewis,1996; S. Rizvy, 2010*).

The quantitative parameters should lead to separation performance under:
(i) maximal driving forces,
(ii) minimal material expenditure,
(iii) minimal energy consumption,
(iv) minimal equipment (investment) costs,
(v) minimal time of separation.

The **most efficient alternatives** fulfil both technical (technological) and economic demands (*see e.g. S.M. Roberts, 1964; F.I. Petlyiuk, V.M. Platonov, D.M. Slavinskii, 1965; A.S. Michaels, 1968; P.R. Rony, 1972; R. W. Thomson, C.J. King, 1972; J.E. Hendry, R.R. Hughes, 1972; V. V. Kafarov, V.N. Shelgov, N. Dorokhov, 1976; C. Hanson, 1979; H.R. Null, 1980; V.N. Schrodt, A.M. Saunders, 1981*).

The goals themselves must be rationalized; the technological tasks of "waste-free" technologies and purity limits such as "as low as reasonably possible" could be ruinously expensive to achieve, with implied costs of energy and reagents that produce other wastes elsewhere. A cost-benefit analysis of the separation subsystems becomes an absolute necessity.

Specific demands occur in **analytical separations** where an adequate, but reproducible yield is the principal aim, whereas the removal of species interfering with the analytical response signal, and the preconcentration to reach instrumental sensitivity levels are tasks related to the qualities of the analytical equipment (*Yu.A. Zolotov, N.M. Kuzmin, 1971; B.L. Karger, L.R. Snyder, C. Horvath, 1973; J. Khym, 1974; J. Michal, 1974; J.M. Miller, 1975; Yu.A. Zolotov, 1978, J. Minczewski, J. J.Chwastowska, R.Dybczynski, 1982; J. Inczedy, 1982; J.P. Gosling, L.V. Basso. 1994; K. Valko, 2000*).

Specific problems arise when scaling-up a laboratory process to pilot and technological conditions and the set of similitude criteria become rather complicated *(P.W. Bridgman,1950; J.H. Rushton, 1951; H.E. Huntley, 1967; S.E. Charm, C. C. Matheo, 1971; R.L. Lacey, S. Loeb, 1972; A.M. Rozen, V.S. Krylov, 1974; B.J. McCoy, 1986; T. Gu, 1995)*.

2.1 ELEMENTARY SEPARATION AND SEPARATION UNIT

The **separation unit (separation module)** is the smallest and self-operating device which produces a partial or full separation of components of a mixture.

To characterize the function of the separation unit, it is sufficient to investigate the behavior of a binary mixture whose total variance — eqn. 1.274 — is $V = 4$; (as usual, temperature, pressure and the masses of two coexisting phases, say P and W). Multicomponent mixtures can also be considered as two-component: a selected component represents the first component (A) while the balance of the mixture can be considered the second component (B) (*E.J. Henley, E.M. Rosen, 1969; H.R. Null, 1970; V. Hlavacek, V. Vaclavek, M. Kubicek, 1979; S. Peter, 1979; J.H. Prausnitz, 1981; S.M. Walas, 1985*).

Various **parameters based on universal chemical potential** can be used to describe the dynamics of interphase transfer in a separation unit, as was demonstrated for heterogeneous isotope exchange in eqns. 1.703–1.707. The driving force of separation in a separation unit is usually defined not directly through the difference of chemical potentials (J mol^{-1}), eqn. 1.39, but through its entropic term. The difference θ (mol m^{-3}) between the instantaneous and equilibrium concentrations, c and c^{eq}, respectively, the **concentration pressure**, is widely used. For phase P the values for components A and B can be defined as

$$(\vartheta_A)_y = c_A - c_A^{eq} = (c_t^P)_P(y_A - y_A^{eq}) \qquad (2.1a)$$

$$(\vartheta_B)_y = c_B - c_B^{eq} = (c_t^P)_P(y_B - y_B^{eq}) \qquad (2.1b)$$

where $(c_t^P)_P = \frac{P}{V^P}$ is the total concentration (mol dm^{-3}) of the components in phase P — see eqn. 1.51. Examples of the driving forces have been discussed for crystallization from supersaturated solution and for adsorption — eqns. 1.565 and 1.572.

The convenience of the values of the concentration pressures is substantiated — see the discussion of eqns. 1.399–1.403 — by the value of the resulting cross fluxes (mol m^{-2}s^{-1}) of the separated substances between the fractions P and W in the separation unit:

$$j_A^P = 10^3 (K_A)_y (\vartheta_A)_y \tag{2.2a}$$

$$j_B^P = 10^3 (K_B)_y (\vartheta_B)_y \tag{2.2b}$$

where K's (m s^{-1}) are the mass transfer coefficients of the components A and B in phase P, while their concentrations are y's.

The separation unit can be either
(i) an **open (dynamic) system** in which feed flows (usually one or two) enter and at least two partial fluxes, containing "fully" or partially separated components, leave the unit, or
(ii) a **closed (static) system** in which a batch of feed is separated into two fractions.

Equilibrium separation can be achieved only in the second case, and for the open system a steady-state operation is the obvious mode. The process of unequal distribution of components between the inputs of the separation unit is called elementary separation. Some examples of separation units are in the Table:

unit	input(s) (F, G)	outputs	
		product (P)	waste (W)
filter	suspension	filter cake	filtrate
thickener	pulp	sediment	supernatant
absorber	gas, sorbent bed	sorbent with adsorbate	lean gas
distiller	beer, crude oil etc.	distillate, condensate	distillation residue
crystallizer	crude solution, syrup, etc.	crystals	mother liquor
precipitator	solution and precipitant	precipitate	supernatant liquid
electrolyzer	electrolyte solution	electrolytic deposit	lean electrolyte
mixer-settler	aqueous and organic solvent	extract	raffinate
permeator (membrane module)	gaseous or liquid solution	rich gas, permeate	lean gas, retentate
percolator	solid, and leaching solution	leachate, extract	extract residue

For example, the elementary separation of two components with different volatility can occur when
(a) liquid and vapor phase remain in a closed vessel at thermodynamic equilibrium corresponding to a given temperature and pressure — see Fig. 18;
(b) a batch of liquid is heated, the vapors are cooled and the condensate is continuously removed from the system, i.e. batch distillation;

(c) liquid is fed continuously into a system (b) with a rate corresponding to the distillation of liquid — continuous distillation mode;

(d) the liquid flow (reflux) is directed countercurrent-wise to the vapor flow in a series of contacting units, i.e. the feeds and outputs are both liquid and gaseous, i.e. fractional distillation or rectification (*E.V. Murphree, 1925; T.P. Carney, 1949; C.S. Robinson, E.R. Gilliland, 1950; A. Weissberger, 1951; E. Krell, 1952; S.A. Bagaturov, 1961; D.W. Tedder, D.F. Rudd, 1978; H.Z. Kister, 1992*).

The designation of outputs as product and waste is purely conventional and depends entirely on the behavior of the most valuable component in the separation procedure. The value of "waste" is never negligible, because it represents just a less-complex and therefore more difficult-to-obtain mixture. **Waste-free separation** is possible in closed but not isolated systems.

The composition of the outputs is primarily determined by thermodynamic and rate factors, but material balance is also a consideration. A separation unit in equilibrium or steady state is described by a set of distribution parameters and material balance equations (*E.J. Henley, E. M. Rosen, 1969*).

2.1.1 Equilibrium processes

If components A and B are present in binary mixture(s) with mole fractions in feed(s) F (and G) of the separation unit:

$$x_A^F + x_B^F = 1 \tag{2.3a}$$

(and

$$x_A^G + x_B^G = 1) \tag{2.3b}$$

for outputs P and W of the separation unit, the following balances should be fulfilled:

$$x_A^P + x_B^P = 1 \tag{2.4a}$$

$$x_A^W + x_B^W = 1 \tag{2.4b}$$

Further, in a closed separation system, the gross amount of the components remains constant, i.e.

$$F + G = P + W \tag{2.5}$$

where F and G are the moles of feed mixtures and P and W are the moles of outputs, or for a particular component

$$Fx_A^F + Gx_A^G = Px_A^P + Wx_A^W \tag{2.6a}$$

$$Fx_B^F + Gx_B^G = Px_B^P + Wx_B^W \tag{2.6b}$$

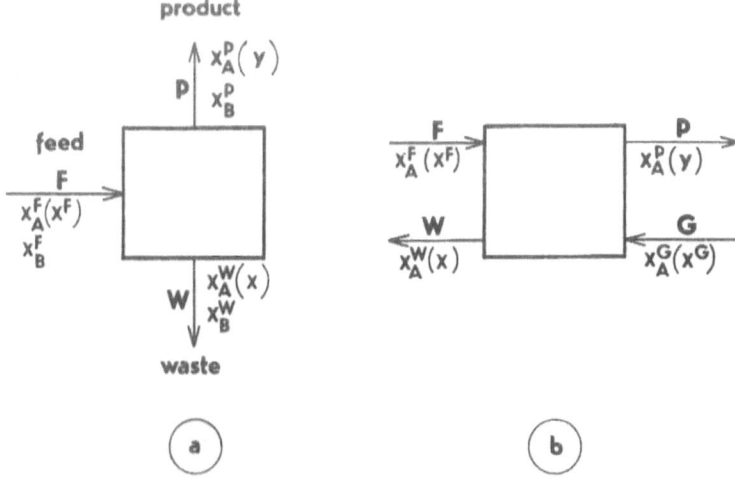

Figure 52. Parameters of concentration unit: a) with one input (F). b) with two inputs (F, G)

These equations also contain the total balance (2.5) as can be proved easily after their summation and consideration of the equation set 2.4., and are applicable to more complicated systems (see e.g. *W.H. Walker, W.K. Lewis, W.H. McAdams, 1927; G.G. Brown, 1951; E.J. Henley, E.M. Rosen, 1969; V. Hlavacek, V. Vaclavek, M. Kubicek, 1979; E.J. Henley, J.D. Seader, 1981*).

There is no necessity to involve the chemical changes of components in this type of calculation because on chemical interaction the amount of resulting bound species can be easily recalculated from the stoichiometry of initial and resulting components — see part 2.2.1.

To avoid more complicated indexing, the description of binary mixtures can be simplified in accordance with the simple balance of mole fractions, eqn. 2.4. Hence, the data relating to x_B are redundant and simple symbols can be introduced to define more clearly the composition data; consequently the concentrations in fraction W can be replaced by

$$x_A^W \equiv x \tag{2.7a}$$

$$x_B^W \equiv 1 - x \tag{2.7b}$$

and for fraction P

$$x_A^P \equiv y \tag{2.7c}$$

$$x_B^P \equiv 1 - y \tag{2.7d}$$

By analogy, for the feeds, we have

$$x_A^F \equiv x^F \tag{2.8a}$$

$$x_B^F \equiv 1 - x^F \tag{2.8b}$$

$$x_A^G \equiv x^G \tag{2.9a}$$

$$x_B^G \equiv 1 - x^G \tag{2.9b}$$

which corresponds to univariancy of the binary system in the separation unit, y being a function of x or vice versa, $y = f(x)$ or $x = \varphi(y)$. If the components form an ideal mixture, Raoult's law, eqn. I.78, or the Berthelot-Nernst distribution law, eqn. 1.308, would be valid for the interphase distribution of the components. Thus, the constant distribution ratio, eqn. 1.309, can be used for characterization of the linear equilibrium relations between x and y which occur in the separation unit:

$$\frac{y}{x} = D_A \qquad (2.10a)$$

$$\frac{1-y}{1-x} = D_B \qquad (2.10b)$$

Now the balance, eqn. 2.6a, will be considered for a single feed F $(G = 0)$, and rewritten as

$$Fx^F = Py + Wx \qquad (2.11)$$

The **yield of separation** in the fraction, $0 \leq R \leq 1$, for component A will be

$$R_A = \frac{Py}{Fx^F} \qquad (2.12a)$$

and for component B of the binary mixture

$$R_B = \frac{P(1-y)}{F(1-x^F)} \qquad (2.12b)$$

With respect to the linear relations 2.10, from eqns. 2.11 and 2.12 it can be shown that the yields are connected with distribution ratios through

$$R_A = \frac{rD_A}{1+rD_A} \qquad (2.13a)$$

$$R_B = \frac{rD_B}{1+rD_B} \qquad (2.13b)$$

where

$$r = \frac{P}{W} \tag{2.14a}$$

is the **stage cut**, i.e. the volume or molar ratio of the product and waste fractions. The **stage cut** could also be defined through the fraction θ of product or waste to feed:

$$P = \theta F \tag{2.14b}$$

$$W = (1 - \theta)F \tag{2.14c}$$

i.e.

$$r = \frac{\theta}{1-\theta} \tag{2.14d}$$

The stage cut is an important characteristic of the separation unit operation and one of its controlling parameters. A further characteristic of the separation unit, which combines the equilibrium distribution resulting from the chemical composition of the phases, and the phase ratio r, is the **mass distribution ratio (capacitance ratio capacity factor)** D_m:

$$D_m = \frac{m^P}{m^W} \tag{2.15a}$$

where m^P and m^W are the amounts (in coherent units) of a component in the product and the waste fractions. For components A and B, particularly,

$$D_m^A = \frac{Py}{Wx} = \frac{PD_A}{W} = rD_A \tag{2.15b}$$

$$D_m^B = \frac{P(1-y)}{W(1-x)} = \frac{PD_B}{W} = rD_B \tag{2.15c}$$

The yield of separation is related to D_m as follows:

$$R = \frac{D_m}{1+D_m} \tag{2.16}$$

and on the logistic (s-shaped) curve, reflecting R as a function of log D_m (Fig. 53), the characteristic value $R = 0.5$ (50% yield of separation) at $D = 1$.

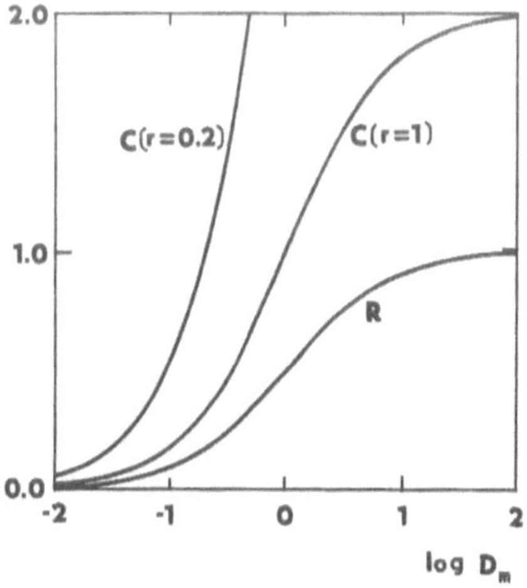

Figure 53. Yield of separation (R) and preconcentration factor (C) at various stage cuts (r) as a function of mass distribution ratio D_m (logD_m).

Such a point of inflexion is critical for both manifold and countercurrent separation processes because at this point a critical change in the operation regime occurs — part 2.2.2.

The main purpose of the separation unit can be either the absolute separation (preconcentration) of a component from the matrix or its relative preconcentration (separation of A from other minor components) — see part 1.1. Fulfilment of each task must be quantified by different relations between the compositions of the feeds and outputs.

The **preconcentration factor** C is the ratio of the concentrations of the concentrated component A in the product to that in the feed:

$$C = \frac{y}{x^F} \tag{2.17}$$

Eqns. 2.6, 2.12a and 2.14 result in

$$C = \frac{F}{P}R_A = \frac{r+1}{r}R_A = \frac{R_A}{\theta} \qquad (2.18a)$$

If $F \approx W$, i.e. the solute separated is the minor component transferred from feed to carrier phase P

$$C_0 = \frac{1}{r}R_A \qquad (2.18b)$$

At the defined stage cut r, a single parameter (R, D, or C) is sufficient to describe the results of the operation of the separation unit. A rapid orientation among the characteristics is provided as follows:

	D	R	D_m	C	C_0
D	$\equiv$	$\dfrac{R}{r(1-R)}$	$\dfrac{D_m}{r}$	$\dfrac{C}{1+r(1-C)}$	$\dfrac{C_0}{1+r}$
R	$\dfrac{rD}{1+rD}$	$\equiv$	$\dfrac{D_m}{1+D_m}$	$\dfrac{rC}{1+r}$	rC_0
D_m	rD	$\dfrac{R}{1-R}$	$\equiv$	$\dfrac{rC}{1+r(1-C)}$	$\dfrac{rC_0}{1+rC_0}$
C	$\dfrac{(1+r)D}{1+rD}$	$\dfrac{(r+1)R}{r}$	$\dfrac{(1+r)D_m}{r(1+D_m)}$	$\equiv$	N.A.
C_0	$\dfrac{D}{1+rD}$	$\dfrac{R}{r}$	$\dfrac{D_m}{r(1+D_m)}$	N.A.	$\equiv$

Separation of the components (A and B) can be specified as the ratio of the distribution ratios, i.e. the **separation factor** α,

$$\alpha = \frac{D_A}{D_B} \qquad (2.19)$$

which can be rewritten as

$$\alpha = \frac{R_A}{1-R_A}\frac{1-R_B}{R_B} \qquad (2.20)$$

or

$$\alpha = \frac{y}{1-y}\frac{1-x}{x} \qquad (2.21)$$

Theoretically, α may achieve any positive value, but conventionally the pair of separated components A and B is defined in such a way that $\alpha > 1$ (i.e. A is the key component which concentrates preferentially in the product).

The relative preconcentration of component A in the product is characterized by the **enrichment factor** S, defined as

$$S = \frac{R_A}{R_B} \qquad (2.22)$$

For the binary mixture and material balance, eqn. 2.11. the ratio is identical to the expression

$$S = \frac{y}{1-y}\frac{1-x^F}{x^F} \qquad (2.23)$$

In practice, various definitions of enrichment separation factors are used and some care should be taken in their interpretation and application. Thus, in isotope separation, where $\alpha \cong 1$, the term "enrichment factor" is given to the parameter $\varepsilon = \alpha - 1$.

Combining eqns. 2.11, 2.12a, 2.19 and 2.20, the following relation between the separation and enrichment factors is obtained:

$$S = R_A + \alpha(1 - R_A) = \alpha + R_A(1 - \alpha) \qquad (2.24a)$$

i.e. $R_A \leq S \leq \alpha$. It is seen that if the yield R_A increases ($R_A \to 1$), the enrichment factor decreases to unity ($S \to 1$) regardless of the value of the separation factor (Fig. 54a). At small R_A, or at $\alpha \cong 1$, S is approximately the same as α, i.e.

$$1 \leq S \leq \alpha \qquad (2.24b)$$

As an example, we can consider the separation of A and B in two systems which differ in chemical composition and consequently in the distribution ratios D_A and D_B:

	System 1	System 2
D_A	100	1.00
D_B	33.0	0.33

This indicates that "occasionally" the separation factors in both systems are equal, = 100/33 = 1/0.33 ≅ 3.03. However, their separation in a separation unit may be substantially different, e.g. for the phase ratio $r = 1$ where

	System 1	System 2
α	3.03	3.03
R_A	0.99	0.50
R_B	0.97	0.25
S	1.02	2.0

In System 1 there is a high yield of A (99%) but its purity does not differ considerably from that of the feed, because B is also separated in high yield (97 %) which is reflected in the low value of the enrichment factor, $S = 0.99/0.97 = 1.02$. In this case the separation factor a better reflects the waste composition (1% of A and 3% of B) in relation to the feed mixture. In System 2, in spite of the low yield of recovery of A (50%), the yield of B is twice as low (25%) and the enrichment factor is substantialy higher, $S = 0.50/0.25 = 2.0$, than in the previous case. This is a general feature of operation with a low yield of the principal component — see Fig. 54 — which can also be controlled by the stage cut: the maximum enrichment factor is $s \cong 3.03$ at r (or P fraction) close to zero. The disadvantage of low yield can be compensated by multi-fold separation without significant loss of product and lessening of its purity — part 2.3. The optimum functioning of the separation unit will depend on whether a high yield, preconcentration factor, purity, or a combination of the factors is to be considered. From a mathematical point of view, the separation factor α is a practical tool to describe the operation of interconnected separation units and separation columns (part 2.4), while the enrichment factor figures as a mandatory factor for multi-stage separations (part 2.3.) and is well suited for characterization of the final results of the operation.

Eqns. 2.4, 2.10, 2.24 and 2.23 permit evaluation of the composition of outputs of the separation unit when the mixture obeys Raoult's law. For example, from eqn. 2.20 a relation exists between the concentration of component A in the product as a function of its concentration in the waste

$$y = \frac{\alpha x}{1+(\alpha-1)x} \tag{2.25}$$

which is the equation of the **equilibrium line** (Fig. 55). In the region of low x (separation of a microcomponent, $x \ll 1$)

$$y \cong \alpha x \tag{2.26a}$$

and

$$\alpha \cong D_A \tag{2.26b}$$

as can be seen directly from eqns. 2.21 (when $x \ll 1$, then $y \ll 1$) and 2.10a. This is, however, valid only for absolute preconcentration of the microcomponent, when B is the matrix.

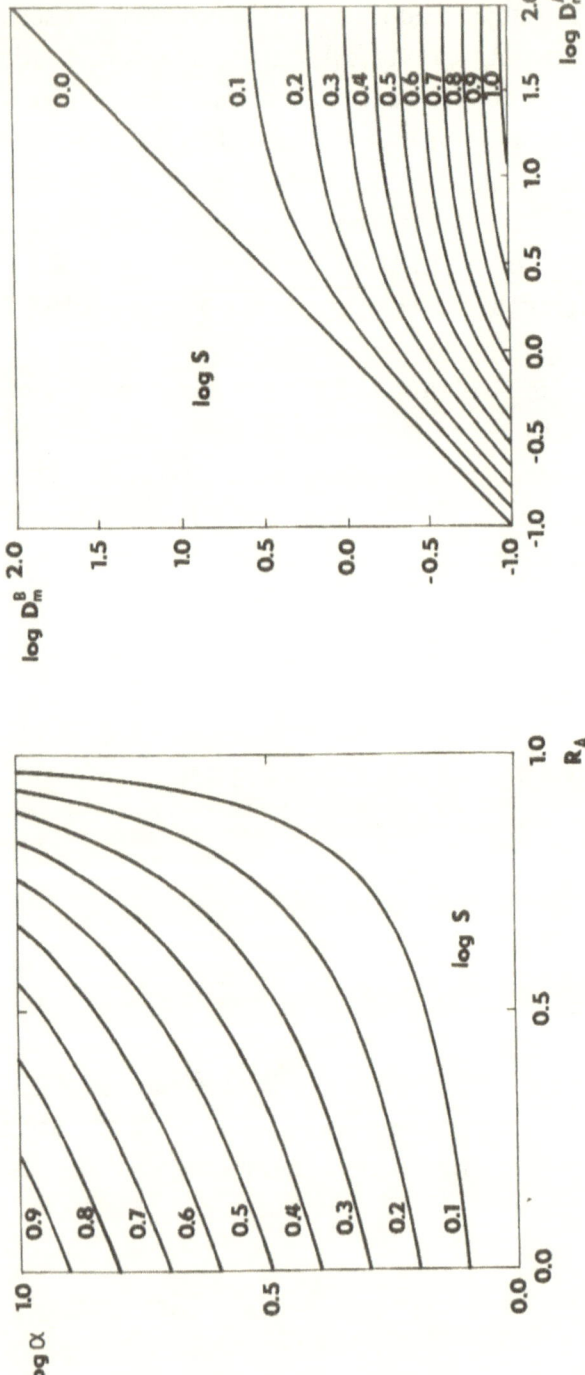

Figure 54. 3-D contour plot of enrichment factor (S) as a function of: a) yield of product A (R_A) and separation factor (α); b) mass distribution ratios of A and B (D_m^A and D_m^B, respectively).

At relative preconcentration when both A and B are **microcomponents** ($x_A + x_B \ll 1$), the separation factor α can be calculated on the basis of eqn. 2.4. The equilibrium lines are linear both for A and B and the separation and enrichment factors should be calculated by eqns. 2.19 (or 2.20) and 2.22 (or 2.24).

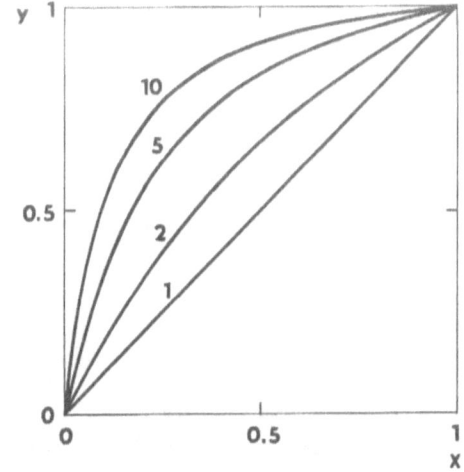

Figure 55. Equilibrium line, a function of concentration y vs. concentration r, at various separation factors α which value is indicated at the curves.

The concentration of the component in the product may be expressed as a function of the initial concentration in the feed, $y = f(x^F)$ in various ways. Under the linearity of distribution, eqn. 2.10a and the balance, eqn. 2.11, a linear relation occurs,

$$y = \frac{(1+r)D_A}{1+rD_A} x^F \qquad (2.27)$$

i.e. y varies from x^F (at $r = \infty$) to $D_A x^F$ (at $r = 0$). If the mixture is ideal and the separation factor does not depend on the composition of the mixture, the separation factor given by eqn. 2.21 is constant, and from eqns. 2.21 and balance 2.11 the result is the quadratic equation

$$y^2 - y \frac{1}{1-\alpha}\left[\frac{1}{r} + \alpha + \frac{r+1}{r}(\alpha-1)x^F\right] + \frac{\alpha(r+1)}{r(\alpha-1)} x^F = 0 \qquad (2.28)$$

For a non-ideal distribution, the empirical equilibrium data should be used to find the composition of the separation unit outputs. Allowance for these variables is traditionally made by a graphical method (*W.H. Walker, W.K. Lewis, W.H. McAdams, 1927*).

In part 1.3.1. we have seen that the lever rule, eqn. 1.283, can be used for finding the composition of the conjugate phase in the phase diagram; the case given by the material balance, eqn. 2.11, is rewritten in the form

$$\frac{P}{W} = \frac{x - x^F}{x^F - y} \tag{2.29}$$

or, in the case of two inputs, from eqn. 2.6a,

$$\frac{P}{W} = \frac{x - x^G}{x^F - y} \tag{2.30}$$

which are, however, more suitable for finding the stage cut at certain points given by the concentrations x and y. For example, if the equilibrium line corresponds to Freundlich's isotherm, eqn. 1.321,

$$y = ax^b \tag{2.31a}$$

for given x in the waste, the result is

$$\frac{P}{W} = \frac{x - x^F}{x^F - ax^b} \tag{2.31b}$$

For $b \neq 1$ and for a given P/W and x^F, the equilibrium value of x can also be found from reciprocal data or by an iterative procedure. If the opposite problem arises, the solution must follow the distribution isotherm (Fig. 23) as an equilibrium line.

The mass balance 2.11 in the y vs. x coordinate is known as the **operating line**

$$y = \frac{F}{P} x^F - \frac{W}{P} x \tag{2.32}$$

or

$$y = x^F - \frac{1}{r}(x - x^F) \qquad (2.33)$$

According to the definition for the separation unit, $x < x^F$ and the point with coordinate (x^F, x^F) is a common intercept of all operating lines.

Over wide concentration ranges, bilogarithmic coordinates, $\log y$ vs. $\log x$ are used for construction of the equilibrium line; the latter is then a straight line and the operating line becomes non-linear. In fact, the solution methods remain the same as before.

In comparison, for a unit with two inputs of immiscible phases where, after contact in the separation unit, the phases become the product and waste respectively, $F \equiv P$ and $G \equiv W$ (see part 2.4), and the operating line is given by eqns.

$$y = \frac{W}{P}x^G + x^F - \frac{W}{P}x \qquad (2.34)$$

or

$$y = x^F - \frac{1}{r}(x - x^G) \qquad (2.35)$$

The operating line will be a straight line with a slope equal to the negative reciprocal value of the stage cut, $-W/P$ (or $1/r$) — Fig. 57. It can be used when the concentration in the product (or waste) is defined and a stage cut should be found. Because the composition of the outlets must fulfil both the equilibrium conditions and material balance, it is found graphically as the intercept of the equilibrium and operating line — the **McCabe-Thiele diagrams** — Figs. 56 and 57 (W.L. McCabe, E.W. Thiele, 1925; W.L. McCabe, J.S. Smith, 1976).

In real time, when the real composition of the outlets — point (x', y') in diagram 56 — differs from those at equilibrium, an

efficiency of the separation unit (stage efficiency) according to Murphree (*E. V. Murphree, 1925*) is defined as having been derived from data for phase *W*

$$\eta_x = \frac{x^F - x'}{x^F - x} \qquad (2.36a)$$

or for phase *P*

$$\eta_y = \frac{y^F - y'}{y^F - y} = \frac{y'}{y} \qquad (2.36b)$$

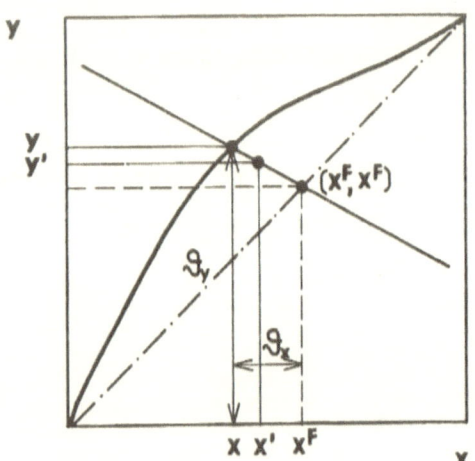

Figure 56. McCabe-Thiele diagram for separation unit with one input F (explanation in text).

(because the initial concentration of A in the product is zero, $y^F = 0$). It may be noted that the difference $x^F - x$ (or $y^F - y$) is the driving force of separation in accordance with eqn. 2.1.

When a separation unit with two feeds has two invariants, x^F and x^G, the equilibrium composition should be sought in another way. The greater the differences in slopes of the operating line (W/P) and the equilibrium line with respect to the phase composition, the greater the values of the driving forces. Conversely, optimal practice usually requires that the lines have roughly the same slope. Such a unit would work in

(i) the enriching regime, when component A passes from fraction $G(\equiv W)$ to fraction $P(\equiv F)$; then $x < x^G$ and $y > x^F$ and the operating line lies below the equilibrium curve;

(ii) the exhausting mode when product A transfers from F ($\equiv P$) to W ($\equiv G$); then $y < x^F$ and $x > x^G$; the operating line will lie above the equilibrium curve.

In both cases **the operating lines of a separation unit with two inputs,** $-x^F = f(x^G - x)$, would be as follows,

$$y - x^F = \frac{W}{P}(x^G - x) \qquad (2.37)$$

have a positive slope $+W/P$ (Fig. 57), and pass through points (x, x^F) and (x^G, y) where x and y are equilibrium concentrations. If, at given input parameters x^F and x^G, output parameters are specified, the stage cut is determined by eqn. 2.30. Conversely, at a given ratio of W/P, one of the output parameters (x or y) can be found if the second is specified, or both of them by means of a trial-and-error procedure, until a right-angle triangle is found whose hypotenuse is parallel to slope W/P and whose apices have coordinates (x, x^F), (x^G, y) and (x, y) — Fig. 57.

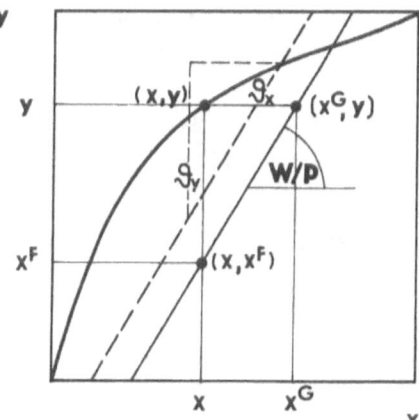

Figure 57. McCabe-Thiele diagram for separation unit with two inputs, F and G (explanation in text).

In large-scale processes, the **mechanical energy** and **heat balance** of a separation unit connected with weak and chemical interactions of solutes and phase transformations (absorption, distillation, drying, melting, etc.) is important for the overall operation of the unit (S.A. Bagaturov, 1961). In some cases, the quantity of heat liberated or consumed in the separation unit is so great that its removal or introduction become the controlling factor of unit design and performance. The terms of mechanical energy of materials are usually negligible when compared to the heat content. In this case the **heat balance** of the separation unit becomes the **enthalpy balance**:

$$\Delta \overline{H}^F F + \Delta \overline{H}^G G + Q_{in} = \Delta \overline{H}^P P + \Delta \overline{H}^W W + Q_{out} \qquad (2.38)$$

where $\overline{\Delta H}'s$ are molar or specific enthalpies (J mol^{-1} or J kg^{-1}) of mixtures (fractions). Q_{in} and Q_{out} are respectively the heat (J) introduced or removed from the separation unit. In a non-equilibrium state, the character of the thermal energy distribution depends on the heat transfer between the fractions, across the separation unit surface and surroundings or specially constructed heat exchangers. Because enthalpies cannot be measured directly, the relation between molar (specific) enthalpy and molar (specific) heat capacity at constant pressure, $\bar{C}_p$ (J mol^{-1} K^{-1}) is applied,

$$\bar{C}_p = \frac{\partial Q_p}{\partial T} = \left(\frac{\partial \overline{H}}{\partial T}\right)_p \qquad (2.39)$$

It follows that

$$\overline{H} = \int_{T_2}^{T_1} \overline{C_p}\, dT \qquad (2.40)$$

and if $\overline{C_p}$ is constant over a temperature range $\langle T_1, T_2 \rangle$ the enthalpies are proportional to the temperature difference

$$\overline{H} = \bar{C}_p(T_2 - T_1) \tag{2.41}$$

Because enthalpies of real mixtures are not ideally additive functions of the individual components, the functions $\overline{H}(x)$, from enthalpy diagrams for actual mixtures, should be known to perform calculations.

For example, if 1 kg of 20% ethanol at $20°C$ ($\bar{C}_p = 4.375$ kJ kg^{-1} K^{-1}) is to be partially distilled to leave 0.75 kg of 10% ethanol at $20°C$ ($\bar{C}_p = 4.320$ kJ kg^{-1} K^{-1}) and yields 0.25 kg of 50% ethanol at $20°C$ ($\bar{C}_p = 4.152$ kJ kg^{-1} K^{-1}) in an adiabatic distillation unit with a full recycle of condensation heat ($Q_{out} = 0$), the heat deficit Q_{in}=(4.125+0.25×4.320×0.75-4.375×1)293 = -28.6 kJ, (to the very rough approximation that the heat capacity is constant).

The condition $Q_{out} = 0$ indicated the assumption that the heat of evaporation was fully recycled by condensation; this, however, does not correspond to the third law of thermodynamics. Even in a full **adiabatic regime**, in an actual separation unit, entropic changes must be taken into account:

$$\overline{\Delta S}^{sep} = \overline{\Delta S}^F F + \overline{\Delta S}^G G - \overline{\Delta S}^P P - \overline{\Delta S}^W W \tag{2.42}$$

In the distillation unit, for example, ΔS^{sep} is positive due to the actual temperature difference between the evaporator (T_0) and condenser (T_H) which is, at molar a heat of evaporation L (J mol^{-1})

$$\overline{\Delta S}^{sep} = L\left(\frac{1}{T_H} - \frac{1}{T_0}\right) \tag{2.43}$$

and therefore additional heat should be introduced to decrease the Gibbs energy — see eqn. 1.13. The entropic effect is significant when an extremely high preconcentration factor C (eqn. 2.17) at low feed concentration s to be achieved. Because in this case $x^F \ll 1$ and $x^P \cong 1$

$$Fx^F \cong P \text{ and } Wx \cong 0 \tag{2.44a}$$

and according to eqn. 1.29

$$\Delta S^{sep} \cong \Delta S^F - \Delta S^P \cong$$

$$\cong PR\frac{1}{x^F}[x^F \ln x^F + (1 - x^F) \ln (1 - x^F)] \cong PR(\ln x^F - 1) \quad (2.44b)$$

E.g. the separation of 1 kg of heavy water ($P = 50$ mol of D_2O) from natural water ($x^F = 1.5 \times 10^{-5}$) at $T = 300K$ needs minimal energy due to this entropic factor,

$E_{min} = 300 \times 8.31 \times 50(\ln 1.5 \times 10^{-5} - 1) = 1508$ kJ.

High energy demands for maintaining concentration and heat driving forces are typical for separation units operating in a continuous mode and in cascades — see part 2.4. The **thermodynamic efficiency** calculated from the ratio of minimal (entropic) energy input, E_{min}, and real energy demands,

$$\eta = \frac{E_{min}}{E} \quad (2.45)$$

is very rarely above 10% and for many separation units is a minute fraction of unity, due to the energy spent on heating, cooling, and pumping accessories of the separation unit.

2.1.2 Rate and local equilibrium processes

We have previously encountered the rate-controlled separation process, particularly in part 1.4.4. for the case of co-crystallization —see eqn. 1.568. In such processes, the **separation factor** is not given by the ratio of thermodynamic but kinetic (transfer coefficient) parameters,

$$\alpha = \frac{K_A}{K_B} \quad (2.46)$$

Examples of such constants can be found in the equations of co-crystallization (eqn. 1.570), electrolytic decomposition (eqn.

1.578), effusion, Knudsen diffusion (eqn. 1.629), electrodialysis (eqn. 1.662), membrane permeation (eqn. 1.676), and photochemically-induced separation (eqn. 1.700). All of these processes have a more-or-less irreversible nature. However, even such processes as distillation, where a local equilibrium exists at least at the liquid-vapor interface, can proceed in separation units where either equilibrium is not reached — eqn. 2.2 — or the unit is an open system with a perpetual inward and outward flux of material, e.g. batch distillation or continuous distillation, as discussed at the beginning of part 2.1.1. (*P.C. Wankat, 1944, 1990, 2012; M. Valcarel, M.D. Lugue de Castro, 1991*).

As regards to the actual arrangement of the separation unit, generally with a single inlet flow $\dot{F}$ (mol s^{-1}) and one or two outlet flows $\dot{W}$ and $\dot{P}$ (mol s^{-1}), there exist three specific and practically important situations (Fig. 58):

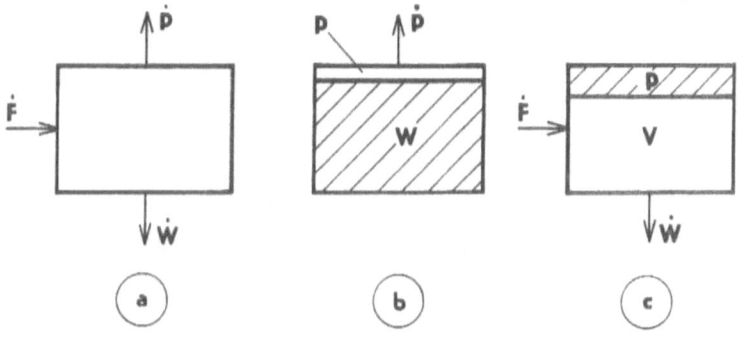

Figure 58. Actual arrangement of product and waste fractions in a rate separation unit: a) continuous product and waste outlet, b) portion of product P and entire waste W retained in the unit, c) portion of waste (feed) phase contacted with bulk of phase P.

(i) the constant inlet is split continuously into outlets $\dot{P}$ and $\dot{W}$;
(ii) the inlet is loaded batch-wise into a separation unit, i.e. $\dot{F}$ is zero, the product is withdrawn continually, and in the separation unit equilibrium exists between a small

fraction of product P and the rest of the separated mixture representing the waste fraction, $P \ll W$;

(iii) the feed is introduced at a constant speed into the separation unit where part V is contacted with fraction P resting in the unit ($\dot{P} = 0$) and then leaves the unit as fraction W with a rate $\dot{W}$.

The attribution of real fluxes to P and W in Figs. 58a, c can be easily converted as seen in part 2.1.1.

In the general mass balance of the units, the **outer parameters** of the separation unit, the flows (mol s^{-1}) of feed ($\dot{F}$), product ($\dot{P}$) and waste ($\dot{W}$) and the **inner parameter**, an **accumulation term** of the total amount of component in moles per time unit ($dn/dt = \dot{n}$) can be distinguished and considered. The fluxes are positive when they enter the separation unit and negative on leaving. Thus the balance is

$$\dot{F} + \dot{P} + \dot{W} + \dot{n} = 0 \tag{2.47}$$

The material balance of component A is, correspondingly

$$\dot{F}x^F + \dot{P}\langle y \rangle + \dot{W}\langle x \rangle + \dot{n} = 0 \tag{2.48}$$

where $\langle x \rangle$ and $\langle y \rangle$ are the area-averaged concentrations in the flows and $\dot{n}$ is the rate of accumulation of the component during the operation of the separation unit.

A dynamic relation between $\langle y \rangle$ and $\langle x \rangle$ results from the transfer of components by outer fluxes P and W equal to the inner cross-diffusion flux j (mol m^{-2} s^{-1}) of components through the interface area A(m^2) at the steady state established in the separation unit:

$$\dot{P}\langle y \rangle = A\, j_A^P \tag{2.49}$$

$$\dot{W}\langle x \rangle = A\, j_A^W \tag{2.50}$$

Considering eqn. 2.2 and the total molar concentrations of phase P and W,

$$c_t^P = \dot{P}t/V^P \tag{2.51a}$$

$$c_t^W = \dot{W}t/V^W, \tag{2.51b}$$

the result is

$$\frac{\langle y \rangle}{\langle x \rangle} = \frac{K_A^P(y^{eq}-\langle y \rangle)a^P}{K_A^W(\langle x \rangle - x^{eq})a^W} \tag{2.52}$$

where

$$a^P = A/V^P \tag{2.53a}$$

$$a^W = A/V^W \tag{2.53b}$$

are the specific surface areas with respect to phases P and W. It indicates that the **dynamic distribution ratio** $\langle y \rangle/\langle x \rangle$ is formally equivalent to the thermodynamic distribution ratio, eqn. 2.10a, in the equilibrium separation unit, and depends not only on the ratio of mass transfer coefficients but on the geometry of the phases and the driving forces operating in the corresponding phases (fractions). The driving forces are given by the deviations of the actual concentrations $\langle x \rangle$ and $\langle y \rangle$ from the equilibrium values, x^{eq} and y^{eq}, attainable in the separation unit. If the unit is not in a steady state, the values are time-dependent. An approach to the steady- or equilibrium state will be discussed in part 2.2.2.

The separation factor α, defined through the ratio, eqn. 2.46, and the balance equations 2.47 and 2.48, in spite of their formal uniformity with eqns. 2.5 and 2.6, should be treated in their differential forms.

In cases where the feed is introduced at a constant speed (Fig. 58a, c) or batch-wise, we have $\dot{F}x^F = \text{const}$, or

$$d(\dot{F}x^F) = 0 \tag{2.54}$$

In the **steady state**, when there is no change in the outlet fluxes and no accumulation, i.e. the feed simply splits into two outlets — Fig. 58a — by a constant stage cut, $d\dot{P} = 0$ and $dW = 0$, and from eqns. 2.53 and 2.54 the result is

$$\frac{d\langle y \rangle}{d\langle x \rangle} = -\frac{\dot{W}}{\dot{P}} \tag{2.55}$$

which is the equation for the **operating line under steady state** conditions.

Differential separation (Fig. 58b) starts with a batch F of fraction W and the product flux $\dot{P}$ is generated by the change of the batch amount and composition, i.e. $\dot{n} \equiv dW/dt$. The particular case of the balances, eqn. 2.47 and 2.48, (the symbols of average concentrations are omitted for simplicity) is

$$\dot{P} + \frac{dW}{dt} = 0 \tag{2.56}$$

while that for a single component A results in

$$\dot{P}y + \frac{d(Wx)}{dt} = 0 \tag{2.57}$$

(accumulation of fraction P inside the separation unit is considered negligible). From these equations it follows that

$$dP = -dW \tag{2.58}$$

and

$$ydP = -d(Wx) \tag{2.59}$$

After elimination of P, the differential balance takes the form

$$\frac{dW}{W} = \frac{dx}{y-x} \tag{2.60}$$

which is integrable as

$$\int_W^F \frac{dW}{W} = \int_{\langle x \rangle}^{x_F} \frac{dx}{y-x} \tag{2.61}$$

where x^F is the concentration of component in the feed (when $W \equiv F, x = x^F$). When **local equilibrium** exists at the contact of phases W and P in the separation unit, equilibrium y vs. x data can be used for the integration: for actual mixtures the integration should be carried out using the empirical curve $1/(y - x)$ vs. x. In case of the linear isotherm, $y = \alpha x$.

$$\frac{1}{y-x} = \frac{1}{(\alpha-1)x} \tag{2.62}$$

and after substitution in eqn. 2.61 it gives on integration

$$\frac{1}{\alpha-1} \ln \frac{x^F}{\langle x \rangle} = \ln \frac{F}{W} \tag{2.63a}$$

or

$$\frac{x^F}{\langle x \rangle} = \left(\frac{F}{W}\right)^{\alpha-1} \tag{2.63b}$$

where $\langle x \rangle$ is the average (resulting) concentration in fraction W. From the equilibrium line of the **ideal mixture**, eqn. 2.25, the result is

$$\frac{1}{y-x} = \frac{1}{\alpha-1}\left(\frac{1}{x} + \frac{1}{1-x}\right) + \frac{1}{1-x} \tag{2.64}$$

and integration of eqn. 2.61 with respect to eqn. 2.64. produces the **Rayleigh equation** (J.W. Rayleigh, 1902) **for the differential separation** of an ideal binary mixture:

$$\frac{1-\langle x\rangle}{1-x^F}\left(\frac{x^F}{\langle x\rangle}\right)^{\frac{1}{\alpha}} = \left(\frac{F}{W}\right)^{\frac{\alpha-1}{\alpha}} \tag{2.65}$$

This equation was derived from differential distillation (a simple evaporation of the batch) but can be applied for any other model of differential separation (*C.J. King, 1980*), e.g. gaseous diffusion in conventional permeator (*S.T. Hwang, K.H. Yuen, J.M. Thorman, 1980*) or electrolysis and co-crystallization.

When the differential separation of matrix A is used for the **preconcentration** of B, assuming that $1 - x = x_B$ and $x \cong 1$, the equation 2.65 becomes

$$\frac{\langle x_B^W\rangle W}{x_B^F} = \left(\frac{F}{W}\right)^{\frac{\alpha-1}{\alpha}} \tag{2.66}$$

where the ratio of the mole fractions can be replaced by the yield of B in concentrate W, i.e. the "waste" of the matrix is actually the "product" of the concentrate:

$$R_B^W = \frac{\langle x_B^W\rangle W}{x_B^F F} = \left(\frac{W}{F}\right)^{\frac{1}{\alpha}} \tag{2.67}$$

which is identical with eqn. 2.63b.

If a sufficiently low ratio W/F is achieved, the yield of preconcentration is sufficiently high even at low values of α. Deuterium was obtained by the simple evaporation of liquid deuterium (*H.C. Urey, F.L. Brickwedde, L.M. Murphy, 1932*), and the electrolysis of water to a small residue preconcentrates deuterium (*G.N. Lewis, 1933; A.M. Brodskii, 1934*) and especially tritium (*M.L. Eidinoff, 1947; M. Saito, 2008*). For example, on electrolytic preconcentration of tritium in natural water, the separation factor is $\alpha = 14$ (tritium is preconcentrated in the liquid phase); at 90% electrolysis, $W/F = 0.1$ and, according to eqn. 2.66, the concentration of tritium increases $(1/0.1)^{(14-1)/14} = 8.5$ times and so that its yield, according to eqn. 2.67 is $(0.1/1)^{1/14} = 0.85$ (85%).

The concentration of the product during differential separation cannot be expressed explicitly; in these cases graphical methods are used, or an iteration method of solution of the transcendental equations

$$x_{i+1} = (\frac{1-x_i}{b})^\alpha, \quad \text{if } b > 1 \tag{2.68a}$$

or

$$x_{i+1} = 1 - bx_i^{1/\alpha}, \quad \text{if } b < 1 \tag{2.68b}$$

where

$$b = \left(\frac{F}{W}\right)^{\frac{\alpha-1}{\alpha}} (1 - x^F)(x^F)^{1/\alpha} \tag{2.68c}$$

are applied. The zero approximation of b for eqn. 2.68a is $x_0 = 1/b$ and for eqn. 2.66b $x_0 = 1 - b$. For example, when $F = 1$, $x^F = 0.5$, $W = 0.1$ and $\alpha = 3$, equation 2.68c gives $b = 2.92$ and $x_0 = 1/2.92 = 0.34$; the first approximation, according to eqn. 2.68a is $x_1 = 0.040$ and yet the third trial gives the correct value, $x_3 = 0.036$.

During the differential separation, the first fractions of the product obtained and the residuals of the waste fraction differ significantly from the feed — therefore these fractions are usually collected separately and segregated from the middle fractions for which, however, the enrichment factor remains much higher than for integral equilibrium separation. For example, if the separation of a binary equimolar mixture ($x^F = 0.5$) occurs with a separation factor $\alpha = 3$, performed once at equilibrium and another time in differential mode at three various stage cuts, $P/W = 0.01:0.99$, $0.90:0.10$ and $0.99:0.01$, then the fraction compositions and separation factors will be as follows:

	stage cut ($P:W$)					
	0.01 : 0.99		0.90 : 0.10		0.99 : 0.01	
	operation mode					
	equil.	differ.	equil.	differ.	equil.	differ.
$\langle x \rangle$	0.498	0.498	0.269	**0.036**	0.253	**0.00005**
$\langle y \rangle$	**0.700**	**0.698**	0.527	0.551	0.503	0.505
S	2.33	2.31	1.11	1.23	1.012	1.02

Most remarkable is the extremely low content (0.036–0.00005) of the more readily separable component remaining in W by the differential method at high cuts, which is several orders lower than during the equilibrium mode of operation (0.269–0.253). This is a most significant feature of the non-equilibrium differential method for preparation component B in a very pure state.

Very often separation (sorption, washing, leaching) proceeds with one **immobile phase**, i.e. a solid or liquid bed (sorbent, filter cake, extractant), which comes into contact with the second phase which is flowing through the separation unit (filter, small packed column, agitated vessel, etc.) — Fig. 58c. The material balances, eqn. 2.47 and 2.48, take the form

$$\dot{F} + \dot{W} + \frac{dP}{dt} + \frac{dV}{dt} = 0 \tag{2.69}$$

and

$$\dot{F}x^F + \dot{W}x + \frac{d(P\langle y \rangle)}{dt} + \frac{d(Vx)}{dt} = 0 \tag{2.70}$$

where $\langle y \rangle$ is the average concentration in product P and V is the portion (mol) of waste fraction retained but continuously replaced by the F stream in the separation unit and contacted with immobile phase P. Because after loading the separation unit with a batch of feed ($V \equiv F$) the equality

$$P + V = \text{const} \tag{2.71}$$

holds, and the outer fluxes are related through

$$\dot{F} = -\dot{W} \tag{2.72a}$$

and

$$\dot{F}dx = -\dot{W}dx \tag{2.72b}$$

then the balances, eqns. 2.69 and 2.70, are reduced to a differential form of the material balance

$$x^F dF + x dW + P d\langle y \rangle + V dx = 0 \tag{2.73a}$$

or the integral

$$x^F F = W\langle x \rangle + P\langle y \rangle + Vx \tag{2.73b}$$

where $\langle x \rangle$ is the average concentration in the total waste fraction. At the **local equilibrium** or pseudo-equilibrium (steady) state between the fractions contacted in the separation unit (V and P), a linear isotherm relation exists

$$\langle y \rangle = Dx \tag{2.74}$$

where D is either the thermodynamic distribution ratio, eqn. 1.309, or the dynamic distribution ratio, eqn. 2.52. As a result, the balance, eqn. 2.73, is ready for integration:

$$\int_0^W dW = (DP + V) \int_{x_0}^x \frac{dx}{x - x^F} \tag{2.75a}$$

where x_0 is obtained from the material balance of an initial batch-wise loaded portion V of the feed,

$$V x^F = x_0 V + y_0 P = x_0 V + D x_0 P \tag{2.75b}$$

The resulting equation linking the current concentration x in the outlet flow, which is identical to the concentration in volume V, with that in the feed (x^F), is

$$\frac{x}{x^F} = 1 - \frac{DP}{V+DP} \exp\left(-\frac{W}{V+DP}\right) \tag{2.76}$$

(compare this form with eqn. 2.138 in section 2.3). The average concentration $\langle x \rangle$ of component A in W is easily obtained by eqn. 2.74 for local equilibrium and the global material balance (2.73b):

$$\frac{\langle x \rangle}{x^F} = 1 - \frac{DP}{W}\left[1 - \exp\left(-\frac{W}{V+DP}\right)\right] \tag{2.77}$$

which is a yield of separation and which for small W's gives $\langle x \rangle / x^F \cong V/(V+DP)$. If such differential separation is performed by equilibrium contact of the whole volume of feed ($F = V + W$) with product P batch-wise (instantaneously), the ratio

$$\frac{\langle x \rangle}{x^F} = \frac{F}{F+DP} = \frac{V+W}{V+W+DP} \tag{2.78}$$

under static conditions is always higher, though at high distribution ratios and low F/P values, when $D \gg F/P$, the difference between "static" and "dynamic" operations is negligible.

At $DP/W < 1$ (which indicates, according to eqn. 2.15, that $D_m < 1$) the separation unit in Fig. 58c operates to remove component A from phase P, as in washing of a filter cake, leaching of a solid batch or stripping of an extract.

When considering variations in the operational mode of a separation unit, the following rule applies to the moving phase: the infinitesimal portions which are continuously contacted with the immobilized phase (differential contact) possess a greater ability to take up the distributed solutes than under static conditions (bulk contact). This is particularly valid when the separation unit is in reality a non-structured device, i.e. the contacted flows are sufficiently randomly oriented and the unit acts as a one-stage separator. Practical experience has shown that models of actual one-stage or multi-stage separators (see part 2.3) become confused and that either the efficiency of the one-stage unit can be overrated or that of multi-stage devices is underestimated. These factors are of prime importance for the overall fraction of solutes transferred from the streaming feed to the product immobilized in a bed, or conversely in separation units which use rate and local equilibrium distributions.

2.2 CHEMICAL CONTROL OF SEPARATION

In part 1.1.2. we concluded that the chemical potential gradient for separation of substances can be influenced by such chemical compositions of systems that result in the formation of individual phases or in the desired distribution between existing phases. Operation of the separation unit can be adjusted by the amounts of inputs and outputs, characterized by operating lines, and the thermodynamic and rate parameters linked to the equilibrium (or steady-state) lines, as was shown in the previous section. However easily the changes of temperature and pressure can be to realize, they are usually narrow as a practical range, while the amounts of phases are limited by the means of mechanical handling. Thus, one of the most important **controlling tools** of chemical separation techniques remains the wide spectrum of interactions between mixture components which cause changes in shape and charges of the molecules, association, solvation, and compound formation as well as aggregation and structuring.

If, for example, a tenfold change of mass distribution ratio, eqn. 2.15, is required it can be achieved by a tenfold change
(i) in the product-to-waste molar ratio P/W or
(ii) of the activity coefficients of the distribuend in non-ideal mixtures which takes place, e.g.

—for a uni-uni valent electrolyte by a change of ionic strength of about unity (according to eqs. 1.108 and 1.217),

—on introduction of 1-2 methylene groups in a component molecule (see eqn. 1.249) or

—by stoichiometric excess of about 4×10^{-4} mol dm^{-3} of a ligand with coordination bond strength 25 kJ mol^{-1} (according to eqn. 1.139), etc.

The realization of such a change depends on the character of the mixture and further treatment of separated fractions. (*G.K. Boreskov, M.G. Slinko, 1965; V. Hlavacek, V. Vaclavek, M. Kubicek, 1979*). A rich palette of chemical additives, influencing 41 types of molecular interactions, is available and its use is mostly limited only by the reversibility of their interactions. A classic example in the use of weak interactions to effect a change in the chemical composition of organic mixtures is the so-called **extractive distillation**, by which the equilibrium line of the binary mixture (A + B) is perturbed by addition of a third component (C). If interaction of the latter with A and B is characterized by the interaction constants u_{AC} and u_{BC}, the theory of regular solutions — section 1.2.5. — results in a change in the separation factor from α_0 to α_C (*A.M. Rosen, 1951*):

$$RT \ln \alpha_C/\alpha_0 = (u_{AC} - u_{BC})x_C + u_{AB}(x_B - x_A)/(1 - x_C) \quad (2.79)$$

where the constants u can be found from solubility parameters. In the same way, additives are employed in zone melting (*V.N. Vigdorovich, V.V. Marichev, 1965; Y. Waseda, M. Isshiiki, 2002; W.L.F. Armarego, C.L.L. Chai, 2013*) and in using the hydrotropic effect in solvent extraction (*V.G. Gaikar, 1986*).

The irreversible oxidation of impurities promotes their removal during distillation (*H. Bradley, R.M. Pankhurst, 1965*) or at sorption (*H.C. Teurerer, 1960*).

In section 1.2.3., complex equilibria were discussed as being the best-developed and universal model of reversible chemical interactions. Accordingly, it is expedient to begin a discussion of complex equilibrium systems with the important feature that the same distribution ratio D can be achieved by different solution compositions, as was substantiated in detail for liquid-liquid systems using the theory of **corresponding biphasic systems** (F. Macasek, 1974) and **isodistribution systems** (F. Macasek, 1980). The idea behind corresponding biphasic systems is the two-phase analogue of Bjerrum's corresponding systems (J. Bjerrum, 1941) where systems with various metal: ligand ratios can be obtained with equal absorbance of the complex; these systems are characterized by the same equilibrium concentrations of their complex-forming components. The same idea can be applied to systems in which a two-phase distribution of such a complex occurs, the salient points as follows.

The principal controlling complex-forming reaction can often be reduced to equilibria of type eqn. 1.343b, which is sufficiently universal because it also includes solvates with ligand acid (HL) formation (J. Rydberg, 1955):

$$(M^{n+})^I + n(HL)^{II} \rightleftarrows [ML_n(OH)_p]^{II} + (n+p)(H^+)^I \quad (2.80)$$

Generally, as the concentrations of the three reactants are varied, the question is, which system composition will lead to a particular distribution ratio determined by the general ratio, eqn. 1.346? When polynuclear complexes $[M_m L_n (OH)_p]$ are neglected from consideration, the full differential of log D depends only on the concentration of HL in phase II and of H^+ in phase I:

$$\mathrm{d}\log D = \left(\frac{\partial \log D}{\partial \log [HL]_{II}}\right)_{[H^+]} d\log [HL]_{II} + \left(\frac{\partial \log D}{\partial \log [H^+]_I}\right)_{[HL]} d\log [H^+]_I \quad (2.81)$$

Taking into account eqns. 1.347, 1.193, 1.203, and 1.204, derivatives are obtained as differences in the average ligand numbers in phase I and II, respectively

$$\frac{\partial \log D}{\partial \log [HL]_{II}} = \frac{\partial \log X^{II}}{\partial \log [HL]_{II}} - \frac{\partial \log X^I}{\partial \log [HL]_{II}} = \langle n^{II} \rangle - \langle n^I \rangle = \Delta n \quad (2.82a)$$

$$\frac{\partial \log D}{\partial \log [H^+]_I} = \frac{\partial \log X^{II}}{\partial \log [H^+]_I} - \frac{\partial \log X^I}{\partial \log [H^+]_I} = \langle p^{II} \rangle - \langle p^I \rangle = \Delta p \quad (2.82b)$$

Non-integer, non-zero values of the differences Δn and Δp indicate the coexistence of various $ML_i(OH)_j$ species, but not vice versa. In isodistribution systems, an equal distribution ratio occurs in all the systems, so that

$$d\log D = 0 \qquad (2.83)$$

Combining eqns. 2.81–2.83, the distribution ratio will remain constant throughout the systems when the equilibrium concentrations fulfil the condition

$$\frac{[HL]_{II}^{\Delta n}}{[H^+]_I^{(\Delta n + \Delta p)}} = \text{const} \qquad (2.84)$$

To achieve the condition of initial (analytical) concentrations, eqn. 1.356 should be used to link the concentrations of the metal atom (c_M) and the ligand (c_L) to the phase ratio r. Thus, the material balance equation 1.356 is adjusted so that

$$rc_L = \frac{[HL]_{II}}{\alpha_L} + \langle n^t \rangle c_M \qquad (2.85a)$$

where the function α_L indicates the ratio of the amounts of all forms H_iL of the ligand to the basic form HL in phase II and is connected with the similar function Π^t, discussed previously, (eqn. 1.356) as follows

$$\alpha_L = \frac{[HL]_{II}}{\Pi^t [L]_I} \qquad (2.85b)$$

If form HL dominates strongly, $\alpha_L \cong 1/r$. Then, by means of eqn. 2.85a, the isodistribution system is linked by the equation

$$\frac{\alpha_L (rc_L - \langle n^t \rangle) c_M}{[H^+]^{\frac{\Delta n + \Delta p}{\Delta n}}} = \text{const} \qquad (2.86a)$$

where the constant is derived from one of the corresponding systems when c_L, c_M, $[H^+]$, and $\langle n^t \rangle$ are known. The simplest approximation for $\langle n \rangle$ is obtained from eqn 1.353a as $\langle n^t \rangle \cong nR$, and

$$\frac{\alpha_L (rc_L - nRc_M)}{[H^+]^{\frac{\Delta n + \Delta p}{\Delta n}}} = \text{const} \qquad (2.86b)$$

For example, in a system where $r = 0.5$, $c_M = 10^{-4}$ M, $c_L = 1$ M, and at $[H^+] = 10^{-5}$ M when $\alpha_L \cong 1/r = 2$, the distribution ratio D is 20 ($R = 0.91$). Next, another isodistribution system is to be found at the same phase ratio r, but at a different concentration: the first step, if the direct experimental estimation of derivatives, eqn. 2.82, is unavailable, is to propose an adequate model of chemical equilibrium to approximate the average ligand numbers. For example, if in phase I a complex M(OH) is assumed to dominate and in phase II only the single species ML_3 exists, then this substantiates the approximation $\Delta n = 3.0 - 0 = 3.0$, $\Delta p = 0 - 1.0 = -1.0$ and $\langle n^t \rangle = nR = 3 \times 0.91 = 2.73$. Then the constant in eqn. 2.86b can be evaluated: const $= 2(0.5 \times 10^{-3} - 3 \times 0.91 \times 10^{-4})/(10^{-5})^{2/3} = 0.978$. At a different concentration of metal, for example $c_M = 3 \times 10^{-4}$M, an identical constant (and consequently an identical distribution of metal in the separation system) will be achieved at for example

c_L	5×10^{-3} M	1.7×10^{-3} M	2.0×10^{-3} M	4.0×10^{-3} M	...
$[H^+]$	unreal	5.05×10^{-7} M	7.12×10^{-6} M	1.19×10^{-4} M	...

(when $\alpha_L \cong 2$). This indicates that the same yield of separation of metal, limited by the condition $rc_L - nRc_M > 0$ can be achieved at an arbitrary number of combinations of the ligand concentration c_L and acidity [H^+].

In practice, deviations from calculations made according to eqn. 2.86 can be expected due to incorrectly established and inconstant values for Δn and Δp, as well as to the non-ideality of the phases and the presence of other components. Hence, the ultimate choice of isodistribution system will depend primarily on the behavior of the minor species (impurities) in such a system (*F. Macasek, 1980, F. Macasek, D. Vanco, 1981; R. Kopunec, J. Kovalancik, 1989*).

The theory of biphasic corresponding and isodistribution systems emphasizes the distribution ratio as a **dimensionless invariant of separation systems** of various chemical compositions, although considerable latitude is possible in selecting reagents and media. In developing a new **separation system**, however, the following criteria are useful to be applied and proved (*Y. Marcus, 1976*):

(i) Do satisfactory separation methods already exist to deal with the problem in hand (that provide the same separation of the components)?

(ii) Is the proposed method (system) indeed superior to the existing methods (systems) according to measurable parameters (yields, separation and enrichment factors, cost, speed, etc.)?

Although maximization of the distribution ratio is important for all pre-concentration techniques, it is the behavior of minor components that is crucial with respect to the mutual separation and purity of the product. Usually there is no problem in reaching the desired yield of a single component by adjustment of the chemical composition, but difficulties appear in connection with suppressing the distribution of the matrix and impurities. This is what makes the field of chemical separation so broad and colorful.

2.2.1 Initial and equilibrium parameters

In general practice, the phase diagrams (section 1.3.1., Figs. 18–20) enabling the determination of both composition and amounts of coexisting phases are well suited to separation science. If electrochemical sensors are used to examine the process, the diagrams of composition vs. equilibrium activity data (part 1.3.2., Figs. 21–22) are more appropriate. In addition, the theoretical prediction of the heterogeneous distribution (section 1.3.2., eqns. 1.335–1.362) becomes more convenient when equilibrium rather than the initial ("total", "analytical") parameters are used.

In addition to those in section 1.2.3., some further relations between initial and equilibrium parameters will now be discussed with respect to complex equilibria as a guide to reversible, multimolecular reactions.

The separation enhanced by chemical interactions (S. Kulprathipanja, 2002) can be modelled by a reaction of type 1.126 proceeding in a two-phase system, such as chemisorption, bioaffinity sorption, ion exchange or chelate extraction, in a separation unit with two feeds and two outputs, all of which are represented by phase I and phase II (Fig. 52b):

$$\nu_1(A_1)_I + \nu_2(A_2)_{II} \rightleftarrows (A_3)_{II} \tag{2.87a}$$

where A_3 is identical with the compound (complex, associate):

$$A_3 \equiv (A_1)_{\nu_1}(A_2)_{\nu_2} \tag{2.87b}$$

The above equilibrium can be characterized by the heterogeneous equilibrium constant

$$K^H = \frac{y_3}{(a_1 x_1)^{\nu_1}(a_2 y_2)^{\nu_2}} \tag{2.88}$$

where x's are equilibrium concentrations in phase I and y's are those in phase II, and α's are the dimensionless (Ringbom) coefficients ($\alpha < 1$) characterizing the equilibria of component side reactions — cf. eqns. 1.165 and 1.344b.

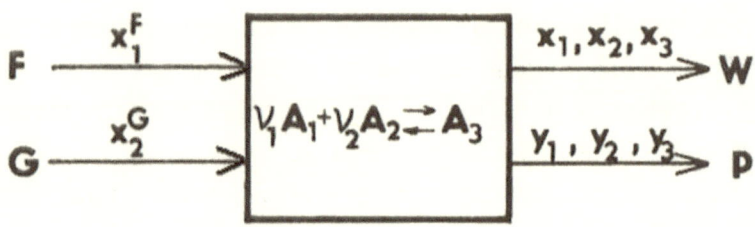

Figure 59. Separation unit with chemical reaction between two input components, A_1 and A_2.

With respect to stoichiometric relations, the material balance of reactants A_1 and A_2 will be

$$Fx_1^F = W(x_1 + v_1 x_3) + P(y_1 + v_1 y_3) \qquad (2.89a)$$

$$Gx_2^G = W(x_3 + v_2 x_3) + P(y_2 + v_2 y_3) \qquad (2.89b)$$

For the individual distribution of reactants A_1, A_2 and product A_3, a purely weak-interaction distribution and linear isotherms can be assumed:

$$y_1 = D_1 x_1 \qquad (2.90a)$$

$$y_2 = D_2 x_2 \qquad (2.90b)$$

$$y_3 = D_3 x_3 \qquad (2.90c)$$

i.e. there is no influence of one component on the physical distribution of the other.

Further, for the equilibrium line of A_2, which is also a component of the product A_3, the following values (total concentrations) should be determined:

$$y \equiv y_1 + v_1 y_3 \qquad (2.91a)$$

$$x \equiv x_1 + v_1 x_3 \qquad (2.91b)$$

The total variance of the system — eqn. 1.274 — is $V = 3 + 2 = 5$ which, at constant pressure and temperature, can be the amounts of phases (W, P) and one of the input concentration. The task is to formulate the distribution ratio of the species present as a function of this controlling concentration.

Under the above conditions, a **transcendental equilibrium line equation** is obtained from eqns. 2.88–2.90 (if $G \equiv P$ and $F \equiv W$, $P/W = r$) and by means of **coordinates normalized** to one input concentration, namely x_2^G, and **normalized efficient constant** (see eqn. 1.169):

$$K = (v_1/v_2)^{v_1+v_2-1} K^H (\alpha_1)^{v_1} (\alpha_2)^{v_2} (r x_2^G)^{v_1+v_2-1} \qquad (2.92a)$$

The equation is thus (when neither D_2 nor D_3 equals zero)

$$K = \frac{v - D_1 q}{v_1 \left[\frac{q - v(r+1/D_3)}{1 - D_1/D_3}\right]^{v_1} \left\{\frac{v_2[1-(v-D_1 q)(r+1/D_3)]}{v_1(r+1/D_2)}\right\}^{v_2}} \qquad (2.92b)$$

where

$$q = \frac{v_2 x_1^F}{v_1 r x_2^G} \qquad (2.92c)$$

and

$$v = \frac{v_2 y}{v_1 r x_2^G} \qquad (2.92d)$$

are the normalized coordinates of the feed concentration x_1^F and the equilibrium value y in the product respectively.

When $K \gg 1$, the limiting values $v = q$ (if $q \leq 1$) and $v = 1$ (if $q > 1$) are obtained. Generally, the two asymptotes of the values of v in the coordinates log v vs. log q are log $v = 0$ as $q \to \infty$, and $\log v = \log K + v_2 \log(v_2/v_1) + v_1 \log q$ as $q \to 0$.

The **operating line**, according to eqns. 2.34, 2.89 and 2.91, is simply a vertical line. When necessary, the equilibrium value x in phase W can be found as

$$\frac{x}{x_2^G} = \frac{v_1}{v_2} r(q - v) \qquad (2.93)$$

Obviously, application of the isotherms is easier when compared with Fig. 57 for two inputs, because the number of stages is found when using the differences between the lines $\log v = \log q$ and $\log v = f(\log q)$ — Fig.60.

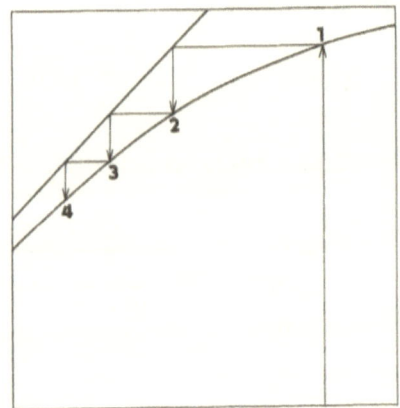

Figure 60. McCabe-Thiele diagram for separation units with chemical reactions; normalized concentration (v) of component A_1 in flow G vs. normalized concentration of the component (q) in feed F.

Fig. 61 shows a few examples of the diagram calculated for representative stoichiometries:

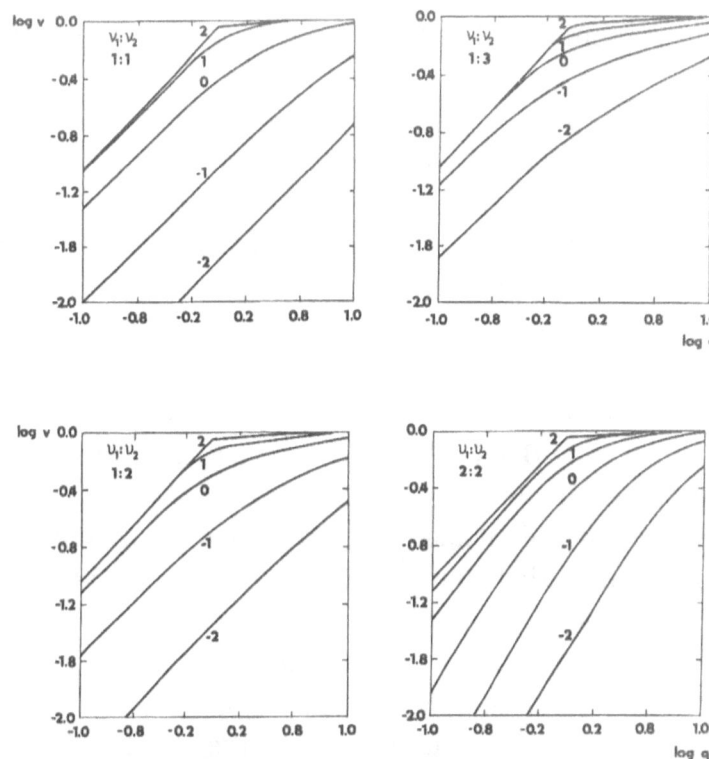

Figure 61. Separation isotherms at chemical equilibrium of A_1 and A_2 at stoichiometry $v_1:v_2$ in normalized coordinates v (normalized concentration in flow G) as a function of q (normalized concentration component A_1 in feed F), log K values are indicated at the curves.

For example, for the reaction $2A_1 + 2A_2 \rightarrow A_3 (v_1 = v_2 = 2)$ with the heterogeneous constant $K^H = 10^{10}$, in the separation unit with stage cut $r = 0.5$ and concentrations of reactants A_1 and A_2 in feeds $x_1 = 8.40 \times 10^{-4}$, $x_2^G = 2 \times 10^{-3}$ (arbitrary concentration units), the normalized constant is $K = (2/2)^{2+2-1} \times 10^{10} \times (10^{-3})^{2+2-1} = 10$. When equilibrium concentrations are to be found, for $q = 2 \times 8.4 \times 10^{-4}/2 \times 10^{-3} = 0.84$, from Fig. 61d it is determined that $\log v = -0.36$, i.e. $v = 0.437$. This indicates that the equilibrium concentration

$y = 4.37 \times 10^{-4}$ and the normalized equilibrium concentration $x/x_2^G = (2/2) \times 0.5 \times (0.84 - 0.44) = 0.20$, i.e. the equilibrium concentration $x = 4 \times 10^{-4}$. The distribution ratio of the A_1 component (both bound in A_3 and free) is $y/x = 4.37 \times 10^{-4}/4 \times 10^{-4} = 1.09$.

Conversely, if the equilibrium concentration $y = 8 \times 10^{-4}$ is achieved, the appropriate concentration of reactant A_1 can be obtained as follows: The value of y is normalized to the value of $v = 8 \times 10^{-4}/(0.5 \times 2 \times 10^{-3}) = 0.80$ and reached (according to the isotherm of Fig. 61d at $q = 1.11$, i.e. about an 11% stoichiometric excess of reactant A_1 is needed. The equilibrium value $x = 0.5(1.11 - 0.80)2 \times 10^{-3} = 3.1 \times 10^{-4}$ can be found according to eqn. 2.93, at which time the distribution ratio will increase to $y/x = 8 \times 10^{-4}/3 \times 10^{-4} = 2.67$ as compared with the previous example.

A semi-empirical calculation approach was used to determine the changes in the distribution ratio as a function of the system's quantitative composition and applied to the separative-analytical method of **concentration-dependent distribution (CDD)** in liquid-solid and liquid-liquid systems (*M. Kyrs, 1965, J. Tőlgyessy, M. Kyrs, 1989*). The distribution ratio D is related to the ratio D_0 in a reference system.

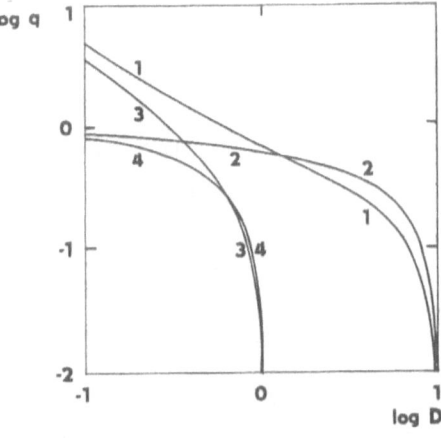

Figure 62. Inverse function of concentration dependent distribution: stoichiometric ratio q calculated as a function of distribution ratio D: curve 1: $D_0 = 10$, $\langle n_{II} \rangle = \Delta n = 3$, curve 2: $D_0 = 10$, $\langle n_{II} \rangle = 3$, $\Delta n = 2$, curve 3; as curve 1 for $D_0 = 1$, curve 4: as curve 2 for $D_0 = 1$.

For example, for the case of liquid-liquid extraction of complexes ML_n, which arise from the reaction of metal M and ligand L, $M + nL \to ML_n$, and where its distribution obeys eqn. 1.346, D was derived as a function of the system composition expressed through the stoichiometric ratio

$$q = \frac{nc_M}{rc_L} \qquad (2.94a)$$

(see Fig. 10), which is a particular case of the normalized parameter, eqn.2.92c. The system with a large excess of reagent L ($q \approx 0$) was chosen as the reference, so that

$$\left[1 - \left(\frac{D}{D_0}\right)^{\frac{1}{n}}\right]\frac{1+rD}{rD} = q \qquad (2.94b)$$

This indicates that D can be found from the inverse function $q = q(D)$ (Fig. 62). It is remarkable that no explicit equilibrium constants are necessary for the calculation: they are hidden in the value of D_0 estimated at a similar composition and under other thermodynamic conditions. When function D is linearized in the vicinity of D_0 through a Taylor series, and with respect to the possible existence of various complexes in phase I and phase II as characterized by eqn. 2.82a, the result is

$$\log D = \log D_0 + (\langle n_{II}\rangle - \langle n_I\rangle)(\log [L_0] - \log [L]) \qquad (2.95)$$

Substituting for concentration [L] in the material balance equation 1.356 results in the more general equation (*F. Macasek, 1974*) including possible **decrease of method sensitivity**:

$$\left[1 - \left(\frac{D}{D_0}\right)^{\frac{1}{\Delta n}}\right]\frac{n(1+rD)}{\langle n_{II}\rangle(1+rD)-\Delta n} = q \qquad (2.96)$$

where Δn is the difference of the average ligand numbers between phase II and phase I (eqn. 2.82a). Eqn. 2.94 is the same as eqn.

2.96 when there is no complexation in phase I, $\langle n_1 \rangle = 0$, and in phase II only one complex exists, $\langle n_{II} \rangle = n$, i.e. $\Delta n = \langle n_I \rangle = n$.

The general solution of the equations with many reactants corresponds to the scheme discussed in part 1.2.3., eqn. 1.146; however the absence of reliable equilibrium constants and the non-ideally of real systems should not be overlooked. To determine the appropriate mathematical treatment, large universal programs such as HALTAFALL (*N. Ingri et al., 1967; F. Gaizer, 1979; M. Meloun, J. Havel, E. Hogfeldt, 1988*) have been written. In particular cases, a specific solution for the single-equilibrium eqn. 2.92 can be obtained. For example, the free ligand A_2 concentration can be found when A_3 represents the complex ML_n in extraction equilibria: $A_1 \equiv M$, $A_2 \equiv L$, $A_3 \equiv ML_n$; $v_1 \equiv 1$, $v_2 \equiv n$; $F \equiv W$; $G \equiv P$; $K^H \equiv \beta_n^H$; $\alpha_1 = 1$; $\alpha_2 = 1/\Pi_t$ (for the side-protonation reaction of the ligand, $L + iH^+$).

The equilibrium value of the **free ligand concentration** can be found from the resulting equation

$$v^{n+1} + v^n(q-1) + v - u = 0 \qquad (2.97)$$

where $u, v,$ and q are normalized coordinates

$$v = [L]\sqrt[n]{\beta_n^W} \qquad (2.98a)$$

$$u = \frac{rc_L}{\Pi}\sqrt[n]{\beta_n^W} \qquad (2.98b)$$

where β_n^W is the weighted biphasic stability constant, eqn. 1.349. The polynomial, eqn. 2.97, can be solved by an iteration algorithm and normalized values are also tabulated for systems with $n \leq 4$ and the stoichiometric ratios q met in practice (*F. Macasek, D. Vanco, 1981; F. Macasek, J. Klas, 1993*).

The dependence of the distribution ratio on the chemical composition of the inlets is rather important in controlling the operation of the separation unit. The greater the values of $D/\partial x^F$, $\partial D/\partial \alpha$, etc., the more sensitive is the regulation of the separation process (see section 2.2.2.). As for the yields of separation, changes in large distribution ratios are irrelevant, because the yield varies within a fraction of a percent; of crucial importance to the product purity are changes in the distributions of impurities. From this standpoint, the systems which possess some **buffering properties** are important.

Least convenient for control are systems with high equilibrium constants at stoichiometric equivalence, when $Fv_2 x_1^F = Gv_1 x_2^G$ — see eqn. 1.136 — and equilibrium concentrations of reactants which are very sensitive to changes in feed concentration. For example, a change of q in systems with the chelate ML_2, characterized by eqn. 2.97, within the ratio 0.98-1.02 (± 2 %) may lead to a change of equilibrium ligand concentration [L] of more than three orders of magnitude, so that distribution of n-valent minor elements may be changed of the order of 10^{3n}! (*F. Macasek, D. Vanco, 1982*).

Conversely, the operation at a stoichiometric (equivalence) deficit of the reagent, abbreviated as **substoichiometric** (*J. Ruzicka, J. Stary, 1963*) or **subequivalence** (*J. Klas, J. Tolgyessy, J. Lesny, 1977*) separation systems, in the case of metal-ligand equilibria are called **ligand buffers** (*M. Tanaka, 1963*). The limiting, buffering value, $[L]_{buf}$ of the free ligand concentration is reached at high *pH* and can be derived from eqn. 2.97 when $u(q-1) \gg v$:

$$[L]_{buf} = (q-1)\sqrt[n]{\beta_n^W} \qquad (2.99)$$

Hence, the **buffering properties of substoichiometric systems** ($q > 1$) can be used successfully in analytical practice when the separation of constant amounts of metal must be ensured (*N. Suzuki, 1959; J. Ruzicka, J. Stary, 1963*), if the antibody is determined after its reaction with substoichiometric amounts of antigen by measuring the distribution of free and bound antigen (*G. Scatchard, 1949; A. Berson, R.S. Yalow, 1959; R.S. Yalow, 1960; K.O. Smith et al., 1977*), or when the operational stability of industrial chelate extraction units was designed (*V. Bumbalek, V. Horak, J. Haman, 1985*). When the high enrichment factors in these systems, due to moderate distribution ratios of the major component, are taken into account —see part 2.1. — it becomes clear that for optimal separation of component A_1 it is unnecessary to have an excess of A_2, but rather the contrary.

2.2.2 Actual and measured parameters; cybernetic principles

The practical performance of separation units takes place under actual conditions when fluctuations and changes in flow, composition and heat exchange occur. These changes act as perturbations that exert certain influences on the output parameters and therefore on the efficiency of the separation process as well. The separation unit represents a particular case of a system transforming an input function (pressure, concentration, *pH*, temperature, flow, etc.) $U(t)$ at time t to a function $Y(t)$ on the output — Fig. 63.

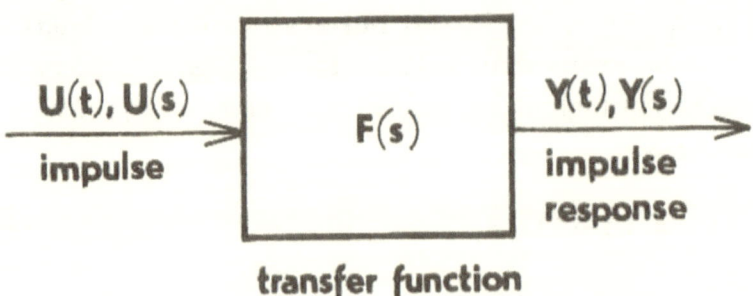

Figure 63. Transformation of input impulse to impulse response in a separation unit possessing in Laplace transform a transfer function $F(s)$.

Initially such general questions arise as:
(i) how does the form of the transfer function depend on the physical parameters of the separation unit, and
(ii) what operations must be applied to stabilize the outlet parameters of the separation unit?

Both of these problems are solved by means of **chemical cybernetics** (*D.R. Coughanowr, 1964; V. V. Kafarov, 1976*).

Control of the separation unit is ensured according to the general network of **automatic regulation** with a **feedback** — Fig. 64. An output parameter (flow, concentration, pressure, temperature, etc.) is measured by an analytical device, consisting of a sensor and a converter (*S. Siggia, 1959; W.R. Seitz, 1984; C. Nylander, 1985; J. Janata, R.J. Huker, 1985*). The response y_p of the sensor to the outlet y is sent to a converter (transducer) that returns the suitably formed signal y_m. The latter is compared with a set-point variable w in a comparator and in the controller or computer, the signal difference is transformed to an **advice impulse** that executes a correcting action on the input u (manipulated variable) by means of an active device: a valve positioner, doser, mixer, heater, etc. Such control is of great importance, especially for a separation device operating in a continuous regime to operate efficiently, because the less the deviations from equilibrium (steady state), the greater the thermodynamic efficiency of the separation (see e.g. equation 2.36).

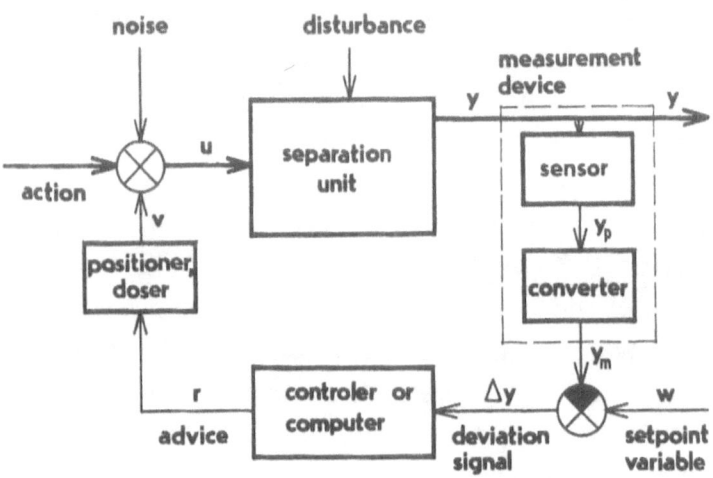

Figure 64. Signal processing and regulation of separation unit: u — manipulated (controlling) variable, y – process (resulting) variable.

In section 2.2.1. it was shown that output composition is mostly a complicated, non-linear function of the input composition. Hence it is desirable to linearize the transfer functions; a universal approach uses Taylor series linear terms in the vicinity of a standard state of the function $Y(u_1, u_2, \ldots, u_n)$. Both equilibrium or steady-state values (u_{is}) are well suited for the standard state. Therefore,

$$Y(u_1, u_2, \ldots, u_n) = Y_s + \sum_i^n \left(\frac{\partial Y}{\partial u_i}\right)(u_i - u_{is}) \qquad (2.100a)$$

where

$$Y_s = Y(u_{1s}, u_{2s}, \ldots, u_{ns}) \qquad (2.100b)$$

is the **response function** at the standard state. The differences

$$u_i - u_{is} \equiv \vartheta_i \qquad (2.100c)$$

are the appropriate **difference variables**, which in separation science are identical to the driving forces, eqn. 2.1. Because $d\vartheta_i = du_i$, function 2.100 can be rewritten as

$$\vartheta_Y = Y - Y_s = \sum_i^n \left(\frac{\partial Y}{\partial \vartheta_i}\right)_s \vartheta_i \qquad (2.101)$$

Because the differentials, i.e. the values $\partial Y/\partial \vartheta_i$ in the steady state, are some constants, both Y and θ_Y become a linear function of the difference variables ϑ_i. Each ϑ_i introduces its specific contribution (its weight being determined by $\partial Y/\partial \vartheta_i$) to the output value Y. Generally, each variable ϑ depends upon time and may have various characteristics (short impulse, jump, waves, etc.).

Let us see, for example, how the simplest dynamic separation unit (Fig. 58a) will respond to a change in chemical composition, i.e. the concentration in feed, x^F, if in particular the unit has a finite size and the volume of the product fraction retained in the unit is V. If the balance, eqn. 2.48, is identical with the function $Y = 0$, then $\vartheta_Y = 0$. The flows F, W, and $\dot{P}$ can be considered constant. According to eqn. 2.101, for the response function Y the result is

$$\vartheta_Y = \dot{F}\vartheta_x + \dot{W}\vartheta_x + \dot{P}\vartheta_y + \vartheta_n = 0 \qquad (2.102)$$

Considering the balance of flows, eqn. 2.47, $\dot{F} = -(\dot{W} + \dot{P})$, and according 2.2, the interphase fluxes are equal,

$$K_x \vartheta_x + K_y \vartheta_y = 0 \qquad (2.103)$$

i.e.

$$\vartheta_x = -K\vartheta_y \qquad (2.104a)$$

where

$$K \equiv K_y/K_x \qquad (2.104b)$$

—see eqn. 1.529. For the sake of simplicity, $\theta_{sw}m$ be given as the dimensionless differences of the mole fractions; further, the total molalities c_ℓ will be included in the corresponding constants and the K's may have the dimensions mol s^{-1} or m^2 mol^{-1}s^{-1}. The difference parameter of the accumulation factor $\theta_{\dot{n}}$ is simply

$$\vartheta_{\dot{n}} = \dot{n} \qquad (2.105)$$

because in the steady state $\dot{n}_s = 0$. The amount accumulated n on ideal mixing is connected with volume V and the total molarity of product c_t^P as $n = Vc = V \cdot y \cdot c_t^P$, and therefore

$$\dot{n} = c_t^P V \frac{dy}{dt} = c_t^P V \frac{d\vartheta_y}{dt} \qquad (2.106)$$

Eqn. 2.102 is transformed into the **standard form** (with a unit coefficient in ϑ_y) of the **response function equation**:

$$\tau \frac{d\theta_y(t)}{dt} + \vartheta_y(t) = K_s \vartheta_x^F(t) \qquad (2.107a)$$

where

$$\tau = \frac{c_t^P V}{\dot{P} + \dot{W}K} \qquad (2.107b)$$

represents the **relaxation time** (s) and

$$K_s = \frac{\dot{P} + \dot{W}}{\dot{P} + \dot{W} + K} \qquad (2.107c)$$

is the **static amplifying constant** (dimensionless). Because the equation contains the first derivation of ϑ_y, the system is classified as a first-order system.

Although integration of eqn. 2.107 does not present any difficulties, a more general solution is found — cf. eqn. 1.432 — by means of the Laplace transformation (R.V. Churchill, 1950; V.A. Ditkin, A.P. Prudnikov, 1965),

$$L[\vartheta(t)] = \vartheta(s) \tag{2.108}$$

If the change of concentration at input, ϑ_x^F is realized as a **Dirac impulse (shock)** $\vartheta_x^F(t) = \vartheta_x^F \cdot \delta(t)$, i.e. an extremely sharp change (Fig. 65a), its Laplace transform is unity, $[\delta(t)] = 1(s)$, and eqn. 2.107a gives

$$\tau s \vartheta_y(s) + \vartheta_y(s) = K_s \vartheta_x^F \cdot 1(s) \tag{2.109}$$

and

$$\vartheta_y(s) = \frac{K_s}{\tau s + 1} \vartheta_x^F \tag{2.110a}$$

or

$$\vartheta_y(s) = \frac{K_s}{\tau} \frac{1}{s + 1/\tau} \vartheta_x^F \tag{2.110b}$$

The original of the Laplace transform, eqn. 2.110 for the Dirac impulse is

$$\theta_y(t) = \frac{K_s \vartheta_x^F}{\tau} e^{-t/\tau} \tag{2.111}$$

At a **unit concentration jump**, ϑ_x^F, when $L[1(f)] = 1/s$, eqn. 2.107 takes the form:

$$\tau s \vartheta_y(s) + \vartheta_y(s) = K_s \vartheta_x^F \frac{1}{s} \tag{2.112}$$

which gives

$$\vartheta_y(s) = \frac{K}{s(\tau s+1)} \vartheta_x^F \qquad (2.113a)$$

or

$$\vartheta_y(s) = K_s \vartheta_x^F \left[\frac{1}{s} - \frac{1}{s+1/\tau}\right] \qquad (2.113b)$$

The original of this image can be found in Laplace transform tables as

$$\vartheta_y(t) = K_s \vartheta_x^F (1 - e^{-t/\tau}) \qquad (2.114)$$

This signifies that if there is a sharp concentration pulse in the input of the separation unit, the output concentration changes gradually in time, according to the response function, examples of which are represented by eqns. 2.111 and 2.114 (Fig. 65). A new equilibrium (steady-state) value is generally reached after a time of $3 - 4\tau$.

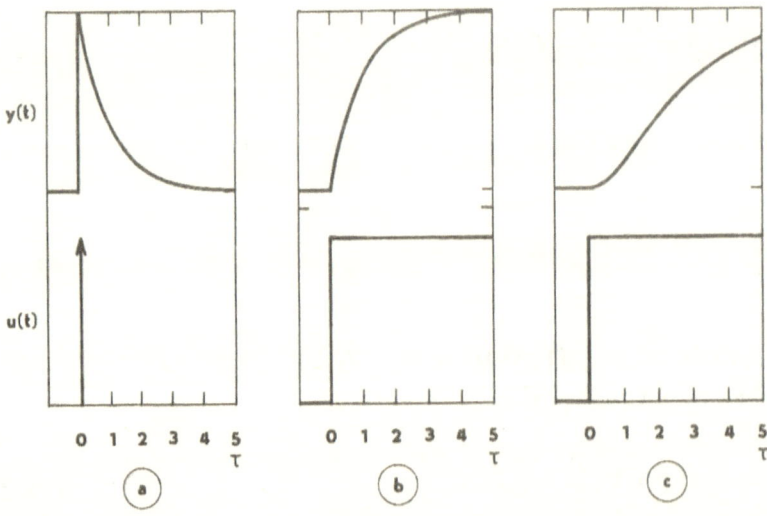

Figure 65. Concentration impulse response $y(t)$ as a function of relaxation time τ for a) infinitely short (Dirac) concentration impulse, b) concentration jump, c) concentration jump in series of units, $\tau_2 = \tau_1/2$.

The fraction $K_\varepsilon/(\tau s + 1)$, which appears before ϑ_x in eqn. 2.110a, plays the role of the **transfer function** between impulse ϑ_x and its response ϑ_y. Generaly, the transfer function is defined as the ratio of the impulse response and the impulse itself, both presented in the form of a Laplace transform image:

$$F(s) = \frac{Y(s)}{U(s)} \tag{2.115}$$

The convenience of such a definition consists in the easy application of the **superposition principle**: when impulse $U(s)$ is a linear combination of partial impulses $U_i(s)$, including noise and disturbances, we have

$$U(s) = \sum a_i U_i(s) \tag{2.116}$$

which has a response

$$Y(s) = F(s) \sum a_i U_i(s) \tag{2.117}$$

Further, if the systems are structured as a series (e.g. cascades of separation units, see section 2.4.1., or an in-line sensor of the separation unit parameter) where an impulse input of the $(i + 1)$ th unit is identical with the response output of the i-th unit,

$$F(s) = \frac{Y_n(s)}{U(s)} = F_1(s)F_2(s) \dots F_i(s) \dots F_n(s) \tag{2.118a}$$

This again underlines the expediency of using Laplace transforms in describing the system's dynamics. For instance, if

$$F_1(s) = \frac{K_1}{\tau_1 s + 1} \tag{2.118b}$$

and

$$F_2(s) = \frac{K_2}{\tau_2 s + 1} \tag{2.118c}$$

($\tau_1 \neq \tau_2$) and the image of the concentration pulse is

$$U(s) = \vartheta^F/s \tag{2.118d}$$

then the resulting response will be, according to eqn. 2.118a

$$Y_2(s) = \frac{\vartheta^F}{\vartheta} \frac{K_1}{\tau_1 s+1} \frac{K_2}{\tau_2 s+1} \tag{2.119a}$$

or, after decomposition to the elementary fractions (if $\tau_1 \neq \tau_2$):

$$Y(s) = \vartheta^F K_1 K_2 \left(\frac{1}{s} - \frac{\tau_1^2}{\tau_1 - \tau_2}\frac{1}{\tau_1 s+1} + \frac{\tau_2^2}{\tau_1 - \tau_2}\frac{1}{\tau_2 s+1}\right) \tag{2.119b}$$

Thus, there is a resulting response on the time scale, according to the Laplace transform:

$$\vartheta(t) = \vartheta^F K_1 K_2 \left[1 - \frac{\tau_1}{\tau_1 - \tau_2}e^{-t/\tau_1} + \frac{\tau_2}{\tau_1 - \tau_2}e^{-t/\tau_2}\right] \tag{2.120}$$

This is illustrated by Fig. 65c in the case when the relaxation time of unit 2 is one half of that of unit 1 ($\tau_2 = \tau_1/2$).

Pulse-response analysis can be used to characterize both separation devices and their mixture behavior, for example in countercurrent extraction cascades (*A.M. Rosen, 1958; V.V. Kafarov et al. 1967*) pulse-chromatographic and tracer-chromatographic techniques (*S.H. Hyun, R.P. Danner, 1982*).

Disturbances, unless they are unavoidable e.g. before a steady-state operation is achieved, must be avoided because they complicate both the operation and evaluation of the separation process.

On the other hand, a smooth course of chromatographic eluent concentration in time is the result of short relaxation constants which are achieved by a suitable mixing regime in multi-component gradient-forming devices at the input of chromatographic columns (*R.S. Alm, 1952; R.S. Alm, R.J.P. Williams,*

A. Tiselius, 1952; C. Liteanu, S. Gocan, 1974). Pulsing of the flow in high-pressure, high-performance chromatography might cause unwanted deflections of the detector and a loss of resolution (O. Mikes, 1979).

When the response curve represents a measurement of some parameter of a separation device, the transfer function provides the relation between the actual and measured values of an evaluated parameter. The relaxation properties of both separation device and sensor are included in the resulting transient function. Analytical chemical sensor systems are usually of the first order and often have a sample transport delay (t_d) that is caused by mechanical (hydrodynamical) transport, chemical treatment of samples, etc. Because the Laplace transform results in $L[t - t_d] = \exp(-t_d s)$, the **transient characteristics of sensors** are usually of the type

$$F(s) = \frac{Y_m(s)}{Y(s)} = \frac{Ae^{-st_d}}{\tau s + 1} \qquad (2.121)$$

or, for a unit concentration jump $L[1(t)] = 1/s$, particularly

$$Y_m(t) = K_s(1 - e^{-t/\tau})(t - t_d) \qquad (2.122)$$

The above relations illustrate the importance of response characteristics both for separation devices and their analytical sensors, because under higher momentum the predictability of the separation process and timely regulation of the input parameters, which is based on incomplete and delayed response data, is a complicated task. Modern controlling systems include a computer element between the sensor and controlling device in a closed loop (S.C. Gates, J. Becker, 1989). A computer receives measured data, evaluates them, makes optimization calculations and then either resets the parameters of the regulators — **on-line digital supervisory control (DSC)** or directly gives commands to a controller — **on-line digital direct control (DDC)** *(L.C. Smith, 1972; P.G. Barker, 1983; M.S. Nardone, 1999)*.

2.3 MANY-FOLD SEPARATION

A separation unit can rarely fulfil the demands of the separation process, which is to obtain more-or-less pure components of mixtures. In fact, in a unit with one input and two outputs, the complete separation of a binary mixture is achieved only at an infinite separation factor, and this is possible only during extreme and selective intermolecular interactions of one of the component with one of the phase matrices.

However, each output of the separation unit may serve as the feed of the next unit to

(i) increase the yield of product, reprocessing the waste output, or

(ii) improve the purity of product, subjugating the product output to an additional separation process, or

(iii) achieve separation of both outputs for all available components,

and therefore the separation is repeated in the same (or the same type of) separation unit or in another one *(E.J. Henley, J.D. Seader, 1981)*. If the number of components (k) is above 2, the number of connection modes to achieve their separation using identical separation units with one feed and two outputs is

$$N = \frac{[2(k-1)!}{k!(k-1)!} \qquad (2.123)$$

and if u various types of unit are used,

$$N = \frac{[2(k-1)]!}{k!(k-1)!} u^{k-1} \qquad (2.124)$$

The number of combinations increases rapidly for a multicomponent system ($k > 4$). For example, if $u = 2$ unit types are to be used for the separation of a six-component mixture ($k = 6$), they can be inter-connected by $N = 1344$ combinations. Therefore, it is necessary to use the system approach to analyze and design the separation process as a whole (*F.B. Petlyiuk et al., 1965; J.E. Hendry, R.R. Hughes, 1972; R.W. Thomson, C.J. King, 1972; P.L. Thibaut Brian, 1972; D.F. Rudd, G.J. Powers, J.J. Siirola, 1973*). There still exists a lack of analytical approaches for complicated systems. Initially, a mixture usually consists of easily-separable groups and the processes with a low fractionation capacity and high productivity (phase separation, etc.) are useful. On terminal separation, the components with only slightly differing properties usually remain and techniques with higher fractionation capacity are preferred — see section 1.4.5. Most general **heuristic rules** for industrial design, and also for small-scale separation, contain the following recommendations for the optimal operation of the separation process:

(a) direct, one-by-one separation of the pure components,
(b) the first steps involve separation of the prevailing, hazardous or corrosive components,
(c) finally, difficult-to-separate components are separated.

For preparative scale one is recommended as follows (*J.M. Douglas, 1988; R.M. Kelly, 1987; Y.A. Liu, 1987*):
- always generate more than one alternative separation method or sequence,
- favor separation using energy separation agents (physical separation: distillation or crystallization, electrophoresis, centrifugation),
- prefer process with minimum of separating agent (solvent, carrier gas etc.),
- avoid solids and if they are present remove them first,
- favor schemes which give the smallest output sets (components of the same product or waste should not be separated).

Some examples of optimization will be discussed in section 2.5. They can be conferred as chemical process steady state simulation and mathematical optimization techniques. Many problems can be overcome with a good choice of a non-linear mathematical program (*M.J.D. Powell, 1964; A.I. Boyarinov, V.V. Kafarov, 1975; L.T.Biegler, 1985; D.M. Himmelblau, 1972*).

Multi-stage separation is performed by
I. a sequence of separation steps via the connection of separation units which does not depend on the time factor, i.e. a **batch** or **periodical separation**, or by
II. a series of inter-connected separation units — **continuous separation**.

Batch (periodical) separation is often used in laboratory analytical procedures but continuous separation is usually preferred for pilot preparatory purposes and in industrial processes (*A. Weissberger, 1951; E. Cremer, T. Davies, 1956*).

2.3.1 Stage separation

The classical problem of manifold separation is well known: if one wants to increase the overall yield in the separation process which is performed in the separation unit by the stage cut and results in the product amount P_1, which is better? (i) to double the stage cut, or (ii) to repeat the same separation with the waste. In both cases, the total amount of product is the same ($= 2P_1$) but the efficiency is determined by the resulting composition of the product fraction(s). Generally, the problem is solved by manifold **separation with transfer of one phase** (*E.L. Smith, 1928; A. Rose, 1941*) (Fig. 66). The waste is conditionally designed as the transferring phase so that the feed of the i-th unit consists of the waste from the previous unit:

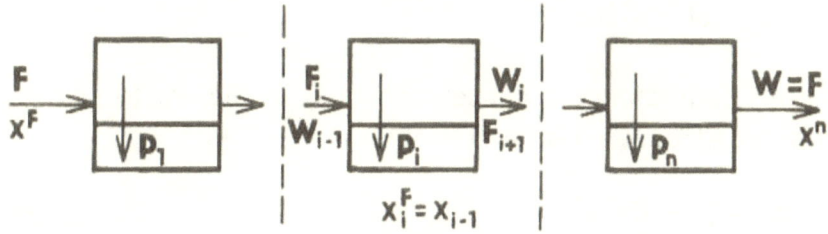

Figure 66. Multi-stage separation with transfer of one phase (flow $F \equiv W$).

$$F_i = W_{i-1} \tag{2.125}$$

and the material balance is

$$W_{i-1}x_{i-1} = F_i x_{i-1} = P_i y_i + F_i x_i \tag{2.126}$$

When the recovery at the i-th stage is

$$R_i = \frac{P_i y_i}{F_i x_{i-1}} \tag{2.127}$$

and there are no losses in the flow, i.e.

$$F_i = \text{const} \equiv F \tag{2.128}$$

Substitution into eqn. 2.126 gives

$$x_i = (1 - R_i)x_{i-1} \tag{2.129}$$

If the yield is constant at each stage, i.e.

$$R_i = \text{const} \equiv R \tag{2.130}$$

the series, eqn. 2.128 is

$$x_1 = (1 - R)x^F \tag{2.131a}$$

$$x_2 = (1-R)x_1 = (1-R)^2 x^F \qquad (2.131b)$$

and by induction, the concentration x_1 in the waste fraction emerging from the n-th unit is obtainable

$$x_n = (1-R)^n x^F \qquad (2.132)$$

The **total yield of n-fold separation** resulting from a difference in the input and output concentrations in stream W, is therefore

$$R_n = 1 - \frac{x_n}{x^F} \qquad (2.133)$$

With respect to eqns. 2.131 and 2.13, the yield expressed through the one-step yield or distribution ratio D is

$$R_n = 1 - (1-R)^n = 1 - \frac{1}{(1+P_i D/F)^n} \qquad (2.134)$$

From eqns. 2.132 or 2.134, an evaluation can be made — through various parameters, see section 2.1.1. — that the number (n) of stages necessary to obtain a given yield R_n or to decrease the concentration in feed from x down to x_n:

$$n = \frac{\log(x_n/x^F)}{\log(1-R)} = \frac{\log(x^F/x_n)}{\log(1+D_m)} = \frac{\log(1-R_n)}{\log(1-R)} \qquad (2.135)$$

The dependence of the yields on the stage cut can now be clarified more precisely. The yield of the single separation procedure can be expressed through the total amounts P and W, as follows,

$$R_1 = 1 - \frac{1}{1+PD/F} \qquad (2.136)$$

and by splitting the amount P of the product phase into n parts and performing an n-fold separation, the final total yield becomes, in accordance with eqn. 2.134,

$$R_n = 1 - \left(\frac{1}{1+PD/nF}\right)^n = 1 - \left(\frac{1}{1+D_m/n}\right)^n \qquad (2.137)$$

where D_m is the mass distribution ratio (eqn. 2.15) for one-unit operation. The conclusion is that the yield R_1 is always less than R_n, because $(1 + PD/nF)^n > 1 + PD/F$. Further, on the division of the phase P into infinitesimal portions, $n \to \infty$, (equivalent to differential separation by flow F — see eqns. 2.69–2.78), $\lim (1 + a/n)^n = e^a$ and the **maximal yield of multifold separation** becomes

$$R_\infty = 1 - \exp(-D_m) \qquad (2.138)$$

For example, if a single operation is realized by the mass distribution ratio $D_m = 1$, the yield is $R_1 = 0.5$ (50%) while dividing phase P into many parts and performing multifold separation, the maximum yield of $R_\infty = 1 - 1/2.718 = 0.63$ (63%) is reached by expenditure of the same total material.

Unfortunately, in practice such contact of the total feed transfer with very small parts of the fresh second phase is physically hard to perform.

Eqn. 2.135 is useful for evaluating the composition of multicomponent systems, when $R_B \neq 1 - R_A$, because it can be applied for both A and B components simultaneously. The **overall enrichment factor** in n-fold separation is

$$S_n = \frac{R_{nA}}{R_{nB}} = \frac{1-(1-R_A)^n}{1-(1-R_B)^n} \qquad (2.139)$$

Evidence can be given that the over-all separation factor is always less than that for the single operation; however, the latter is less favorable for small yields and the total effect should be considered. In any case, manifold separation is used generally for higher yields, and when the product purity plays a minor role.

The situation is reversed, however, **when phase P is disposed as concentrating impurities**, e.g. filtrates from filter washing solutions. The yield is considered with regard to the composition of the transferred flow, i.e. by calculating the fraction in the waste, $R' = 1 - R$; and because the operating equation 2.132 is valid, the fraction of any component retained after passage of n portions of the exhausted phase is

$$R'_n = (R'_1)^n \qquad (2.140)$$

where R' is the fraction retained after a single procedure. The enrichment factor for component B in fraction W is

$$S'_n = \frac{R'_{nB}}{R'_{nA}} = \left(\frac{R'_B}{R'_A}\right)^n \qquad (2.141)$$

Thus the number of necessary operations can be derived from eqn. 2.141 as

$$n = \frac{\log S'_n}{\log R'_B - \log R'_B} \qquad (2.142)$$

For example, when a separation unit gives yields $R_A = 0.50$ and $R_B = 0.25$, for 99% yield of A, it is necessary, according to eqn. 2.125, to perform $n = \log (1 - 0.99)/ \log (1 - 0.5) = 6.7 \cong 7$ separations. Because simultaneously the yield of B is $R_{nB} = [1 - (1 - 0.25)^7] = 0.866$ (86.6%), the enrichment factor decreases from $S_1 = 2.0$ to $S_7 = [1 - (1 - 0.5)]^7/[1 - (1 - 0.25)]^7 = 1.14$. Noteworthy is the fact that even such an enrichment factor is still higher than that in a system where both distribution ratios are much higher, e.g. when $R_A = 0.99$ and $R_B = 0.97$ for a single step — see section 2.1. If multifold separation is used to decrease the content of impurity A in the waste fraction, say for two orders of magnitude ($S' = 100$), the goal is achieved in $n = \log 100/(\log 0.75 - \log 0.5) = 11.4 \cong 12$ steps. This is achieved, however, at the cost of a very low yield of purified B in the "waste", namely $R' = 0.75^{12} = 0.001$ (0.1%).

Therefore, multifold separation makes it possible to obtain
(i) a less pure component (A) in higher yield than by single separation; and, simultaneously,
(ii) a small fraction of the pure second component (B) in the waste (transferred phase).

There is evidence that the flexibility of manifold separations achieves better separation for the same amount of chemicals than during a single operation.

If the separation isotherm is non-linear, i.e. yield R_i is not constant at each stage, the McCabe-Thiele diagrams (Fig. 67) should be used step-wise: the product serves as the initial point for establishing the next unit distribution (*A.J.P. Martin, R.L.M. Synge, 1941*).

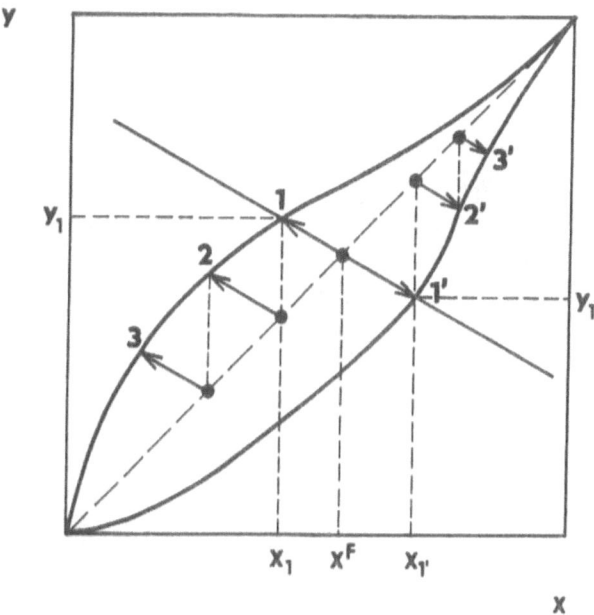

Figure 67. McCabe-Thiele diagram for multifold separation for enriching the product phase (stages 1, 2, 3...) or exhausting it (stages 1', 2', 3' ...).

It can be seen in the diagrams that the uniform splitting of the receiving phase into equal fractions is evident only from linear isotherms, and otherwise an optimal cut should be determined with respect to the isotherm function $= f(x)$. We shall demonstrate such an approach in the latter case when it becomes necessary to decrease concentration from x_i to x_{i+2} in the two steps ($i+1$ and $i+2$) by a minimal total amount of phase P divided into portions P_{i+1} and P_{i+2}; the goal of the optimization consists of finding the proper proportion between the portions. Because x_{i+1} is the output concentration of an earlier step (step i) and at the same time the input concentration for the next step (step $i+2$), the condition of a minimum amount of phase P at fixed $x_i(y_j)$ and $x_{i+2}(y_{i+2})$, and x_{i+1}, y_{i+1} variables, must fulfil the equation

$$\frac{d(P_{i+1}+P_{i+2})}{dx_{i+1}} = 0 \tag{2.143}$$

From material balances of type eqn. 2.32,

$$P_{i+1}y_{i+1} = W_i x_i - W_{i+1} x_{i+1} \tag{2.144a}$$

i.e. in particular, when $W_i = W = \text{const}$,

$$\frac{P_{i+1}}{W} = \frac{x_i - x_{i+1}}{y_{i+1}} \tag{2.144b}$$

and

$$\frac{P_{i+2}}{W} = \frac{x_{i+1} - x_{i+2}}{y_{i+2}} \tag{2.144c}$$

or,

$$P_{i+1} + P_{i+2} = W\left(\frac{x_i - x_{i+1}}{y_{i+1}} + \frac{x_{i+1} + x_{i+2}}{y_{i+2}}\right) \tag{2.145}$$

and after differentiation of eqn. 2.145,

$$0 = W\left(-\frac{1}{y_{i+1}} - \frac{x_i - x_{i+1}}{y_{i+1}^2}\frac{dy_{i+1}}{dx_{i+1}} + \frac{1}{y_{i+2}}\right) \qquad (2.146a)$$

the general **condition for optimal cuts** is obtained in the form:

$$\frac{y_{i+1}}{y_{i+2}} = \frac{x_i - x_{i+1}}{y_{i+1}}\left(\frac{dy}{dx}\right)_{x=x_{i+1}} + 1 \qquad (2.146b)$$

For example, in the particular case of an adsorption which follows the Freundlich isotherm, eqn. 1.321, $y = ax^b$ ($dy/dx = abx^{b-1}$), substitution into eqn. 2.146a leads to an implicit equation determining the equilibrium concentration after the $(i + 1)$th operation,

$$x_{i+1} = \frac{bx_i}{(x_{i+1}/x_{i+2})^b + b - 1} \qquad (2.147)$$

which can be used as an iteration formula. In particular, when $b = 1$, x_{i+1} is given as the geometrical mean, $x_{i+1} = \sqrt{x_i x_{i+2}}$, the ratio of the amounts of product remains uniform, $P_{i+1} : P_{i+2} = 1 : 1$, and does not depend on concentration (otherwise the cut is always proportional to x_i^{1-b}).

For example, if the coefficient of the Freundlich isotherm $b = 0.25$, the concentration after the i-th adsorption is $x_i = 0.15$ and the required concentration after two further steps is $x_{i+2} = 0.01$. Starting from a trial $x_{i+1} = 0.07$ in eqn. 2.147, after 4 iterations, $x_{i+1} = 0.0194$ is obtained as the optimal value in the intermediate step; if the second coefficient of isotherm, $a = 0.09$, according to eqns. 2.144 and 1.321, the optimal values $P_{i+1}/W = 3.89$ in the first step and $P_{i+2}/W = 0.331$ in the next step. Hence, the optimal phase splitting is quite asymmetrical and the amounts of sorbent in the two steps are substantially different (13:1). In the same way, the whole concentration region and an arbitrary number of operations can be designed for efficient absorption recovery of the product.

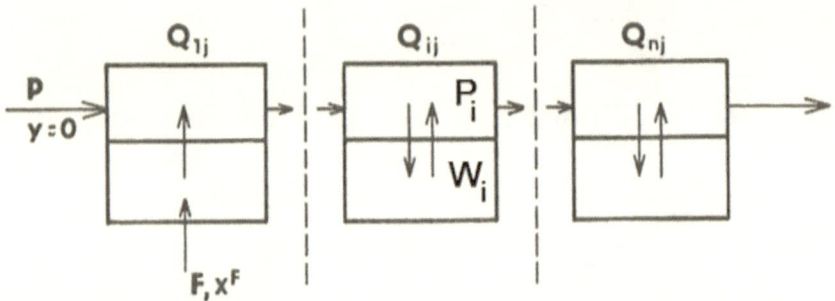

Figure 68. Craig technique of quasi-countercurrent separation with a piston-flow transfer of one phase (flow P) and enrichment of the product phase (explanation in text).

A more sophisticated mode of phase transfer was designed (*L. Craig, 1944*) to enhance both the yield and purity in a **quasi-countercurrent separation** arrangement (see section 2.4.) with transfer of the concentration impulse. The feed is introduced batch-wise as phase W in the first separation unit; phase P is transferred as a "piston-flow" (Fig. 68). To simplify calculation, the total amount in the unit will be lettered as Q's with as subscript the corresponding unit number i:

$$P_i y_i + W_i x_i = Q_i \qquad (2.148)$$

Performed operation steps will be indexed by the subscript j, i.e. $Q_{i,j}$ means amount Q in unit i and step j. Therefore, the feed is $Q_{1,0}$ ($W_{1,0} \equiv y_1 = 0$)

$$F x^F = W_{1,0} x^F = Q_{1,0} \qquad (2.149)$$

After the first portion of phase P, which contains no separated substances, passes through unit 1, the distribution in the unit with yield R in phase P results in the amount $Q_{1,1}$ remaining in the unit

$$Q_{1,1} = (1-R)Q_{1,0} \tag{2.150}$$

and the product transferred into unit 2 by the same step 1 is

$$Q_{2,1} = RQ_{1,0} \tag{2.151}$$

If yield R remain constant while repeating the separation with transfer of products and supplying a fresh phase P in unit 1, the balance at each step j is

$$Q_{i,j} = \quad (1-R)Q_{i,j-1} \quad + \quad RQ_{i-1,j-1} \tag{2.152}$$

a rest in i-th unit   a transformer from the $(i-1)$ th unit

Direct solution of sets of recurrent eqns. 2.150 is non-trivial. In the first steps the ratios Q_{ij}/Q_{10} are as follows:

Step	Stage (i)					
(j)	0	1	2	3	4	...
0	1	0	0	0	0	
1	1-R	R	0	0	0	
2	(1-R)²	2R(1-R)	R²	0	0	
3	(1-R)³	3R(1-R)	2R²(1-R)	R³	0	
...						
j	\multicolumn{6}{l}{terms of binomial $[R+(1-R)]^j$ (their sum =1) the i-th term corresponds to the i-th stage}

(actually, any of the stages can be the ultimate one, i.e. then the concentrations calculated for "further" stages are fictitious).

Using this method of contacts and transfers, even if the substance is strongly retained in phase W, to some extent the substance is transferred to the same phase in the next unit(s). For example, at a regular yield $R = 0.3$, in the first unit ($i = 1$) on the first step ($j = 1$) remains $Q_{1,1} = 1 - R = 0.7$ (70%) of substance introduced by the feed; after 3-fold separations, however, there remains only $Q_{1,3} = (1-R)^3 = 0.7^3 = 0.343$ (34%) and the main part accumulates in unit 2, $Q_{2,3} = 3R(1-R) = 0.63$ (63%).

By induction, the values for certain units at the j-th step is represented by the terms of the binomial expansion $[R-(1-R)]^j$:

$$Q_{ij} = Q_{1,0} \frac{j!}{(j-i)!\,i!} R^i (1-R)^{j-i} \tag{2.153}$$

(B. Williamson, L. C. Craig, 1947).

The greatest fraction of component at step j is located in the stationary phase of unit number

$$i_{max} = Rj \tag{2.154}$$

e.g. at $R = 0.3$ and after a 10-fold transfer ($j = 10$) the concentration culminates in unit 3 ($i_{max} = 0.3 \times 10 = 3$) and, for example in unit 5 this maximum appears after $j = R/i_{max} = 5/0.3 \cong 17$ operations.

Using the **Craig method** it is possible to achieve fractions of pure components from multicomponent mixtures. The distribution of substances A and B for $R_A = 0.5$ and $R_B = 0.25$ is illustrated in Fig. 69.

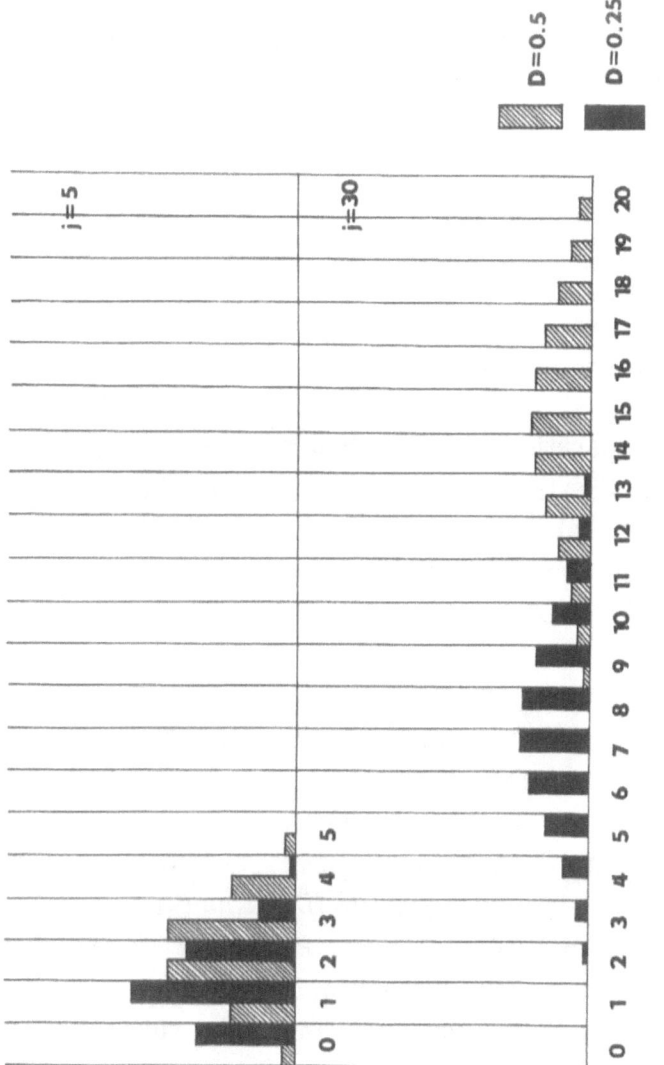

Figure 69. Distribution of substance A (full fields) and B (linked fields) for mass distribution ratios D_m^A and D_m^B after $j = 5$ and $j = 30$ steps of the Craig technique.

The transfer of one phase is best performed in liquid-liquid separation systems in a mode of repeating solvent extraction or piston-flow transfer (*E. Jantzen, 1932; L.C. Craig, 1950; F.A. Metzsch, 1953; O. Wichterle, O. Mikes, 1957; I.G. Fleetwood, 1962*). Adsorption separation units with liquid-solid systems are less appropriate for transfer operation, and their number in series does not usually exceed two in industrial flow-sheets. Gas-liquid and gas-solid systems are better accommodated for use in gas-absorption countercurrent equipment — section 2.4.

Optimization of the stage cut on double transfer of phases, "diamond patterns" can be obtained as well the theory (*M. T. Bush, P.M. Denser, 1948*).

2.3.2 Column separation

Many-fold separation has been characterized as that which proceeds in a series of real stages (absorbers, mixer-settlers, crystallizers, etc.) where each separation unit can be specified by equilibrium or steady-state characteristics. The situation is much less clear in cases where one of the phases passes through a column (tower) packed by a quasi-continuous (granulated solid, etc.) bed (*R. Billet, 1995*). In such a situation solutes in the mobile phase are carried by the flow and distributed between the stationary phase and the fluid; the question remains, which characteristics peculiar to single separation unit and possessing, for the most part, thermodynamic character, can be adopted for the time-space continuous process (*D. DeVault, 1943*).

The general features of convective transport were discussed in section 1.4.5. It will now be convenient to bridge the continuous and discrete (stage) models. The principles of chromatographic column modelling consist of their subdivision into sections labelled **theoretical stages**, i.e. units with equilibrium or pseudo-equilibrium distribution (*A.J.P. Martin, R.L.M. Synge, 1941*) or, with respect to the rate of distribution and transfer processes, the division of the column into sections called transfer units (*A.P. Colburn, 1934*).

The **height equivalent to a theoretical plate (HETP)** is that part (length) of the separation column in which the output concentrations are identical to the corresponding equilibrium in one separation unit, the theoretical plate (a plate with equilibrium concentrations at its outlet). For a column of theoretical plates, therefore, it is possible to represent the relationship between the flow leaving any plate, or flow, and the bed of the plate, by an appropriate equilibrium line. This is discussed as follows.

If the volume of the theoretical plate is V, the volume of the stationary phase is P and its average concentration is $\{y\}$, and the volume of the mobile phase is W with an average concentration of $\{x\}$, then the ratio of the average concentrations should be identical with the equilibrium distribution ratio:

$$\frac{\langle y \rangle}{\langle x \rangle} = D \tag{2.155}$$

and the material balance

$$V\bar{y} = P\langle y \rangle + W\langle x \rangle \tag{2.156}$$

where $\bar{y}$ is the average concentration of component in the volume of the theoretical plate as a whole. The ratio, eqn. 2.155, is reached only if in some cross-sections of the HETP segment the actual ratio $y/x < D$ and the opposite in others, $y/x > D$, i.e. both negative and positive deviations from the true equilibrium value occur.

In this respect the model of theoretical stages is close to the stage model with the transfer of one phase (*A.J.P. Martin, R.L.M. Synge, 1941; J.C. Giddings, 1961, 1963*). **Glueckauf's model** of a chromatographic column (*E. Glueckauf, 1946; J.I. Coates, E. Glueckauf, 1947; E. Glueckauf, 1955*) goes further because it considers the dynamics of separation by an actual bed of columns as consisting of spherical particles of the stationary phase (Fig. 70).

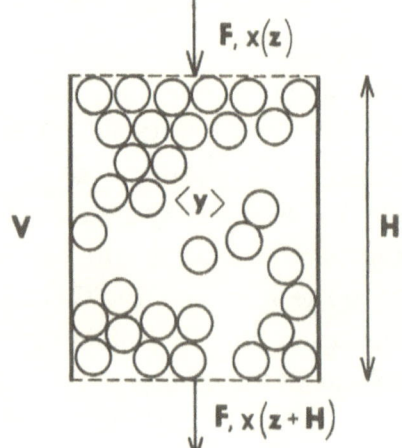

Figure 70. Scheme of theoretical stage in a packed chromatographic column.

If the column is packed with uniform, homogeneous beads of the stationary phase (sorbent or solid carrier) which are much smaller than the column diameter (the condition is fulfilled in fact if the diameter of the beads is less than one-eighth the diameter of the column), then the phase volume relationship in the column can be given by the ratio

$$\varepsilon = \frac{W}{V} = \frac{W}{W+P} = \frac{1}{1+r} \qquad (2.157)$$

known as the **free volume of the column (void fraction, total porosity of the bed)**; its value falls in the narrow range $0.38 \leq \varepsilon \leq 0.40$. Hence, the phase ratio is determined by the packing of the solid bed, $= 1.50 - 1.63$. Though the value of W (mol) in gas-solid systems depends on the pressure and temperature, the column operation should generally be controlled entirely by the chemical character and composition of the stationary and mobile phase, the **eluent**, and hydrodynamic factors (geometry, flow rate, temperature). Because the chemical composition is used to change the distribution ratio D, and the flow rate w (m s^{-1})

can be considered to be the most important dynamic factor, the plate theory for chromatographic columns was derived using these parameters.

If z is the distance coordinate along the column, and if during the time interval Δt the volume of fluid $\Delta F(m^3)$ passes through the volume element of height H then the average concentration (t) changes to $(t + \Delta t)$, the distribuend concentration $x(z)$ in the flow becomes $x(z + H)$ and the material balance of the process is

$$\Delta F x(z) + V(t) = \Delta F x(z + H) + V\bar{y}(t + \Delta t) \qquad (2.158)$$

The interval Δt can be made infinitesimal but H should have a finite value, limited at least by the size of sorbent (package) partides. Hence, the transformation of the difference equation 2.158 to a differential equation is necessary.

The volume flow $\Delta F/\Delta t (m^3 s^{-1})$ can be recalculated from the linear superficial velocity w (m s^{-1}) by means of the void cross-section of the column q (m^2):

$$\frac{\Delta F}{\Delta t} = qw \qquad (2.159)$$

and the relation between H and Δt is simply

$$H = w\Delta t \qquad (2.160)$$

As a result, the **material balance** (eqn. 2.158) **in the column segment** can be rewritten as

$$qH\left[\bar{y}(t + \Delta t) - \bar{y}(t)\right] = wq\Delta t\left[x(z) - x(z + H)\right] \qquad (2.161)$$

and at $\Delta t \to 0$ we find

$$\frac{d\bar{y}}{dt} = w\frac{x(z) - x(z + H)}{H} \qquad (2.162)$$

After expansion as a Taylor series, and using the first linear term only, the result is

$$x(z) - x(z+H) = -\frac{\partial x(z_j)}{\partial z} H \qquad (2.163)$$

where z_j refers to the interval $(z, z + H)$. If coordinate z_j is assumed to be in the middle of the interval, i.e.

$$z_j = z + \frac{H}{2} \qquad (2.164)$$

then eqn. 2.163 becomes, after a second expansion of the difference x:

$$\frac{\partial x(z_j)}{\partial z} = \frac{\partial x}{\partial z} - \frac{H}{2}\frac{\partial}{\partial z}\left(\frac{\partial x}{\partial z}\right) \qquad (2.165)$$

Now the difference-differential equation 2.162 is transformed into the differential equation

$$\frac{d\bar{y}}{dt} = -w\frac{\partial x}{\partial z} + \frac{wH}{2}\frac{\partial^2 x}{\partial z^2} \qquad (2.166)$$

which has universal validity, regardless of an equilibrium down the height H. As regards to the validity of the linear distribution isotherm, considering eqns. 2.15 and 2.157, concentration $\bar{y}$ is related to concentration x through the capacity ratio D_m and the void fraction ε as

$$\bar{y} = (D_m + 1)\varepsilon x \qquad (2.167)$$

and the **equation of linear chromatography** (at quasi-equilibrium distribution) becomes

$$\frac{dx}{dt} = -\frac{w}{(D_m+1)\varepsilon}\frac{\partial x}{\partial z} + \frac{w}{(D_m+1)\varepsilon}\frac{H}{2}\frac{\partial^2 x}{\partial z^2} \qquad (2.168)$$

Comparing this equation with the more general eqn. 1.592, derived from combined convective and interphase transport, the relationship between the parameters of movement specifies that the retention factor R_F would become

$$R_F = \frac{1}{(D_m+1)\varepsilon} \qquad (2.169)$$

—cf. eqn. 1.539 — and the effective dispersion coefficient

$$D = \frac{wH}{(D_m+1)\varepsilon} \qquad (2.170)$$

is related both to the mass distribution ratio D_m and the height H.

The effects of longitudinal mixing have been treated elsewhere (*L. Lapidus, N.R. Amundson, 1952; J.J. Van Deemter, F.J. Zwiderweg, A. Klinkenberg, 1956; J.C. Giddings, 1965*) — see section 1.4.5.

Diffusion in the solid phase is included in a more complicated way (*J.B. Rosen, 1954; E. Glueckauf, 1955; M.J.E. Golay, 1957; A.S. Said, 1981*) and generally it accounts for even more of an increase in spreading of the moving zones of the solutes. For example, diffusion in the mobile phase film adjacent to the solid (diffusion coefficient D') and diffusion in the solid beads of radius r_0 (diffusion coefficient D'') can be inserted in the value H (cm) as

$$H = kr_0 + \frac{D_m}{(D_m+1)^2\varepsilon}\frac{0.142 r_0^2 w}{D^{II}} + \left(\frac{D_m}{D_m+1}\right)^2 \frac{1.33 r_0 w \delta}{D^{I}} \qquad (2.171a)$$

where k is a constant ($k = 1.6 - 8$) depending on the size of the solid beads, and δ is the thickness of the film adjacent to the solid phase, which can be approximated at low flow rates w as

$$\delta = \frac{0.2 r_0}{1+70 r_0 w} \cong 0.2 r_0 \qquad (2.171b)$$

The capacity factor D_m can be altered by the chemical composition of the mobile and anchored phases, and an optimum is achieved when diffusion in the solid phase does not prevail over diffusion in the fluid film, i.e. when the second and third terms of the right side of eqn. 2.171a are of the same order ($D_m \cong D^{II}/D^{I}$). This is important only if the solid phase is porous or swollen, and the diffusivity D^{II} does not exceed D^{I} by more than for one order of magnitude. An optimum is reached at a Péclet number of about unity when the third term (the contribution of diffusion in the fluid film) equals approximately the first, geometrical term (the effective size of the solid phase beads, kr_0), i.e. when the flow rate is $w \cong D^{I}/r_0$ (F. Helfferich, 1959).

For large-scale chromatography a more complete optimization includes non-linearity of adsorption isotherms (G. Guiochon, S. Golshan-Shiraz, 1994; T. Gu, 1995) and economic factors, because for example the flow rate increases the cost of solvent, and the income function needs to be maximized (B.J. McCoy, 1986).

The **height equivalent to a theoretical plate** (HETP), derived from eqns. 2.170 and 2.169, is:

$$H = \frac{2D}{(D_m+1)\varepsilon w} = \frac{2D}{R_F w} \tag{2.172}$$

The effective dispersion coefficient D can be estimated from the spread of the chromatographic band from its original, narrow width d (Fig. 71) (J.C. Giddings, H. Eyring, 1955; E. Grushka, 1972).

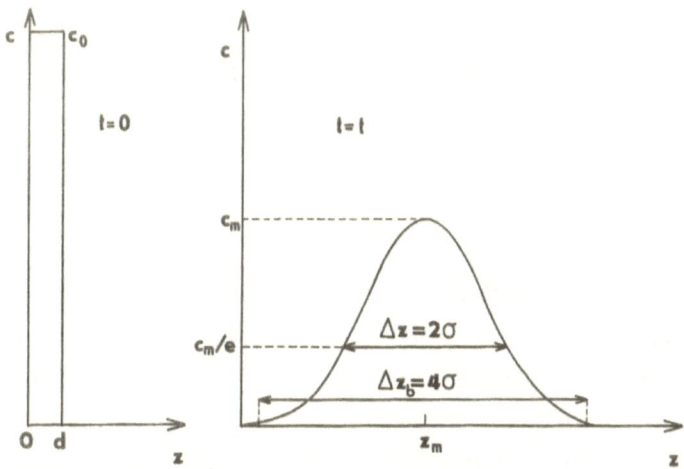

Figure 71. The spread of chromatographic band.

According to eqns. 1.428, 1.427b, 1.436, and 1.437 (or 1.623 and 1.624, Fig. 29a), and 1.662, the equation for a chromatographic band ($c \equiv x$) at migration length $z = wtR_F$ is

$$c = \frac{c_0 d}{\Delta z}\sqrt{\frac{2}{\pi}}\exp\left[-2\left(\frac{z-z_m}{\Delta z}\right)^2\right] \tag{2.173a}$$

where $z_m (m)$ is the linear coordinate at the **peak maximum path** and

$$\Delta z = 2\sqrt{2Dz/wR_F} = 2\sigma \tag{2.173b}$$

when $d \ll \Delta z$. The latter value, according to the general property of the Gaussian function 1.621, is the width of the chromatographic band measured at a height $c_m/e = 0.38c_m$, i.e. at about one third of the maximal concentration. If the peak width is measured at the base (usually between the intercepts of the linear extrapolation of the sides of sharp peaks with the abscissa), this value is

$$\Delta z_b \cong 4\sigma = 2\Delta z \tag{2.173c}$$

Additional broadening of the peaks proceeds outside the column, in open tubes interconnecting the columns with collectors or detectors on the output (*J.G. Atwood, M.J.E. Golay, 1981*).

Thus, HETP can be determined from eqns. 2.172 and 2.173b as

$$H = \frac{(\Delta z)^2}{4z_m} \tag{2.174}$$

This means that each component passing through the column may have its own height equivalent of a theoretical plate, corresponding to its retention factor, R_F value. The transport of a solute through a column of the total length L is equivalent to its transfer through the **number of theoretical stages** (separation units) of the column,

$$n = \frac{L}{H} \tag{2.175}$$

expressive of the column efficiency. A band that moves with linear velocity wR_F appears at the end of column (when $z_m = L$) after an elution time (eq. 1.590)

$$t_m = \frac{L}{wR_F} \tag{2.176}$$

or, after passing an **elution volume**

$$V_m = \frac{Lq}{R_F} \tag{2.177}$$

If the band width Δz is expressed in volume units, the number of theoretical stages can be evaluated from one of the following relations,

$$n = 4\left(\frac{z_m}{\Delta z}\right)^2 = 4\left(\frac{V_m}{\Delta z}\right)^2 = 16\left(\frac{V_m}{\Delta z_b}\right)^2 \tag{2.178}$$

HETP is an adequate measure of column efficiency for a single component, but separation of a pair A and B could better be measured by the resolution characterizing the relative positions and shape of both chromatographic bands. Because the elution volume is proportional to the length of the column L (eqn. 2.176) and the distance between them is proportional to $\sqrt{L}$ (eqns. 2.178 and 2.175), an optimum of migration length (column length) may be achieved to distinguish moving bands of the two different solutes. The **chromatographic resolution** (R_s) is defined as the distance between two peak maxima, expressed in units of the mean standard deviations, i.e.

$$R_S = \frac{z_A - z_B}{(\Delta z_A + \Delta z_B)/4} \qquad (2.179)$$

where z_A and z_B are distances from the start to the peaks of solutes A and B, respectively (Fig. 72), and Δz_A and Δz_B are their widths (*P.H. Purnell, 1960; J. Knox, 1961; M. Kubin, 1965; J.F.K. Huber, J.A.R. Hulsman, 1965; F. Helfferich, G. Klein, 1970; Ya.I. Yashin, 1976*).

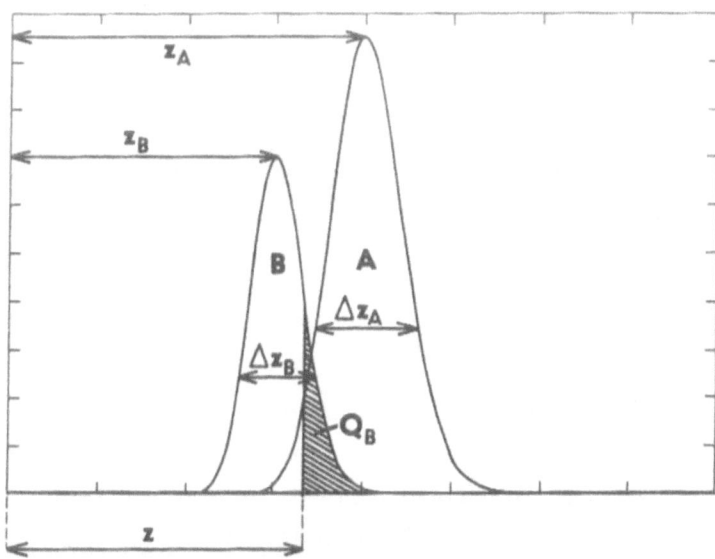

Figure 72. Chromatographic resolution of two bands ($D_B > D_A$), Q_B — contamination of peak A with component B at disconnection of the peaks at volume z (elution curve is a mirror reflection).

With respect to eqns. 2.174–2.177, from eqn. 2.179 it can be shown that

$$R_S = \frac{\sqrt{n}}{2} \frac{z_A - z_B}{z_A + z_B} \tag{2.180}$$

where z's are separation paths (elution volumes) and n is the number of theoretical plates. Because elution volumes are inversely proportional to mass distribution ratios, capacity factors D_m — eqns. 2.169 and 2.177

$$z = \frac{wt}{(D_m + 1)\varepsilon} \tag{2.181}$$

the chromatographic resolution, eqn. 2.180, can be expressed through the separation factor $\alpha = D_A/D_B$ (eqn. 2.19) as follows

$$R_S = \frac{\sqrt{n}}{2} \frac{\alpha - 1}{\alpha + 1} \frac{\bar{D}_m}{\bar{D}_m + 1} \tag{2.182a}$$

where $\bar{D}$ is the **mean capacity factor**,

$$\bar{D}_m = (D_m^A + D_m^B)/2 \tag{2.182b}$$

At small separation factors ($\alpha \leq 1.2$), the approximate relationship

$$R_S = \frac{\sqrt{n}}{2}(\alpha - 1) \frac{D_m^B}{D_m^B + 1} \tag{2.183}$$

is sufficiently correct. In accordance with eqn. 2.180, the simplest approach to improve column separation consists in increasing the number of theoretical plates by diminishing the flow rate and increasing the length of the column. Doubling the length of the column will only increase the resolution by a factor of $\sqrt{2}$.

For the majority of analytical applications, a peak resolution $R_S \geq 1.5$ is sufficient to reach an adequate proportionality of the height or area of the peak to the amount of component (*D. Abbott, R.S. Andrews, 1969; A.B. Littlewood, 1970; W. Rieman III, H.F. Walton, 1970; J.A. Khym, 1974; J. Michal, 1974; J.M. Miller, 1975; J. Novak, 1975; S. Hala, M. Kures, M. Popl, 1981; A. Braithwaite, F.J. Smith 1985*).

High efficiency is important for complicated mixtures of hydrocarbons and biomacromolecules; it is achieved by spherical particles of very small size ($5 - 10\mu m$) either in gas or liquid chromatographic systems. Because the microparticles need a high pressure for the sufficient flow rate of eluent, the chromatographic technique was called high-pressure gas or liquid chromatography, and now the liquid chromatographic systems are called **HPLC, high-performance liquid chromatography** (*P.R. Brown, 1973; J.C. Berridge, 1985; A. Fallon, R.F.G. Booth, L.D. Bell, 1987; D. Ishii, 1988; L.R. Snyder, J.L. Glajch, J.J. Kirkland, 1989; I.N. Papadoyanis, 1989; J.A. Adamovics, 1990; W.J. Lough, I.W. Wainer, 1995*).

The contamination of a fraction by another component can be calculated as an integral part of the chromatographic band, e.g. as the part of component B which is included in the band of solute A when the segregation of the peaks is completed at a separation path z. The content ΔQ of each component in the fraction of eluate in the range $0 \leftrightarrow z$ is given by the integral of the peak equation, eqn. 2.173, i.e. when the total amount is $Q = \varepsilon_0 d$,

$$\frac{\Delta Q}{Q} = \int_z^\infty \sqrt{\frac{2}{\pi}} \frac{1}{\Delta z} \exp\left[-2\left(\frac{z-z_m}{\Delta z}\right)^2\right] dz \qquad (2.184)$$

or, considering eqn. 2.178,

$$\frac{\Delta Q}{Q} = \frac{1}{2}\text{erfc}\left[\sqrt{\frac{n}{2}}\left(\frac{z}{z_m} - 1\right)\right] \qquad (2.185)$$

(see eqn. 1.435b). Because $\Delta Q/Q$ is the yield of separation R, in accordance with eqn. 2.22, the enrichment factor, characterizing the purity of peak A disengaged at coordinate (elution volume, separation path) z, is

$$S = \frac{R_A}{R_B} = \frac{\text{erfc}\left[\sqrt{\frac{n}{2}}\left(\frac{z}{z_A}-1\right)\right]}{\text{erfc}\left[\sqrt{\frac{n}{2}}\left(\frac{z}{z_B}-1\right)\right]} \qquad (2.186)$$

When coordinate z is normalized to the value z_A (elution volume of component A), the ratio z_A/z_B is close to the separation factor α ($z_B = z_A D_A/D_B = z_A/\alpha$) – Fig.73.

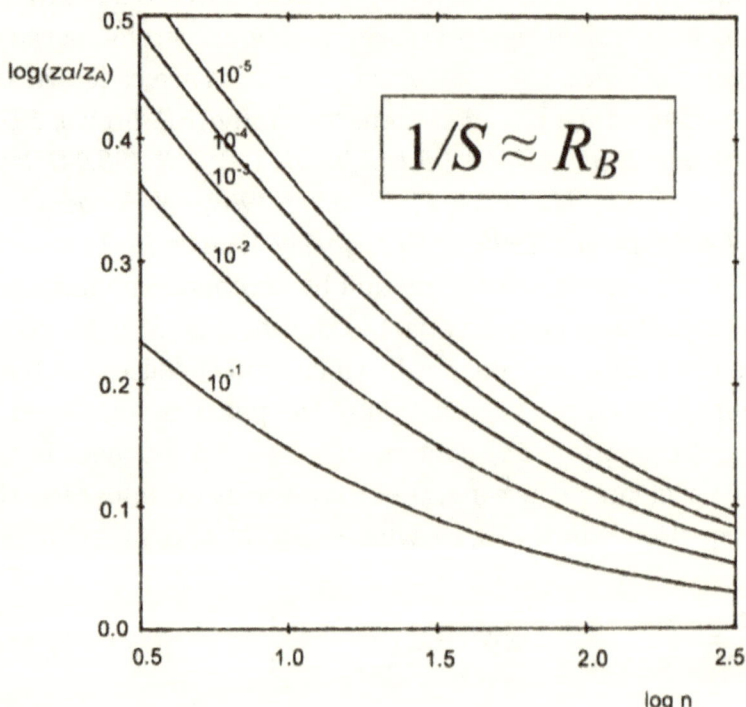

Figure 73. 3D contour plot of a peak contamination as a function of number of theoretical stages (n), separation factor (α) and elution volume of peak $A(z_A)$ at a certain volume of disconnection of the peaks (z). For example, at disconnection of peaks for which separation factor is $\alpha = D_B/D_A = 2$ at $\approx -0.74 z_A$, i.e. log ($z\alpha/z_A$) = 0.16, contamination R_B below 10^{-2} (1%) will occur at n above 32 (log n = 1.5).

Further, under reasonable conditions, the prevailing fraction of the principal component A appears in the peak, $R_A \approx 1$; and to calculate [3] the enrichment factor S it is sufficient to have data on z_A, a and n:

$$S = \frac{2}{\mathrm{erfc}\left[\sqrt{\frac{n}{2}}\left(\frac{z_A}{z_B}-1\right)\right]} \qquad (2.187)$$

For many preparative purposes a purity of fractions 90-95 % is reasonably high with respect to the time devoted to separation in the entire effort (*B.A. Bidlingmeyer, 1987*). In other instances, however, a much higher purity is required; for example, a should be above 1.1 to reach 99.99% purity [$\log(1-R) = -4$] using suitable equipment and optimal performance. Conversely, when the retention factors differ too greatly, in **gradient chromatography**, instead of constant - **isocratic** - mobile phase composition its change is designed to decrease the distribution ratios of the components gradually (*R.S. Alm, R.J.P. Williams, A. Tiselius, 1952; C. Liteanu, S. Gocan, 1974; P. Jandera, J. Churacek, 1985*) or, if using gas chromatography, this can be achieved by a programmable change of temperature in the columns (*E. Kovats, W. Simon, E. Heilbronner, 1958; W.R. Melander, K.B. Chen, C. Horvath, 1979*).

However, in the fields of molecular biology, recombinant DNA production of proteins and allied biotechnologies, high resolution does not provide a panacea for separating complex mixtures of biomacromolecules because their secondary and ternary structure may be changed and the optimization of preparative biopolymer fractionation should therefore include

---

[3] For large arguments $(x > 3)$ of the function $\mathrm{erfc}(x) = 1 - \mathrm{erf}(x)$, there exists a suitable approximation:

$$\mathrm{erf}(x) \approx 1 - \frac{1}{x\sqrt{\pi}} \exp(-x^2)\left(1 - \frac{1}{2x^2}\right)$$

the bioactivity balance (*C.J.O.R. Morris, P. Morris, 1976; M.T.W. Hearn, 1982; 1983; 1984; G. Piljac, V. Piljac, 1986; M.T.W. Hearn, B. Grego, 1988; P.A. Belter, E.L. Cussler, W.S. Hu, 1988; J-C. Janson, L. Ryden, 1989; K.M. Gooding, F.E. Regnier, 1990; M.R. Ladisch, 1991; C. Horvath, L.S. Ettre, 1993*).

The mutual interference of components is much more difficult to treat, and multicomponent preparative chromatography requires a more in-depth approach (*F. Helfferich, G. Klein, 1970; C. Horvath, A. Nahnum, J.H. Freuz, 1981; A.R. Mansour, 1986; P.J. Schoenmakers, 1986; H. Rhee, A. Rutherford, N.R. Amundson, 1989*).

Quite often, a packed (or sprayed) column is used for **recuperative adsorption** of minor components (impurities, hazardous or rare components) from a stream of fluid, e.g. on dehumidification of gases and liquids, deparaffinization and desulphurization of oil, etc. The column is used until its sorption capacity is exhausted and a threshold concentration of sorbate appears in the output. Eqn. 2.77 shows that when the entire column works as one separation stage, the instantaneous concentration c in the effluent is $1/(1 + D_m) \leq c/c_{max} \leq 1$. However, because sorption proceeds gradually, the concentration of sorbate registered in the effluent as a function of volume of passed-through fluid is called the **output curve** or **breakthrough curve** (Fig. 74).

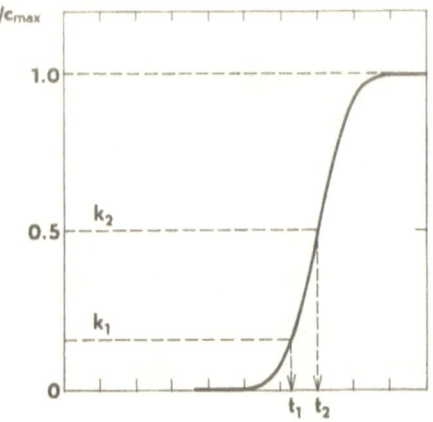

Figure 74. Break-through curve at the waste output of exhausting separation column: k_1, k_2 — coefficients of characteristic relative concentrations, t_1, t_2 — characteristic time values.

Although the change in concentration is rather sudden, the front of the output curve is indistinct due to dispersion phenomena. Various approaches have been attempted to analyze the break-through curve on the basis of step-wise sorption (*M. Dubinin, S. Yavich, 1936; E. Wicke, 1939; E. Glueckauf, 1946, 1955; A.S. Michaels, 1952; J.B. Rosen, 1952, 1954; V.V. Rachinskii, 1954*). Usually, in the output curve certain points are defined when the actual concentration c represents a certain fraction ($k_1, k_2 \ldots$) of maximal concentration c_{max}; as a rule, two points are necessary to derive the total shape of the advanced front.

For example, the time t_1 and t_2 necessary to reach $c_1 = 0.1585\, c_{max}$ and $c_2 = 0.5\, c_{max}$ are related by the number of theoretical stages as follows,

$$n = \frac{t_1 t_2}{(t_1 - t_2)^2} \tag{2.188}$$

and for bed particles of radius r_0 and sorbate inner diffusion coefficient D^{II}, it was concluded (*E. Glueckauf, 1955*) that

$$\frac{c}{c_{max}} = \frac{1}{2} + \mathrm{erf}\left[\frac{15 D^{II}}{2 r_0^2 \sqrt{n}}(t - t_2)\right] \tag{2.189}$$

The influence of the fluid velocity w can be derived from eqns. 2.171 and 2.175: n is directly proportional to w and the bed height L. The fluid velocity is also the explicit parameter of the continuous absorption concept which uses a model of transfer units — see section 2.4 (*J.B. Rosen, 1952, 1954; A.S. Michaels, 1952*). At the linear adsorption isotherm, the breakpoints can be derived from Henry's constant and hydrodynamic parameters (*J.B. Rosen, 1952*).

The time to reach the particular ratio c/c_{max} can be calculated as follows.

$$\frac{c}{c_{max}} = \frac{1}{2} + \frac{1}{2}\,\mathrm{erf}\left[\frac{Y/X - 1}{2\sqrt{V/X}}\right] \tag{2.190a}$$

where X, Y and V are dimensionless parameters related to the bed height L, the radius of the solid beads r_0, the diffusivity D^{II} in solid phase, the distribution constant K^D, the flow rate w and the fluid film transfer coefficient K^I as

$$X = \frac{D^{II} K^D L (1-\varepsilon)}{\varepsilon w r_0^2} \tag{2.190b}$$

$$Y = \frac{D^{II}(t - L/w)}{r_0^2} \tag{2.190c}$$

and

$$V = \frac{D^{II} K^D}{K^I r_0} \tag{2.190d}$$

For example, let us calculate the process when air containing $c_{max} = 0.0027$ wt% of humidity must be dried to 97.5%, i.e. to a ratio of $c/c_{max} = 0.025$ at a flow velocity $w = 0.30$ m s^{-1} on a column with length $L = 0.30$ m packed with a molecular sieve, the particles of which have a radius of $r_0 = 6.3 \times 10^{-4}$ m, the void volume fraction is $\varepsilon = 0.38$ and the dimensionless distribution constant for water molecules between air and sorbent is $K^D = 8.7 \times 10^4$, and the diffusion coefficient of water in the sorbent is $D^{II} = 0.5 \times 10^{-10}$ m^2 s^{-1} (much higher in air, $D^I = 2.64 \times 10^{-5}$ m^2 s^{-1}). First of all, the diffusion film resistance should be evaluated; for given hydrodynamical conditions the Schmidt and Reynolds numbers are, according to eqns. 1.472 and 1.476, $Sc \cong 0.58$ and $Re \cong 9$ (see related values by R.E. Treybal, 1968) and the transfer coefficient $K^I \cong 2 \times 10^{-3}$ s^{-1} corresponds to these; now, the dimensionless parameters of eqn. 2.190a are $X \cong 54$ and $V \simeq 0.065$. The ratio $c/c_{max} = 0.0025$ will be obtained at a value -1.40 of the argument under the error function in eqn. 2.140; therefore $Y/X = [2 \times (-1.40)\sqrt{6.5 \times 10^{-2}/54} + 1]2/3 = 0.603$ ($Y \cong 32.7$); hence, according to eqn. 2.190c, the time for a more humid air break-through will be after $t \cong 2.32 \times 10^5$s $\cong 1.4 \times 10^5$s $\cong 39$ hrs.

In spite of their great separation capacity, the main disadvantage of column separations, with the exception of their superior analytical applications, is their discontinuous, batch-wise mode of operation. Continuous instruments were developed for preparative purposes (*H. Pichler, H. Schulz, 1958; V.I. Sakodinskii, S.A. Volkov, 1972; R.M. Canon, J.M. Begovich, W.G. Sisson, 1980*) and those based on circular or cylindrical columns with a moving bed or a moving port deserve particular attention (*P.E. Barker, R.E. Deeble, 1973; P.C. Wankat, 1984; J.D. Navratil, A. Murphy, D. Sun, 1990*).

2.4 COUNTERCURRENT SEPARATIONS

Examples of manifold separation techniques indicate the usefulness of combining separation units with one feed. Manifold separation, however, is primarily suitable for increasing the yield of separation; the desirable high-level purity can be achieved only for small fractions of components or by applying piston transfer of the product phase when using the Craig method.

Although phase replacement is relative when performing both the techniques discussed in section 2.3. and 2.4., countercurrent operation incorporates the steady dynamic inlet and outlet of both phases at the ends of the separation units (*H.R.C. Pratt, 1967; S. Hartland, 1970; M.Valcarel, M.D. Luque De Castro, 1991*). Counter-current contact is realized as the contact of two phases: usually gas-liquid (distillation, gas adsorption) or two immiscible liquids (solvent extraction) in the:

—fractional distillation (rectifying) columns, a continuous flow or bubbles of vapor (gas) passes upward and the condensed liquid runs down (*T.P. Carney, 1949; C.S. Robinson, E.R. Gilliland, 1950; A. Weissberger, 1951; E. Krell, 1958; S.A. Bagaturov, 1961; H.Z. Kister, 1992; K.Sattler, H.J. Feindt, 1995*);

—**wetted-wall columns**, the liquid layer moves down the walls of the column under gravity (*T.K. Sherwood, R.L. Pigford, 1952*);
—**plated columns**, the gas from lower plates bubbles through the liquid via the openings of caps situated on the plates and the liquid overflows below through the overflow pipes dipped into the liquid of a lower plate (*A.E. Karr, 1980*);
—**packed columns** (*M.R. Fenske, S.O. Tongberg, D. Quiggle, 1934; T.H. Chilton, A.P. Colburn, 1935; K. Cohen, 1940; K. Kroll, 1959; A.M. Rozen, 1960; E.D. Oliver, 1966; R. Billet, 1995*), the liquid flows down the packing surface in thin films and the gas flows upward;
—**spray towers** (*A.P. Colburn, 1930*), a gas passes in at the bottom and a liquid is introduced in the form of the spray from nozzles at the top of tower;
—**bubble-plate absorbers**, the gas stream is dispersed in by a porous plate at the bottom of the liquid-filled column and the gas bubbles rise through the liquid;
—**cascade of solvent extraction mixer settlers** or **extraction columns** (*T.G. Hunter, A. W. Nash, 1932; G.A. Yagodin et al., 1981; V.S. Kislik, 2012*), two liquid counterflows are established as in gas-liquid contactors: the heavy phase (usually the aqueous phase) is fed in at the top while the lighter solvent ascends and one of these phases may be dispersed; due to the smaller differences in density of the phases, in comparison with gas-liquid systems, liquid extraction columns need additional effort to maintain sufficient flow rates and dispersity (*J.B. Lewis, H.R.C. Pratt, 1953; A.E. Karr, 1959, 1980; T. Misek, 1963, 1964; S.A.K. Jeelani, S. Hartland, 1985*);
—**cascade of electrolysis cells**, where the countercurrent flow of electrolyte and gaseous product and their interconversion is achieved by gas-permeable bipolar electrodes (*D.W. Ramey et al., 1980*).

The main advantages achieved in countercurrent separation are based on the following features:
(i) the value of the waste fractions of manifold separation systems is often non-negligible, for the waste may contain even greater amounts of the principal component than the product;
(ii) for a certain number of stages, the counter-current transfer of phases in separation units is equivalent to an increased number of the separation units;
(iii) the countercurrent process is continuous by definition;
(iv) it is possible to maintain strong driving forces throughout the operation, due to the continuous shift of the system from equilibrium;
(v) once established in its steady state, the process is relatively easy to control.

Some limits for the phase ratio in countercurrent processes are established from the hydrodynamic relationships (*R.L. Pigford, 1965; G.B. Wallis, 1969*). As has been mentioned, to establish an efficient interface it is necessary to maintain one phase dispersed in a continuous one. As the phase rate increases, at some point the fractional drag on the drops of dispersed phase causes them to move with the continuous phase and flooding occurs. The flooding point depends on both the flow rates and phase ratio.

The counter-flow of solids is used less often, with the solid being transferred by a **gas lift**, a **bucket elevator** or in **fluidized beds** with intermittent fluidization (*E.D. Oliver, 1966; V.I. Gorshkov, 1983; M.A. Burns, D.J. Groves, 1985; D.M. Ruthven, C.B. Ching, 1989*).

If the feed flow enters the separation cascade as a particular phase, it can leave it with either a decreased or increased concentration of the separated component(s); obviously this depends on the character of the second input, which can be either rich in the component (e.g. a fraction of enriched product), or, conversely, a phase which is poor or even free of the separated component. This is the means by which a more complete separation between two components or two groups of components is achieved. With two outputs, further fractionation of the two groups cannot be achieved but the question arises as to how the two (pseudo-two) component separation factor α of countercurrent devices depends on the operational arrangement of the feeds and elementary units.

The values of the inlets (F, G) and outlets (P, W), and their composition are denoted as the **outer parameters** of the series of separation units and will be used to characterize the work of the **cascade of separation units** as a whole. The parameters of the flows within the cascade between the separation units are denoted as the **inner parameters**, defining the internal distribution of the separated components. The cascades can be regarded as or subdivided into sections where the product accumulates, i.e. the **enriching section (column)**, and those which serve as a source of the product, i.e. the **exhausting section (column)**. It is instructive to use the respective inner parameters of the entire cascade as the outer parameters of the enriching and exhausting sections.

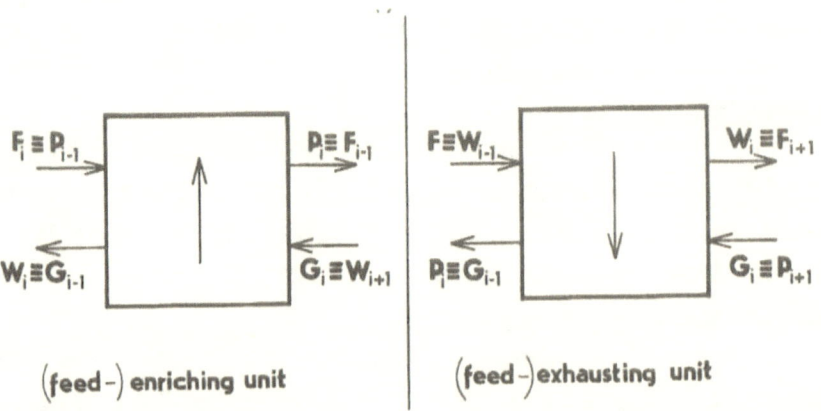

Figure 75. Scheme of separation unit in feed enriching and exhausting cascade, respectively.

If the numbering of the particular stages in the cascade (Fig. 75) ascends from the unit with the outer feed, upwards the enriching section, the feeds of the i-th separation unit consist of the product from the previous stage,

$$F_i = P_{i-1} \tag{2.191a}$$

and the waste from the next stage,

$$G_i = W_{i+1} \tag{2.191b}$$

In the exhausting section the situation is reversed, $F_i = W_{i-1}$ and $G_i = P_{i+1}$, i.e. with respect to phase F, the exhausting section units operate exactly as in manifold separation mode — eqn. 2.125.

2.4.1 Cascades of equilibrium units

To establish the relations between the inner and outer parameters of the cascades and columns, let us consider the material balances between the $(i+1)$ th stage (plate) of an enriching column and the end of the n-th stage (Fig. 76).

This is substantiated when the amounts of materials transferred by both fluxes $F \equiv P$ and $G \equiv W$ are equal at each stage, i.e. in the steady state of separation (compare with equation 2.53):

$$P_i = P_{i+1} = \cdots = P_n = F \tag{2.192a}$$

$$W_i = W_{i+1} = \cdots = W_n = W = G \tag{2.192b}$$

It is evident that, with continuity of the transport, the balances of the molar amounts, eqns. 2.192a and 2.219b, actually represent the equality of the molar fluxes ($\dot{F}, \dot{G}$), mol s^{-1}, in the units. The material balance between the last, the n-th and the i-th stage is

$$P_i y_i + G_n x^G = P_1 y_n + W_{i+1} x_{i+1} \qquad (2.193)$$

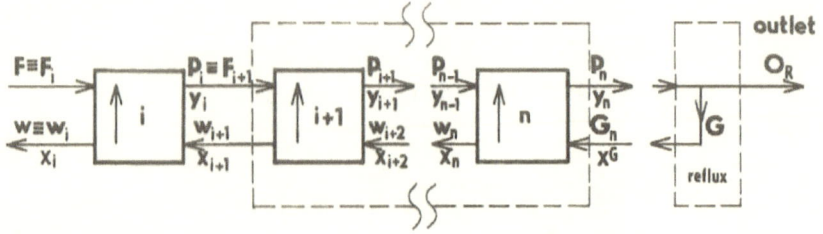

Figure 76. Enriching separation cascade with reflux.

With respect to the equality, eqn. 2.192b, the relationship between the concentrations in the outer streams and those in the product and waste flows between the i-th and $(i + 1)$ th stage, which is obtained as follows

$$y_1 = y_n - \frac{W}{P} x^G + \frac{W}{P} x_{i+1} \qquad (2.194)$$

is the **operating line of the enriching column** — cf. eqns. 2.32 and 2.34; it is important that the operating line does not characterize the work of an isolated stage but the cooperative work of the adjacent stages in the cascade. The operating line is a straight line of slope W/P, and the outer parameters, i.e. the concentration of the component in product (y_n) and in the second phase feed (x^G), determine the concentration y_i in any particular i-th stage.

If the separation cascade is designed for the maximal preconcentration of a sufficiently pure component, i.e. to achieve as high a value of y_n as possible for a given number of stages, the ratio W/P should be maximized; however, this would diminish the productivity of such a column, because the outlet P would be very low compared to the countercurrent flow G which is the principal source of separated component, regardless of its concentration in inlet F. For economical reasons, a part of outlet P is passed backward as a countercurrent feed (flow G) and the operating ratio W/P is controlled by splitting of the outlet P to streams O_R and G,

$$P = O_R + G \tag{2.195}$$

($P \equiv F$) — Fig. 76. The partial return of product is realized, e.g. in distillation as the condensation of vapors when a liquid fraction, called the **reflux** runs down the distillation column. The ratio r_O of the reflux flow G to the product output O_R is called the **reflux ratio** and is related to the stage cut as follows

$$r_O = \frac{G}{O_R} = \frac{W}{P-G} = \frac{1}{P/W - 1} \tag{2.196}$$

($0 \leq r_O \leq \infty$). Further, if there is no other flow joining the reflux, the concentrations are related as

$$y_n = x^G \tag{2.197}$$

and the operating line, eqn. 2.194, becomes a function of the two outer parameters, y_n and x_{i+1}:

$$y_i = \frac{W}{P} x_{i+1} + (1 - W/P) y_n \tag{2.198}$$

We shall see that control of the reflux ratio is of crucial importance for the efficient operation of the entire cascade:

(i) **Separation without reflux** (full outlet)

$$O_R = P_n; r_O = 0; G = W = 0 \tag{2.199}$$

With respect to eqn. 2.194, the concentrations of product in the outer outputs do not differ,

$$y_i = y_n \tag{2.200}$$

i.e. without reflux, the entire column works as a single unit, reaching the efficiency of one separation unit, the i-th, feed stage.

(ii) **Minimal reflux** (r_0^{min}) which is necessary to maintain enrichment in a given sequence of units, i.e. to fulfil the condition of increasing concentration

$$y_{i+1} > y_i \tag{2.201}$$

results from eqns. 2.194, 2.198 and an equilibrium equation. For mixtures that obey Raoult's law, the equilibrium line 2.25 can be used, so that

$$r_0^{min} = \frac{(y_n - y_i)[\alpha - y_i(\alpha - 1)]}{(\alpha - 1)y_i(1 - y_i)} \tag{2.202}$$

Because y_i is close to the concentration in the feed ($y_i \cong x^F$), it is evident that the more this concentration differs from that in the product the greater should be the reflux ratio; the latter becomes very high for mixtures which are difficult to separate, when $\alpha \cong 1$.

(iii) **Total reflux** (zero outlet)

$$O_R = 0; r_0 = \infty; G = W = P \tag{2.203}$$

is the only operation mode when the ideal steady state of separation can be achieved. According to eqns. 2.203 and 2.198, the equilibrium at each stage results in equal concentrations in countercurrent outputs:

$$y_i = x_{i+1} \tag{2.204}$$

In this steady state, the equilibrium is not disturbed by the consumption of product and the product flow leaves the cascade without any visible separation effect. However, each stage is enriched by the separated component and its concentration will increase from one stage to another. Taking into account the equilibrium line eqn. 2.109 and the operating condition, eqn. 2.203, at the opposite end of the cascade we have

$$y_{n-1} = x_n \tag{2.205}$$

The internal parameters of the column will generally be given by eqn. 2.21 which, applied to the arbitrary i-th stage is

$$\frac{y_i}{1-y_i} = \alpha \frac{x_i}{1-x_i} \tag{2.206}$$

Thus, the relationship for concentrations in two adjacent units is

$$\frac{x_n}{1-x_n} = \frac{y_{n-1}}{1-y_{n-1}} = \alpha \frac{x_{n-1}}{1-x_{n-1}} \tag{2.207}$$

Following this procedure for the unit which is situated below the n-th unit, the relation is expressed by

$$\frac{x_n}{1-x_n} = \alpha^j \frac{x_{n-j}}{1-x_{n-j}} \tag{2.208}$$

and, if $n - j = 1$, i.e. the whole column contains n units,

$$\frac{x_n}{1-x_n} = \alpha^{n-1} \frac{x_1}{1-x_1} \tag{2.209}$$

By analogy with the separation factor α of the single separation unit, eqn. 2.21, the ratio

$$q = \frac{y_n}{1-y_n} \frac{1-x_1}{x_1} \tag{2.210}$$

is defined as the **overall separation factor** of the separation cascade (column). It can be considered as another outer parameter of the column and, according to the equilibrium condition 2.206, eqn. 2.209 and the definition, eqn. 2.210, it is connected with the separation factor α (the internal parameter) as follows

$$q = \alpha^n \qquad (2.211)$$

Both eqns. 2.210 and 2.211 are used for calculation of the number of stages (theoretical plates) n of the enriching section of columns for separation of ideal mixtures by **Fenske's equation** (*M.R. Fenske, D. Quiggle, C.O. Tornberg, 1932*):

$$n = \frac{\log q}{\log \alpha} \qquad (2.212)$$

For testing of fractional distillation columns, readily analyzed mixtures with constant, and well-established separation factors are used, e.g. n-heptane —methylcyclohexane ($\alpha = 1.083$), 2,2,4-trimethylpentane—methylcyclohexane ($\alpha = 1.049$) are used. The value of q is evaluated according to eqn. 2.210 from the mole fractions of the more volatile component, which are determined by the refractive index in the column head and in the pot in the steady state (*S.Hala, M. Kuras, M. Popl, 1981*). For small values of the separation factor, $\ln \alpha \cong \alpha - 1$, and eqn. 3.122 can be approximated by

$$n = \frac{2.3 \log q}{\alpha - 1} \qquad (2.213)$$

It is evident that in this case n becomes very large. For example, the rectification column in which the deuterium contents must be increased from natural abundance, $x_1 = 0.0015\%$ to $x_n = 99.8\%$ requires that $= (0.9998/0.002)(0.99985/0.00015) = 3.3 \times 10^6$. For fractional distillation of water at normal pressure $\alpha = 1.026$ and the number of stages $n = 2.3 \log (3.3 \times 10^6)/0.026 = 580$!

For the distillation of liquid hydrogen $\alpha = 1.42$ and the same enrichment is achieved at only $n = \log(3.3 \times 10^6)/\log(1.42) = 43$ stages.

Low values of the separation factor and a resulting greater number of stages in countercurrent separation are also typical for many inorganic substances: distillation of $SiHCl_3 - PCl_3$, the solvent extraction of the pairs Zr-Hf, Nb-Ta, Ni-Co, Am-Cm ($\alpha = 1.1 - 1.4$), etc. Therefore, obtaining for example a separation system with a separation factor $\alpha = 2.5$ in the solvent extraction of lanthanides with di-2-ethylhexylphosphoric acid (DEHPA) was a great success (*D.F. Peppard et al., 1957*).

(iv) An **ideal cascade** should work under conditions of equal concentration in both feeds of the separation unit, i.e. for the i-th unit the equality

$$y_{i-1} = x_{i+1} \tag{2.214}$$

must be fulfilled. The contact of the inputs in the separation unit does not lead to an increase of entropy. This is important for large numbers of stages and for irreversible processes when separation units consume much energy. The theory of these cascades was developed (but not published except for classified reports during World War II) by *A.M. Dirac* and *R. Peierls* (*K. Cohen, 1951; M. Benedict, T.H. Pigford, 1957; A.M. Rozen, 1960*).

Condition, eqn. 2.214, cannot hold unless the total fluxes decrease with an increasing concentration of component on passing from one unit to another. Hence, the operating line does not represent a straight line and the eqns. 2.192 are also invalid. However, considering eqn. 2.106 and an enrichment factor, eqn. 2.23, for the i-th stage, the relation for the enrichment factor S is

$$\frac{y_i}{1-y_i} = S \frac{y_{i-1}}{1-y_{i-1}} \tag{2.215}$$

because y_{i-1} is the concentration in both feeds of the unit. By analogy, for the $(i-1)th$ stage

$$\frac{y_{i-1}}{1-y_{i-1}} = S \frac{x_i}{1-x_i} \tag{2.216}$$

By summation of eqns. 2.215 and 2.216, and comparing the result with the universal eqn. 2.23, condition 2.214 of the ideal cascade operation is transformed to

$$S = \sqrt{\alpha} \tag{2.217}$$

Thus, using $\sqrt{\alpha}$ instead of α in the equations already derived, it is possible, for example, to calculate the minimal reflux ratio, eqn. 2.102. Furthermore, when $\alpha \approx 1$ the overall separation factor is

$$q = (\sqrt{\alpha})^n = \alpha^{n/2} \approx 1 + \frac{n}{2}(\alpha - 1) \tag{2.218}$$

i.e. two stages are necessary to achieve the same enrichment as in one stage of a cascade with a constant reflux ratio. The operating line of the ideal cascade is derived from eqns. 2.114, 2.115, and 2.117 as

$$\frac{y_i}{1-y_i} = \sqrt{\alpha} \frac{x_{i+1}}{1-x_{i+1}} \tag{2.219}$$

By analogy with eqn. 2.202a, the phase ratio at a given stage i of the ideal cascade should be

$$\frac{W_{i+1}}{O_R} = \frac{(y_n - y_i)[\sqrt{\alpha} - y_i(\sqrt{\alpha} - 1)]}{(\sqrt{\alpha} - 1)y_i(1 - y_i)} \tag{2.220a}$$

When $\alpha \approx 1$, $\alpha^{n/2} \approx 1 + n(\alpha-1)/2$ and eqn. 2.220a ($n=1$) turns to

$$\frac{W_{i+1}}{O_R} = \frac{y_n - y_i}{y_i(1 - y_i)} \left(\frac{\alpha+1}{\alpha-1} - y_i \right) \tag{2.220b}$$

As has been shown, the detailed design of the ideal cascade is important. When the inner fluxes are minimized, the size of the cascades and the energy consumption are reduced. This is quite understandable for such an extreme example as the separation of isotopes by gaseous diffusion (*K. Cohen, 1951*). A sophisticated analysis of the cascade indicates that when the concentration changes are very small (because $\alpha \cong 1$) and the component with a linear distribution between the outputs is produced, then the sum of the fluxes between a great number of stages contains an important term called the **value function** or **separation potential**,

$$\Phi(x) = (2x - 1) \ln \frac{x}{1-x} \tag{2.221}$$

where x represents the concentration of the separated component in the mixture. The value function, which was introduced in separation science by *A.M. Dirac* and *R. Peierls* (*K. Cohen, 1950*), is a measure of the equipment necessary to perform the separation of the substance (*V.A. Palkin, V.M. Gadel'shin, O.E. Aleksandrov, V.D. Seleznev, 2014*). For a non-linear equilibrium line, the value function can normally be obtained by numerical integration of the differential equation

$$\frac{d^2 \Phi(x)}{dx^2} = \left(\frac{\alpha - 1}{y - x}\right)^2 \tag{2.222}$$

which results from the general properties of the value function (*K. Stamberg, J. Cabicar, 1984*).

The relationship between the separation potential and the separation work, eqn. 1.31, is

$$\frac{E_{min}}{\Phi(x)} = RTx(1 - x) \tag{2.223}$$

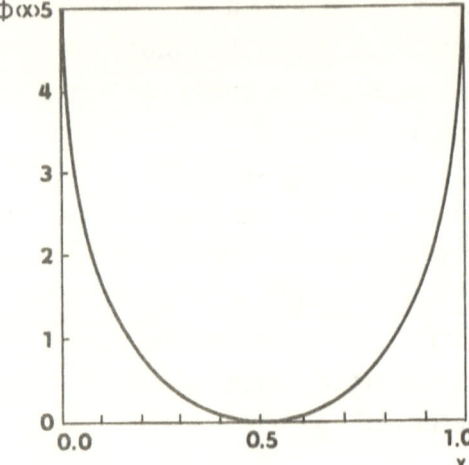

Figure 77. Value function (separation potential 2.221) for various mixture composition.

The separation potential is symmetrical, $d\Phi/dx = 0$ for an equivalent composition of binary (pseudo-binary) mixture, ($x = 0.5$) — Fig. 77. For dilute solutions, when $x_A \to 0$, it tends to infinity, i.e. to obtain a pure component, an infinitely large cascade is required. The separation potential multiplied by the molar fluxes of the outer parameters of the cascade is called the **separation duty** or **separation work**,

$$\Delta U = P\Phi(y_n) + W\Phi(x_1) - F\Phi(x^F) - G\Phi(x^G) \qquad (2.224)$$

The ideal cascade is characterized by those outer parameters whereby an enrichment of the waste which would reach the original level of the component concentration as in the feed would cost as much as the feed itself, i.e. it operates with maximal possible efficiency when the operating and material expenditures are evaluated. A practically important operation of the **exhausting column** (Fig. 78) consists of the **countercurrent flow of the product-free inlet**: the scrubbing a gas by a liquid adsorbent or the stripping of a solute from the liquid extract by an immiscible solvent (*A. V. Izmailov, P.M. Maltsev, Yu. G. Mitskewich, 1975*).

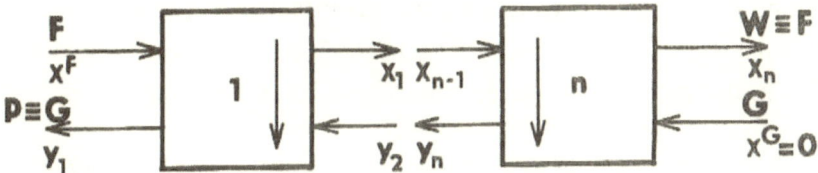

Figure 78. Exhausting separation cascade with product-free inlet G.

The general form of the **operating line** of the **exhausting cascade** is derived from the balance between the first and i-th stages of the column:

$$Wx^F + Py_{i+1} = Wx_i + Py_1 \qquad (2.225)$$

and therefore

$$y_{i+1} = y_1 - \frac{W}{P}x^F + \frac{W}{P}x_i \qquad (2.226)$$

However, the material balance which concerns the inner parameters of the exhausting column is better analyzed from the final, n-th stage, where a waste identical to the exhausted feed ($W \equiv F$) is removed and a fresh inlet that the product from the stream enters ($G \equiv P, x^G = 0$):

$$Wx_{n-1} = Wx_n + Py_n \qquad (2.227)$$

For the linear isotherm, the equation transforms into

$$x_{n-1} = x_n\left(1 + \frac{PD}{W}\right) \qquad (2.228)$$

or

$$x_{n-1} = x_n(1 + D_m) \qquad (2.229)$$

where D is identical to the distribution ratio D_A (eqn. 2.10) and D_m with the mass distribution ratio D_m^A (eqn. 2.15). Because of the above, the mass balance for the $(n-1)$th stage is

$$Wx_{i-2} + Py_n = Wx_{n-1} + Py_{n-1} \qquad (2.230)$$

and results in

$$x_{n-2} = x_{n-1}(1 + D_m) - x_n D_m \qquad (2.231)$$

After considering eqn. 2.229,

$$x_{n-2} = x_n[(1 + D_m)^2 - D_m] = x_n(1 + D_m + D_m^2) \qquad (2.232)$$

By a process of induction, we have

$$x_{n-i} = x_n(1 + D_m + D_m^2 + \cdots + D_m^i) \qquad (2.233)$$

This geometrical progression has the sum of i terms, as follows:

$$x_{n-i} = x_n \frac{D_m^{i+1} - 1}{D_m - 1} \qquad (2.234)$$

For the special case when $D_m = 1$, the direct result from 2.233 is

$$x_{n-i} = x_n(i + 1) \qquad (2.235)$$

At the first stage ($i = n$) of the exhausting column, the value of x_{n-n} is identical to the feed concentration $x_0 \equiv x^F$. Then the concentration in the waste W, as a function of the outer parameter, the concentration in the feed, is obtained as

$$x_n = x^F \frac{D_m - 1}{D_m^{n+1} - 1} \qquad (2.236)$$

which is known as the **Kremser equation** (A. Kremser, 1930). For three-phase systems (con-distillation, slurry adsorption, slurry scrubbers, emulsion liquid membrane systems), a modified Kremser equation that includes two mass distribution ratios has been derived (P.C. Wankat, 1980) — cf. eqns. 1.687.

The corresponding **yield of exhaustion** by countercurrent flow can be calculated from eqn. 2.236 as

$$R_n = 1 - \frac{x_n}{x^F} = \frac{D_m^{n+1} - D_m}{D_m^{n+1} - 1} \qquad (2.237a)$$

—compare with eqn. 2.134 — or, for the particular case when $D_m = 1$, the limiting value of eqn. 2.237 gives

$$R_n = 1 - \frac{1}{1+n} = \frac{n}{1+n} \qquad (2.237b)$$

The number of stages for the necessary recovery of product, with yield R_n, can be calculated from eqn. 2.235 as

$$n = \frac{\log(D_m - R_n) - \log(1 - R_n)}{\log D_m} - 1 \qquad (2.238a)$$

or, for $D_m = 1$,

$$n = \frac{R_n}{1 - R_n} \qquad (2.238b)$$

For non-linear isotherms, the number of equilibrium stages is related to the external parameters in a similar way as we discussed for one-stage separations in units with two inlets — Figs. 57.

Very often it is useful to combine the exhausting with the enriching section to reduce the losses of valuable or other necessary components in the waste stream. In this case the waste W from the enriching section serves as the feed for the exhausting section, (Fig. 79).

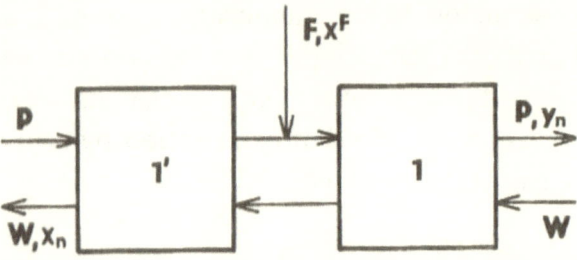

Figure 79. Feeding part of separation cascade with both enriching (1) and exhausting (1') section separation units.

(The numbering of the stages proceeds from an intermediate point where an outer feed is introduced and the parameters of the exhausting sections are indicated by primes). The input concentration of the enriching section is identical to the output concentration of the exhausting section:

$$y_{1'} = x^F \equiv x_1 \qquad (2.239)$$

The equilibrium concentrations in the outputs of the individual stages are obtained from distribution isotherms (equilibrium lines), and operating lines — Fig. 80.

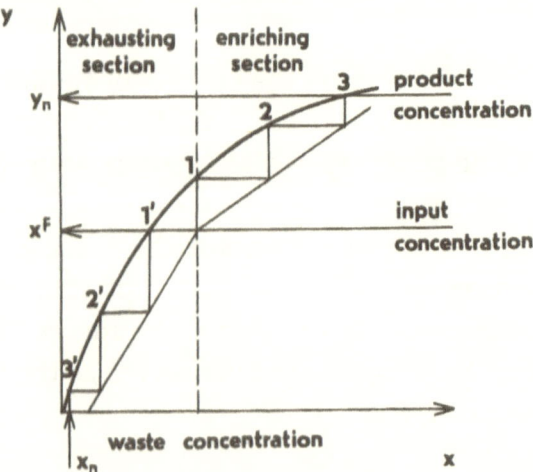

Figure 80. McCabe-Thiele diagrams for separation cascade with both enriching and exhausting sections.

The operating line of the enriching section is given by eqn. 2.194 or 2.198, and that of the exhausting section by eqn. 2.226, respectively. Although the balances are formally identical, it should be recognized that the stage cuts (W/P) in both sections are generally different unless the distribution isotherm is linear along the entire separation path. Because the exhausting section operates at an initial (low concentration) part of the distribution isotherm, which is usually steeper, the stage cut W/P in the exhausting section is also usualy higher than in the enriching section of the cascade. The same level of exhaustion is achieved by a greater number of stages and a smaller counter flow of phase G, or vice versa. For example, a minimum 99% recovery ($R_n \geq 0.99$) may be found at $D_m = 1.0 (R = 0.50)$ only after $n = 99$ stages but on increasing D_m to $D_m = 3.0$ ($R = 0.75$) only $n = 4$ stages are necessary. Under technological conditions, the number of stages and fluxes are optimized because investment costs increase with the number of stages n, and the operational costs are proportional to the material inputs and reflux applied.

The separation of binary mixtures or two groups of components by countercurrent operation for increasing number of stage B will proceed by the separation for two fractions, the product containing mostly the components with $D_m > 1$ and the waste retaining substances with $D_m < 1$. For example, a mixture of lanthanides can be separated in a countercurrent flow of nitric acid solution (F) and tributylphosphate solution (G); the group La-Ce-Nd, having $D_m \leq 0.6$ remains in the aqueous stream and the group Sm-Ln-Y with $D_m \geq 1.2$ is extracted into the organic stream (G.V. Korpusov, 1962).

Limiting recovery fractions in the exhausting cascade depend on the mass distribution ratio, as obtained from eqn. 2.237,

$$\lim_{n \to \infty} R_n = 1 \quad \text{if } D_m \geq 1$$
$$\lim_{n \to \infty} R_n = D_m \quad \text{if } D_m \leq 1 \tag{2.240}$$

As a result, the components with lower distribution ratios will be recovered from the feed with a maximal D_m; if D_m reaches a critical value 1, the yield may approach 100%. In principle, a "full" recovery of every component with $D_m \geq 1$ can be achieved by increasing the number of stages, but its purity is entirely controlled by the lowest distribution ratios of the other components, the limiting values of enrichment factors being

$$\lim_{n \to \infty} S_n = 1 \qquad \text{if } D_m^A \text{ and } D_m^B \geq 1$$
$$\lim_{n \to \infty} S_n = D_m \qquad \text{if } D_m^A \geq 1 \text{ and } D_m^B \ll 1 \qquad (2.241)$$

Such an operation can be achieved by a suitable stage cut, but because of the conflicting demands in purification (large D_m^A and very small D_m^B) it can be met only either at a favorable high distribution of the main component A, or by strong suppression of the distribution of the other components.

For example, in the multifold separation discussed in connection with eqns. 2.138 and 2.142, (i.e. the separation of components having $R_A = 0.50$ and $R_B = 0.25$ in one stage), the mass distribution ratios are $D_m^A = 1$ and $D_m^B = 0.33$. In multi-fold separation, it was enough to have 7 stages for 99% total recovery of ($R_n = 0.99$), but simultaneously the component B was also transferred to the product with a rather high yield (87%). In countercurrent operation, however, for 99% recovery of A, according to the Kremser equation in form eqn. 2.238, $R_n = 0.99$, and $n = 99$ stages are necessary. This is much higher than for the previous mode of separation, but although the distribution of component B is only three times lower, it is recovered with a final yield close to $R_n^B = 0.33$ (33%) and the enrichment factor is 3 times higher, $S_n = 3.0$. Of course, a much superior separation can be designed in systems in which the chemical composition favors a lower distribution of B, and, if possible, a higher distribution of A.

Generally speaking, during countercurrent separation the recovery yields are lower than in multi-fold separation for equal numbers of stages, but the enrichment (purity) of the product stream is higher when the countercurrent system operates at suitable distribution ratios (for the principal component $D_m^A \geq 1$ and for the impurity $D_m^B < 1$).

Obvious complications of the material balance calculations arise through the mutual influence of components and thermodynamic non-ideality (*A. Rose, 1941; E.J. Roche, 1971*). For two-liquid counter streams (liquid-liquid extraction) changes due to the partial miscibility of solvents often need to be taken into account; graphical constructions based on triangular diagrams — section 1.3.1. — have been developed to cope with the problem (*K.A. Varteressian, M.R. Fenske, 1937; A.V. Nikolaev et al., 1967; D.R.F. West, 1982*).

2.4.2 Continuous transfer

In the previous section, the countercurrent cascades were discussed in terms of a series of interconnected units which provide thermodynamic equilibrium or a steady-state distribution between their outlets. For a continuous process, such a distribution may be modelled following the concept of theoretical stages (*A.J.P. Martin, R.L.M. Synge, 1941*) as was demonstrated for column separation (section 2.3.2.). Although the calculations based on models of theoretical stages are rather straightforward, the approach does not take account the dynamics of interphase transfer (*A.M. Rozen, 1956, 1960*). However, for an actual countercurrent arrangement of contact, it is unreasonable to assume that, at some coordinate z down the countercurrent column, the phases would approach equilibrium to a greater extent than the outlet flows at a distance H (Fig. 81).

At each point down the flow path, exists a driving force θ (eqn. 2.1) which causes a crossing flux j from the waste to the product stream (eqn. 2.2). Although both the interfacial area and the mass transfer coefficient K are the subject of much discussion (R.R. Bird, W.E. Stuart, E.N. Lightfoot, 1960; V.V. Kafarov, 1962; T.A. Hatton; E.N. Lightfoot, 1984; Yu.I. Babenko, 1986), their products, the capacity coefficients, represent some constants under given hydrodynamic conditions — see the Sherwood and Schmidt numbers in section 1.4.2. — and can be assumed constant over the entire contact path which varies from $z = 0$ to the column length, $z = L$.

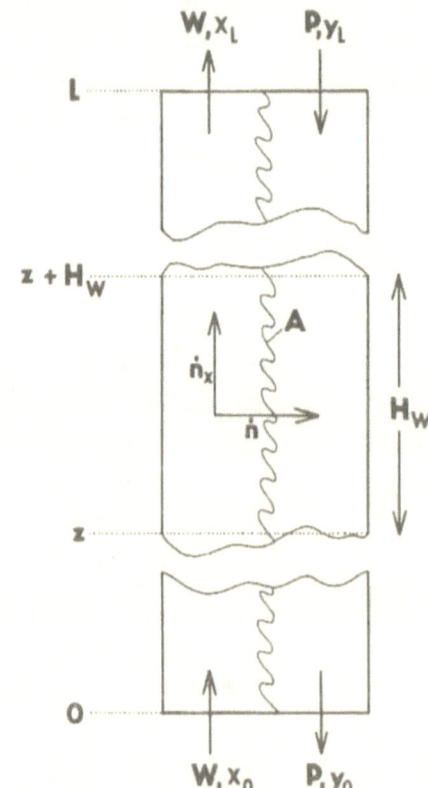

Figure 81. Section of separation countercurrent (pseudo-countercurrent) column representing a transfer unit (explanation in text).

Hence, the total transfer coefficient K (m s^{-1}) for two-film resistance — eqn. 1.530 — should also be constant. The **total fluxes** (mol s^{-1}) of product $\dot{P}$ and waste $\dot{W}$, when the phases are mutually saturated or immiscible (i.e. the transfer of total material between the phases is negligible in this respect) are usually constant as well:

$$K_x \, a_x = \text{const}, K_y \, a_y = \text{const} \, (Ka = \text{const}) \quad (2.242a)$$

$$\dot{P} = \text{const}, \dot{W} = \text{const} \quad (2.242b)$$

In turn, the concentrations, driving forces, and the direct and crossing fluxes of the components are functions of the coordinate z. In the direction of the z-axis, the **fluxes of a component** (mol s^{-1}) are given by the concentrations $x(z)$ in the waste stream and $y(z)$ in the product stream, as that the total fluxes are

$$\frac{dn_x}{dt} = \dot{n}_x = \dot{W}x \quad (2.243a)$$

$$\frac{dn_y}{dt} = \dot{n}_y = \dot{P}y \quad (2.243b)$$

and their change down the column length will be characterized, after differentiation, by

$$d\dot{n}_x = \dot{W}dx \quad (2.244a)$$

$$d\dot{n}_y = \dot{P}dy \quad (2.244b)$$

The driving forces in phases W and P are

$$\vartheta_x = (x - x^{eq})c_{tW} \quad (2.245a)$$

in phase W, with a total concentration of components c_{tW}, (mol dm^{-3}) — eqn. 1.51b — and

$$\vartheta_y = (y^{eq} - y)c_{tP} \quad (2.245b)$$

in phase P with the analogous concentration c_{tP}. When the driving forces are expressed by molar concentrations and the transfer coefficients are K_x and K_y (m s^{-1}), the crossing flux of the component (mol s^{-1}, equal in both phases) through the interphase A(m^2), is

$$\frac{dn}{dt} = \dot{n} = 10^3 K_x \vartheta_x A = 10^3 K_y \vartheta_y A \qquad (2.246)$$

The symmetry of eqns. 2.243-2.246 is evident and therefore it is sufficient to analyze the phase relations in phase P at the concentrations x (in practice, $x \ll y$ and the resistance in the waste phase W usually becomes the single controlling factor for transfer).

In a section of infinitesimal length dz and element volume dV, and through an interphase area dA, the following relations exist,

$$dA = adV = aAdz \qquad (2.247\text{a})$$

where a can be either the specific interface (m^{-1}) with respect to phase W,

$$a_W = \frac{A}{V_W} \qquad (2.247\text{b})$$

or phase P,

$$a_P = \frac{A}{V_P} \qquad (2.247\text{c})$$

The specific surface area in the dispersed two-phase system can be evaluated by the mean Sauter diameter of the dispersed phase (*T. Misek, 1964*) — eqn. 1.684.

The **crossing flux** in the element of column is given by

$$d\dot{n} = 10^3 a_W K_x \vartheta_x A dz \qquad (2.248)$$

The central idea of the dynamic model is that, at some length H of column, the **longitudinal transfer of the component** is the same as the crossing transfer (mol m^{-1}s^{-1}). Its mathematical formulation is

$$\int_z^{z+H} d\dot{n} = \int_z^{z+H} d\dot{n}_x \qquad (2.249)$$

and therefore

$$d\dot{n} = d\dot{n}_x \qquad (2.250)$$

Now the desired relation for change in concentration x vs. the Cartesian coordinate z is obtained from eqn. 2.250 and eqns. 2.245a and 2.244a as:

$$\frac{dx}{dz} = \frac{x - x^{eq}}{H_W} \qquad (2.251)$$

where

$$H_W = \frac{1}{10^3 c_{tw} a_W A} \frac{\dot{W}}{K_x} \qquad (2.252)$$

is a value with the dimension of length (m), which is called the **height of a transfer unit, HTU** (*T.H. Chilton, A.P. Colburn, 1935*).

The above consideration is suitable for one-dimensional separation. For a two-dimensional process, such as membrane separation, the transfer unit does not have a characteristic length but can be characterized by its area, the **area of the transfer unit, ATU**, through which the same amount of a component is transferred as with the stream washing the surface of the membrane (*S. Weller, W.A. Steiner, 1950*),

$$A_W = \frac{\dot{W}}{10^3 c_{tw} K_x} \qquad (2.253)$$

HTU (ATU) is, according to the way it has been derived, the length (area) of the countercurrent device where the same total crossing flux (mol s^{-1}) of a particular separated solute is identical with its stream in one of the current phases. Eqn. 2.253 concerns the waste stream and, by analogy, the same relation can be obtained for the product flow:

$$H_P = \frac{1}{10^3 c_{tP} a_P A} \frac{\dot{P}}{K_y} \qquad (2.254)$$

Generally, the heights of the transfer units in different phases, H_W and H_P respectively, differ; HTU is the greater the smaller the mean driving force in the phase and the greater the longitudinal transfer of the latter.

A more complete approach to the theory of transfer units allows for the complex situation where the flow rates change through the column, the mutable dispersed phase bubble, or drop sizes, and number of them per unit volume, etc. However, even the application of the existing basic theory of physico-chemical hydrodynamics (*M.M. Sharma, P.V. Danckwerts, 1970*) remains topical because the data on HTU's are still unsatisfactory.

From the theory of interphase transfer — section 1.4.4. — two further complications should be mentioned: the additivity of resistances in both flowing phases (*C.J. King, 1964*) and the transfer accomplished by chemical reactions.

The overall transfer units, H_{OW} and H_{OP}, derived from those in particular phases, originate from eqns. 2.252 or 2.254 and the relationships between the transfer coefficients K_x or K_y, and the total transfer coefficient K. When designating phase W as phase I and phase P as phase II, the overall transfer coefficient, eqn. 1.529, in the molarity concentration scale can be written as

$$\frac{1}{K} = \frac{1}{K_x} + \frac{1}{D}\frac{1}{K_y} \qquad (2.255)$$

Dividing this equation by $10^3 c_t a A$, multiplying either by $\dot{W}$ or $\dot{P}$, and taking into account the phase volume ratio r, i.e.

$$\frac{\dot{P}}{c_{tP}} = r\frac{\dot{W}}{c_{tW}} \qquad (2.256)$$

and

$$a_W = ra_P, \qquad (2.257)$$

the **overall HTU related to phase W** becomes

$$H_{OW} = \frac{\dot{W}}{K_x c_{tW} a_W} + \frac{1}{rD}\frac{\dot{P}}{K_y c_{tP} a_P} \qquad (2.258)$$

this and the **overall HTU related to phase P** becomes

$$H_{OP} = \frac{\dot{P}}{K_y c_{tP} a_P} + rD\frac{\dot{W}}{K_x c_{tW} a_W} \qquad (2.259)$$

Comparing these expressions with eqns. 2.252 and 2.254, and knowing that the product of the phase ratio r and distribution ratio D is the mass distribution ratio D_m (eqn. 2.15), it follows that

$$H_{OW} = H_W + \frac{1}{D_m}H_P \qquad 2.260)$$

and

$$H_{OP} = H_P + D_m H_W \qquad (2.261)$$

When the distribution ratio is high (the concentration of solute is much higher in phase P than in phase W) the phase P resistance is small and diffusion in the phase W film is the controlling parameter ($H_{OW} \cong H_W$) and vice versa. Generally, eqn. 2.260 should be used when $D_m > 1$ and eqn. 2.261 when $D_m < 1$ (phase P is the controlling one).

For a continuous system with three phases (two concurrent streams), two overall transfer coefficients should be considered (*P.C. Wankat, 1980; P.C. Wankat, R.D. Noble, 1984*).

When the transfer is accompanied by chemical reactions, the HTU is inversely proportional to the acceleration coefficient E, eqn. 1.542, unless the ss-reaction at the interphase does not take place (section 1.4.4).

Formally, the transfer unit concept greatly resembles the theory of equilibrium plates (section 2.3.2.). Their differentiation, however, is necessary and the "theoretical stage" must always be specified whether or not it concerns the height equivalent to theoretical plate (HETP) or height of transfer unit (HTU). The theory of HTU, as distinguished from HETP, is not limited by thermodynamic equilibrium and takes account of the dynamics of continuous countercurrent separation. For a correct choice of the controlling phase, the HTU characterizes the separation process better than the HETP. At low concentrations and mixtures difficult to separate, i.e. for low driving force, the HTU is obviously larger than the HETP derived from separation data for more concentrated mixtures or the mixtures of easily separable components. Hence, the HTU gives a more complex picture both of the separated mixture and the separation equipment.

By integration of the concentration gradient, eqn. 2.251, within $z = 0$ (the coordinate where the waste stream enters the column) and $z = L$, the end of column with the waste output, if the HTU does not change down the separation path, we can obtain

$$H_W \int_{x_0}^{x_L} \frac{dx}{x - x^{eq}} = \int_0^L dz = L \qquad (2.262)$$

According to the definition, the **number of transfer units**, N, is obtained from eqn. 2.263 as

$$N_W = \frac{L}{H_W} = \int_{x_0}^{x_L} \frac{dx}{x - x^{eq}} \qquad (2.263)$$

In this equation the integral of the dimensionless quantities can be found graphically or numerically by means of the equilibrium and operating lines. If $x^{eq} = 0$ (infinitely high mass distribution ratio D_m), the integration is simple,

$$N_W = \ln(x_L/x_0) \cong -\ln(1 - R_N) \qquad (2.264)$$

i.e. one transfer unit decreases the concentration x down to $1/e = 1/2.718$ (down to about 37%, $R_N = 0.37$).

For linear distribution isotherms and operating lines, i.e. over a narrow concentration interval and for immiscible phases, the HTU can be derived theoretically (A.P. Colburn, 1941). If concentration y is a linear function of concentration over the range (x_0, x_L), the driving force

$$\vartheta_x = x - x^{eq} = \vartheta \qquad (2.265)$$

is also a linear function of x. The result is that the outer parameters of the column (Fig. 81) are related to the driving force through

$$\frac{d\vartheta}{dx} = \frac{\vartheta_0 - \vartheta_L}{x_0 - x_L} \qquad (2.266a)$$

or

$$dx = \frac{x_0 - x_L}{\vartheta_0 - \vartheta_L} d\vartheta \qquad (2.266b)$$

Substituting eqn. 2.265 and 2.266b into the integral eqn. 2.263, results in

$$N_W = \frac{x_0 - x_L}{\vartheta_0 - \vartheta_L} \int_{x_0}^{x_L} \frac{d\vartheta}{\vartheta} \qquad (2.267)$$

and

$$N_W = \frac{x_0 - x_L}{\vartheta_0 - \vartheta_L} \ln \frac{\vartheta_0}{\vartheta_L} = \frac{x_0 - x_L}{\overline{\vartheta}} \qquad (2.268)$$

The value

$$\overline{\vartheta} = \frac{\vartheta_0 - \vartheta_L}{\ln(\vartheta_0/\vartheta_L)} \qquad (2.269)$$

is called the **logarithmic-mean driving force** and gives the mean value of the driving force — the integral, eqn. 2.263 — over the range of the separation path. Eqns. 2.268 and 2.269 are formally the same for the data for the second phase, on replacing concentration x by y. The usefulness of the latter equations consists of their making use of the output parameters, the terminal driving forces (ϑ_0 and ϑ_L) and the composition (x_0 and x_L) which are usually easily determined.

For example, let us consider an extraction (or other exhausting) column with the following outer parameters: at the bottom the phase W with a concentration of extracted component of $x_0 = 0.034$ enters, and counter-currently the solvent stream P with a concentration $y_0 = 0.34$ leaves the column. The phase ratio is $r = P/W = 0.2$, i.e. a 50-fold preconcentration in the product and a recovery of 99.76% are obtained. At the top of the column fresh solvent enters, $y_L = 0$, and raffinate leaves with the concentration decreased to $x_L = 8 \times 10^{-5}$. Linearity between y and x holds over this concentration interval, the distribution ratio being $D = 76$, which means that the HTU should be calculated for phase W. The equilibrium concentration in phase W, which corresponds to the concentration y_0 in the output, is $x_0^{eq} = \frac{y_0}{D} = \frac{0.34}{76} = 4.47 \times 10^{-3}$; thus the driving force in phase W at this end of the column is $\vartheta_0 = 3.4 \times 10^{-2} - 0.45 \times 10^{-2} = 2.95 \times 10^{-2}$. There is zero concentration in stream W at length L, $x_L = 0$ and the driving force $\vartheta_L = 8 \times 10^{-5} - 0 = 8 \times 10^{-5}$. Thus, the main driving force exists at the bottom of the column. The logarithmic-mean of the driving forces, according to eqn. 2.269 is $\vartheta = (2.95 \times 10^{-2} - 8 \times 10^{-5})/(\ln(2.95 \times 10^{-2}/8 \times 10^{-5})) = 5 \times 10^{-3}$ and the number of transfer units in phase W is found from eqn.

2.268 as $N_W = (0.034 - 8 \times 10^{-5})/(5 \times 10^{-3}) = 6.8 \cong 7$ units. Calculation using the simplified formula 2.264 gives $N_W = -\ln(8 \times 10^{-5}/0.034) = 6.03$, which is a reasonable approximation. When, however, the calculation is extended to phase P, where the driving forces are 76-times higher, $\vartheta_0 = 2.27$ and $\vartheta_L = 6 \times 10^{-3}$, the result would be a much smaller value, $N_P = 0.9$, i.e. the efficiency of the column would be underestimated. The overall number of transfer units, with respect to the resistances in both phases, according to eqn. 2.460, gives the value $N_{OW} = 6.8 + [1/(0.2 \times 76)]0.9 = 6.86$ (calculation according to eqn. 2.261 would give a rather overestimated value of $N_{OP} = 517$). If, in comparison, the number of theoretical equilibrium plates is calculated by the Kremser formula, eqn. 2.238, with $D_m = 76 \times 0.2 = 15.2$ and $R_n = 0.9976$, the result is $n = 2.2$. Hence, the evaluation of column efficiency from the concept of equilibrium stages is unfavorable, although with respect to the transfer units, when considering the dynamic properties of column operation, it is reasonably high.

For non-linear isotherms and operating lines, White's method of graphical determination of transfer units is applicable (*G.E. White, 1940*).

Calculated data of HTU are valid for a sufficiently homogeneous flow on the vertical z. Both longitudinal molecular and eddy diffusion and radial in homogeneity of the contact increases the real value of HTU proportionally to the reciprocal value of the Péclet number. This problem has been considered already through combined convective and interphase transport in section 1.4.5.

One further application of the transfer unit theory was proposed (*A.S. Michaels, 1952*) in the treatment of break-through curves in continuous absorption (Fig. 74). The parameters of the curves are related to the numbers of stages N_{13} and N_{23} over the intervals $\langle t_1, t_3 \rangle$ and $\langle t_2, t_3 \rangle$ respectively as follows,

$$\frac{t_2-t_1}{t_3-t_1} = \frac{N_{13}}{N_{23}} \tag{2.270}$$

and the N's are obtainable by integration according to eqn. 2.263.

If the transfer efficiency of a one-dimensional separation unit is to be considered, it is clear that the terminal driving forces are zero, when the complete equilibrium is considered. Because, according to the definition of the equilibrium stage in the unit, the equality

$$L = (\text{HETP}) \tag{2.271}$$

is valid, we have

$$N \cdot (\text{HTU}) = (\text{HETP}) \tag{2.272}$$

However, the number of transfer units N in the equilibrium stage would be infinite because the HTU for zero driving forces is also zero. This problem is overcome by assuming a certain fraction of **approach to equilibrium** in an incomplete equilibrium unit, i.e.

$$y = \varphi y^{eq} \tag{2.273a}$$

and

$$\vartheta_y = (\varphi - 1)y \tag{2.273b}$$

where φ is the arbitrary coefficient, slightly less then unity; though the choice of its value is rather subjective, it is clear that this creates a formal bridge between the concepts of equilibrium and dynamic transfer.

REFERENCES

1. Abbott, D.; Andrews, R.S.: *An introduction to chromatography (Concepts in chemistry).* Houghton Mifflin: Boston, 1969.
2. Abe, M.; Uno, K.: *Synthetic inorganic ion exchange materials. XIX. Ion exchange behavior and separation of alkaline earth metals on crystalline antimonic(V) acid as a cation exchanger.* Sep.Sci.Technol., 14(4) 355-66 (1979).
3. Aboul-Enein, H.Y. (Ed.): *Separation techniques in clinical chemistry.* CRC Press: Boca Raton, 2003.
4. Abraham, M.H.; Liszi, J.: *Calculations on ionic solvation. V. The calculation of partition coefficient of ions.* J.Inorg.Nucl. Chem., 43(1) 143-51 (1981).
5. Abramzon, A.A.; Shchukin, Ye.D. (Eds.): *Poverkhnostno-aktivnye yavleniya i poverkhnostno-aktivnye veshchestva (Surface phenomena and surface-active compounds).* Khimiya: Leningrad, 1984.
6. Adamovics, J.A.: *Chromatographic analysis of pharmaceuticals.* Marcel Dekker: New York, 1990.
7. Adams, B.A.; Holmes, E.L.: *Adsorptive properties of synthetic resins.* J.Soc.Chem.Ind. (London) 54, 1-6 T (1935).
8. Adzumi, H.: *Studies on the flow of gaseous mixtures through capillaries. III. The flow of gaseous mixtures at medium pressures.* Bull.Chem.Soc.Japan, 12, 292-303 (1937).
9. Aguilar, M-I.: *HPLC of peptides and proteins: Methods and protocols.* Humana Press: Totowa, 2004.
10. Aguilar, M.; Cortina J.L. (Eds.): *Solvent extraction and liquid membranes: Fundamentals and applications in new materials,* CRC: Boca Raton, 2010.
11. Aguillela, V.M.; Garrido, J.; Mafe, S.; Pellicer, J.: *A finite-difference method for numerical solution of the steady-state*

Nernst -Planck equations with non-zero convection and electric current density. J.Membr.Sc., 28, 139 (1986).

12. Ahuja, S.: *Chemical derivatization for the liquid chromatography of compounds of pharmaceutical interest.* J.Chrom.Sci., 17, 168-72 (1979).
13. Ahuja, S.: *Chiral separation applications and technology.* American Chemical Society: Washigton DC, 1997.
14. Albertsson, P.A. *Partition of cell particles and macromolecules*, 2nd Edn., Almqvist and Wiksell: Stockholm, 1971.
15. Albertsson, P.A.; Nyns, E.J. *Counter-current distribution of proteins in aqueous polymer phase systems.* Nature 184, 1465 (1959).
16. Alberty, R.A.; King, E.L.: *Moving boundary systems formed by weak electrolyte. Study of cadmium iodide complexes.* J.Am.Chem.Soc., 73, 517 (1951).
17. Albery, W.J.; Burke, J.F.; Leffier, E.B.; Hadgraft, J.: *Interfacial transfer studied with a rotating diffusion cell.* J.Chem.Soc., Faraday Trans. 1, 72(7) 1618-26 (1976).
18. Alfassi, Zeev B.; Wai, Chien M. (Eds.): *Preconcentration techniques for trace elements.* CRC Press: Boca Raton, 1992.
19. Ali, Imran; Aboul-Enein, H.Y.; Gupta, Vinod K.: *Nanochromatography and nanocapillary electrophoresis.* J.Wiley&Sons: New Jersey, 2009.
20. Allen, J.F.: *An improved technetium-99m generator for medical application.* J.Appl. Radiat.Isotopes, 16, 332 (1965).
21. Allen, R.C.; Maurer, H.R.: *Electrophoresis and isoelectric focussing in polyacrylamide gel.* Walter De Gruyter: New York, 1974.
22. Alm, R. S.: *Gradient elution analysis. Ill. Oligosaccharide.* Acta Chem.Scand., 6, 1186-93 (1952).

23. Alm, R.S.; Williams, R.J.P.; Tiselius, A.: *Gradient elution analysis. I. A general treatment.* Acta Chem.Scand., 6, 826-36 (1952).
24. Al-Mutaz, I.S.: *Recent trends to reverse osmosis for desalination in Saudi Arabia.* In: *Chemical separations, Vol. I.*, p. 379. C.J. King, J.D. Navratil, Eds., Litarvan Literature: Denver, 1986.
25. Ambartzumian, R.V.; Letokhov, V.S.: *Multiple photon infrared laser photochemistry.* Chem.Biol.Appl. Lasers, 3, 167-316 (1977).
26. Ames, W.F.: *Numerical methods for partial differential equations.* Nelson: London, 1969.
27. Amidon, G.L; Yalkowsky, S.H.; Anik, S.T.; Valvani, S.C.: *Solubility of non-electrolytes in polar solvents. V. Estimation of the solubility of aliphatic monofunctional compounds in water using a molecular surface area approach.* J.Phys. Chem., 79(21) 2239-46 (1975).
28. Andreev, I.I.: *Ueber die Wachstums und Aufloesungsgeschwindigkeit der Krystalle.* Zh.Russ.Fiz.-Khim.Obshch., 40, 397 (1908).
29. Aniansson, G.: *The radioactive measurement of the adsorption of dissolved substances on liquid surfaces and an application to "impurities" in dodecylsodium sulfate.* J.Phys. Colloid.Chem. 55, 1286-99 (1951).
30. Applegate, L.E.: *Membrane separation processes.* Chem. Eng., June 11, 64 (1984).
31. Aris, R.; Amundson, N.R.: *Longitudinal mixing or diffusion in fixed beds.* Am.Inst.Chem.Eng. Journal, 3, 280 (1957).
32. Armarego, W.L.F.; Chai, Christina Li Lin: *Purification of laboratory chemicals.* 7th Ed. Butterworth-Heinemann: Oxford, 2013.
33. Asenjo, J.: *Separation processes in biotechnology.* Marcel Dekker: New York, 1990.

34. Ashbrook, A.W.: *Chelating reagents in solvent extraction processes. Present position.* Coord. Chem. Rev., 10(4) 283-7 (1975).
35. Astarita, G.: *Gas absorption by first-order chemical reaction in a spherical liquid film.* Chem. Eng. Sci., 17, 708 (1962).
36. Aston, F.W.: *The mass spectra of chemical elements.* Phil. Mag., 39, 449, (1920).
37. Atwood, J.G.; Golay, M.J.E.: *Dispersion of peaks by short straight open tubes in liquid chromatography systems.* J.Chromatogr., 218, 97-122 (1981).
38. Atwood, J.G.; Schmidt, G.J.; Slavin, W.: *Improvement in liquid chromatography column life and method flexibility by saturating mobile phase with silica.* J.Chromatogr., 171, 109-115 (1979).
39. Audinos, R.: *Membranes de microfiltration tangentielle. Traité constants physicochimiques.* K 365. Techniques de l'ingénieur: Paris, 2001.
40. Axen, R.; Ernback, S.: *Chemical fixation of enzymes to cyanogen halide activated polysaccharide carriers.* Eur.J.Biol., 18(3) 351-60 (1971).
41. Axelsen, N.H.: *Quantitative immunoelectrophoresis: new developments and applications.* N. H. Axelsen (Ed.) Universitetsforlaget: Oslo, 1975.
42. Axen, R.; Porath, J.; Ernback, S.: *Chemical coupling of peptides and proteins to polysaccharides by means of cyanogen halides.* Nature, 214, 1302-4 (1967).
43. Babcock, W.C.; Balcer, R.W.; Lachapelle, E.D.; Smith, K.L.: *Coupled transport membranes. I. The mechanism of uranium transport with a tertiary amine.* J.Membr. Sci., 7(1) 71-87 (1980).
44. Babenko, Yu.I.: *Teplomassoobmen. Metod raschota teplovykh i diffuzionnykh potokov. (Heat and mass exchange. Calculation methods for heat and diffusional fluxes).* Khimiya: Leningrad, 1986.

45. Bachmann, K.J.: *The materials science of microelectronics.* VCH Publishers: Weinheim, 1995.
46. Bagaturov, S.A.: *Teoriya i raschet peregonki i rektifikatsii (Theory and practice of distillation and rectificiation).* Gostoptekhizdat: Moskva, 1961.
47. Bamford, C.H.: *Comprehensive chemical kinetics. Vol.23. Kinetics and chemical technology.* Elsevier: Amsterdam, 1985.
48. Barger, G.; Dale, H.H.: *Beta-imidazolathylendiamine, ein blutdrucksenkender Bestandteil der Darmschleimhaut.* J. Physiol., 41, 499-503 (1910).
49. Barker, J.A.: *Cooperative orientation effects in solutions.* J.Chem.Phys., 20, 1526-32 (1952).
50. Barker, J.A.: *Cooperative orientation in solutions. The accuracy of quasichemical approximation.* J.Chem.Phys., 21, 1391-4 (1953).
51. Barker, P.E.; Deeble, R.E.: *Production scale organic mixture separation using a new sequential chromatographic machine.* Anal.Chem., 45(7) 1121-1125 (1973).
52. Barker, P.G.: *Computers in analytical chemistry.* Pergamon Press: Oxford, 1983.
53. Barrer, R.M.: *Hydrothermal chemistry of the silicates. VIII. Low-temperature crystal growth of aluminosilicates and some gallium and germanium analogues.* J.Chem.Soc., 1959, 195-208.
54. Barrer, R.M.; Klinowski, I.: *Ion exchange involving several groups of homogeneous sites.* J.Chem.Soc., Faraday Trans 68(1) 73-8 (1972).
55. Barrer, R.M.; Townsend, R.P.: *Transition metal ion exchange in zeolites. Part 1. Thermodynamics of exchange of hydrated ions in ammonium mordenite.* J.Chem.Soc., Faraday Trans. 1, 72, 661-73 (1976).
56. Barrolier, J.; Watzke, E.; Gibian, H.: *Simple apparatus for carrier-free preparative continuous electrophoresis.* Z. Naturforsch., 13B, 754-8 (1958).

57. Bates, R.G.; Staples, B.R.; Robinson, R.A.: *Ionic hydration and single ion activities in unassociated chlorides at high ionic strengths.* Anal.Chem., 42, 867-71 (1970).
58. Batuner, L.M.; Pozin, M.Ye.: *Matematicheskie metody v khimiicheskoi tekhnike (Mathematical methods in chemical engineering)* 3rd. ed. Gos.Khim.Izd.: Leningrad, 1960.
59. Baudisch, O.: *Quantitative Trennung mit "Cupferron".* Chem.Ztg. 33, 1298 (1909).
60. Bauer, D.; Cote, G.; Pescher-Cluzeau, Y.: *Interfacial properties of 5alkyl-8-quinolinols and rate of phase transfer in Al/ Ga separation.* In: Chemical Separations, vol.1, p.903. C.J. King, J.D. Navratil, Eds., Litarvan Lit.: Denver, 1986.
61. Bautista, R.G. (Ed.): *Emerging separation technologies for metals. II.* Minerals, Metals & Materials Society: Warrendale 1998.
62. Bear, J.: *Dynamics of fluids in porous media.* American Elsevier: New York, 1972.
63. Bechhold, H.: *Kolloidstudien* mit der *Filtrationsmethode.* Z.Physik.Chem, 60, 257 (1907).
64. Beck, M.T.: *Chemistry of complex equilibria.* Akademiai Kiado: Budapest, 1970.
65. Becker, F.E.; Zakak, A.I.: *Recovering energy by mechanical vapor recompression.* Chem.Eng.Prog., 81(7) 45-9 (1985).
66. Becker, F.F.; Wang, Xiao-Bo; Huang Ying; Pethig, R.; Vykoukal, J.; Gascyone, P.R.C.: *Separation of human breast cancer cells from blood by differential dielectric afinity.* Proc. Natl. Acad. Sci. USA, Cell Biology, 92, 860-864 (1995).
67. Beckmann, W. (Ed.): *Crystallization. Basic concepts and industrial applications.* Wiley-VCH Verlag: Weinheim, 2013.
68. Bedzyk, M.J.; Bommarito, G.M.; Caffrey, M.; Penner, T.L.: *Ion distribution profile in liquid layers adjacent to liquid interfacea.* Science, 248, 52 (1990).

69. Belter, P.A.; Cussler, E.L.; Hu, W.S.: *Bioseparations. Downstreams processing for biotechnology.* J. Wiley-Interscience: New York, 1988.
70. Belyaeva, T.V.; Bunkin, F.P.; Savostina, V.M.; Shafeev, G.A.: *Stimulirovannaya nepreryvnym lazernym izlucheniem selektivnaya ekstraktsiya medi i kobalta.* Dokl.AN SSSR 307, 583 (1989).
71. Benedetti-Pichler, A.A.; Rachele, J.R.: *Limits of identification of simple confirmatory tests.* Ind.Eng.Chem., Anal.Ed., 12, 233 (1940).
72. Benedict, M.: *Multistage separation process.* Chem.Eng. Progr. 1, 43(2) 41-60 (1947).
73. Benedict, M.; Pigford, T.H.; Levi, H.: *Nuclear chemical engineering.* 2nd.Ed. McGraw-Hill: New York, 1981.
74. Benes, P.; Majer, V.: *Trace chemistry of aqueous solutions. General chemistry and radiochemistry.* Academia: Prague, 1980.
75. Benes, P.; Paulenova, M.: *Surface charge and adsorption properties of polyethylene in aqueous solutions of inorganic electrolyte I. Streaming potential measurement.* Kolloid-Z.Z.Polym., 251(10) 766-71 (1973).
76. Berezkin, V.G.; Kruglikova, V.S.; Belikova, N.A.; Shirayeva, V.E.: *Pulse chromatographic method for studying the kinetics of bimolecular reactions.* Gaz.Khromatogr. 454-460 (1966).
77. Berezkin, V.G.; Pakhomov, V.P.; Tatarinskii, V.S.; Fateeva, V.M.: *Contribution of adsorption on interphase boundaries to retention volume in gas-liquid solid-phase chromatography.* Dokl.Akad.Nauk SSSR 180(5) 1135-8 (1968).
78. Berezkin, V.G.; Loshchilova, V.D.; Pankov, A.G. et al.: *Khromato-raspredelitelnyi metod (Chromato-distribution method).* Nauka: Moskva, 1976.
79. Berezkin, V.G.: *Khimicheskie metody v gazovoi khromatografii* (Chemical methods in gas chromatography). Khimiya: Moskva, 1980.

80. Bergfors, T.M. (Ed.): *Protein crystallization.* International University Line: La Jolla, 2009.
81. Bernal, J.D.; Fowler, R.H.: *A theory of water and ionic solutions, with particular reference to hydrogen and hydroxyl ions.* J. Chem. Phys., 1(8) 515-48 (1933).
82. Bernhard, M.; Brinckamn, F.E.; Sadler, P.J. (eds.) *The importance of chemical 'speciation' in environmental processes.* Dahlem Konferenzen, Life Science Res. Reports 33. Springer Verlag: Berlin, 1986.
83. Berridge, J.C.: *Techniques for the automated optimization of HPLC separations.* J.Wiley: New York, 1985.
84. Berson, S.A.; Yalow, R.S.: *Recent studies on insulin-binding antibodies.* Ann. N. Y. Aca.dsci., 82, 338-44 (1959).
85. Berstein, R.B.: *Molecular reaction dynamics.* Claredon Press: Oxford, 1974.
86. Berthelot, E.M.; Jungfleisch, E.C.: *Sur les lois qui president au partage d'un corps entre deux dissolvants (Experiences).* Ann. Chim. Phys., 26, 396 (1872).
87. Berthelot, E.M.: *Sur les lois qui president au partage d'un corps entre deux dissolvants (Théorie).* Ann. Chim. Phys., 26, 508 (1872).
88. Berthier, G.: *Determination of the coefficients of self diffusion by the method of heterogeneous isotope exchange.* J. Chim. Phys., 52, 41-7 (1955).
89. Berthod, A.: *Chiral recognition in separation methods: Mechanisms and applications.* Springer Science: New York 2010.
90. Berthoud, A.J.: *Formation of crystal faces.* J.Chim.Phys., 10, 624-635 (1912).
91. Bertrand, J.: *Note sur la similitude en mecanique.* J. Ecole Polytechn., No. 32, (1848).
92. Bidlingmeyer, B.A.: *Preparative liquid chromatography.* Elsevier: Amsterdam 1987.

93. Biegler, L.T.; Hughes, R.: *Feasible path optimisation with sequential modular simulators.* Computers and Chem.Eng., 9(4) 379-394 (1985).
94. Bier, M.: *Electrophoresis. Theory, methods and application.* Academ. Press: New York, 1959.
95. Bikerman, J.J.: *Surface chemistry.* Acad.Press: New York, 1958.
96. Billet, R. *Packed towers in processing and environmental technology.* VCH: Weinheim 1995.
97. Binning, R.C.; James, F.E.: *Now separate by membrane permeation.* Petr.Refin., 39, 214-5 (1958).
98. Bird, R.B.; Stewart, W.E.; Lightfoot, E.N.: *Transport phenomena.* J. Wiley: New York, 1960.
99. Birren, B.; Lai, E.: *Pulsed field gel electrophoresis. A practical guide.* Academic Press: San Diego, 1993.
100. Bjellqkvist, B.; Ek, K.; Righetti, P.G.; Gianazza, E.; Goerg, A.; Westermeier, R.; Postel, W.: *Isoelectric focusing in immobilized pH gradients: principle, methodology, and some applications.* J. Biochem. Biophys., Methods, 6(4) 317-39 (1982).
101. Bjerrum, J.: *Metal ammine formation in aqueous solution.* P. Haase&Son: Copenhagen, 1941.
102. Bjerrum, J.: *A new optical principle for the investigation of step equilibria.* Kgl.Danske Videnselskab.Math-Fys.M., 21(4) 22 (1944).
103. Bjerrum, N.: *Modern views of acid and alkaline reactions and their application to analysis.* Z. Anal.Chem., 56, 13-28 (1916).
104. Bjerrum, N.: *Die Dissoziation der starken Elektrolyte.* Z. Elektrochem., 24, 321-28 (1919).
105. Bjerrum, N.: *Studien uber Chromirhodanide. IV.* Z. Anorg. Allg.Chem., 119, 179-201 (1921).
106. Bjerrum, N.: *Untersuchungen uber Ionenassoziation I. Der Einfluss der Ionnenassoziation auf die Aktivität des Ionen*

bei mittleren Assoziationsgraden. Det. Danske Videsk. Selsk. Math. Med., 7(9) 3 48 (1926).

107. Bloch, J.M.; Yun, W.B.; Yang, X.; Ramanathan, M.; Montano, P.A.; Capasso, C.: *Metal counterion concentrations and structure at liquid interfaces by near-total-external-fluorescence using synchrotron X -rays.* Phys. Rev. Lett., 61, 2941 (1988).
108. Blytas, G.C.; Peterson, D.L.: *Determination of kerosine range n-paraffins by a molecular sieve, gas-liquid chromatography method.* Anal.Chem., 39(12) 1434-7 (1967).
109. Bocek, P.; Gebauer, P.; Dolnik, V.; Foret, F.: *Recent developments in isotachophoresis.* J.Chromatogr., 334, 157-95 (1985).
110. Boltwood, B.B.: *Uber die Radioaktivitat von Thoriumsalzen.* Physik. Zeitschr., 8, 556-561 (1907).
111. Boreskov, G.K.; Slinko, M.G.: *Modeling of chemical reactors.* Pure and Appl. Chem., 10(4) 611-24 (1965).
112. Born, M.: *Volumen und Hydrationswärme der Ionen.* Z.Phys., 1, 45-48 (1920).
113. Born, M.; Mayer, J.E.: *The lattice theory of ion crystals.* Z.Phys., 75, 1-18 (1932).
114. Born, von I.A.: *Métallurgie, on l'amalgation des minéraux metode d'extraire* par le *mercur.* Bern, 1787.
115. Born, von I.A.: *Ueber das Anquicken der gold und silberhaltigen Erze, Rohrsteine, Schwarzkupfer und Hettenspeise.* Bergbaukunde (Leipzig) 1 (1786).
116. Boublik, T.; Fried, V.; Hala, E.: *The vapour pressure of pure substances.* Elsevier: Amsterdam, 1973.
117. Bowen, H.J.M.: *Absorption by polyurethane foams; new method of separation.* J.Chem.Soc. A, 1082 (1970).
118. Boyadzhiev, L.; Lazarova, Z.: *Study of creeping film pertraction. Recovery of copper from dilute aqueous solutions.* Chem. Eng. Sci., 42{5) 1131-5 (1987).

119. Boyadzhiev, L.; Sapundzhiev, T.; Bezenshek, E.: *Modeling of carier mediated transport.* Separ.Sci., 12(5) 541-51 (1977).
120. Boyarinov, A.I.; Kafarov, V.V.: *Metody optimalizatsii v khimii i khimicheskoi tekhnologii. (Optimization theory in chemistry and chemical technology).* Khimiya: Moscow, 1975.
121. Boyd, G.E.; Schubert, J.; Adamson, A.W.: *The exchange adsorption of ions from aqueous solutions by organic zeolites. I. Ion exchange equilibria.* J.Am.Chem.Soc. 69, 2823 (1947).
122. Boyd, T.E.; Cusick, M.J.; Navratil, J.D.: *Ferrite use in separation science and technology.* In: *Recent developments in separation science, Vol. VIII., pp. 201-232.* N. N. Li, J. D. Navratil, Eds., CRC Press: Boca Raton 1986.
123. Bradley, D.: *The hydroclone.* Pergamon Press: New York 1965.
124. Braithwaite, A.; Smith, F.J.: *Chromatographic methods.* 5th Ed., Chapman and Hall: London 1995.
125. Braun, T.: *Trends in using resilient polyurethane foams as sorbents in analytical chemistry.* Fresenius Z. Anal.Chem., 314{7) 652-6 (1983).
126. Braun, T.; Fardy, A.E.: *Polyurethane foams and microspheres in analytical chemistry. Improved liquid-solid, gas-solid and liquid-liquid contact via a new geometry of the solid phase.* Anal. Chim. Acta, 99 (1) 1-36 (1978).
127. Braun, T.; Navratil, J.D.; Farag, A.B.: *Polyurethane foam sorbents in separation science.* CRC Press: Boca Raton 1985.
128. Brezhneva, N.Ye.; Roginskii, S.Z.; Sbilinskii, A.I.: *Catalytic transfer in the isotopic exchange of bromine.* J.Phys.Chem. (USSR) 9, 752-4 (1937).
129. Bridgman, P.W.: *Dimensional analysis. Chemical engineering handbook.* Perry, J.H., Ed., McGraw-Hill: New York, 1950.

130. Brinkley, S.R.: *Calculation of the equilibrium composition of system of many constituents.* J.Chem. Phys., 15(2) 107 (1947).
131. Britton, H.Ts.: *Elektrometrischen Untersuchungen ueber die Fallung von Hydroxiden I-IV.:* J.Chem. Soc., 127, 2110-59 (1925).
132. Brockmann, H.; Schodder, H.: *Aluminium oxide with buffered adsorptive properties for purposes of chromatographic adsorption.* Ber., 74, 73-8 (1941).
133. Brockmann, H.: *New findings in the field of chromatographic adsorption.* Angew. Chem., 59, 199-206 (1947).
134. Brodskii, A.I.; Aleksandrovich, M.M.; Slutzkaya, M.M.; Shelud'ko, M.K.: *Concentrating heavy water.* Doklady AN SSSR, 3, 615-17 (1934).
135. Broomley, L.A.: *Thermodynamic properties of strong electrolytes in aqueous solutions.* Am.Inst.Chem.Eng. Journal, 19(2) 313-20 (1973).
136. Brown, G.G.: *Unit operations.* J. Wiley: New York, 1951.
137. Brown, P.R.: *High pressure liquid chromatography. Biochemical and biomedical applications.* Academic Press: New York, 1973.
138. Brown, P.R.: *Rapid separation of nucleotides in cell extracts using high-pressure liquid chromatography.* J.Chromatogr., 52(2) 257-72 (1970).
139. Brown, P.R.; Grushka, E.: *Structure-retention relations in the reversed-phase high performance liquid chromatography of purine and pyrimidine compounds.* Anal.Chem., 52(8) 1210-15 (1980).
140. Brown, P.R.; Krstulovic, A.M.: *Practical aspects of reversed-phase liquid chromatography applied to biochemical and biological research.* Anal. Biochem., 99(1) 1-21 (1979).
141. Brunauer, S.; Emmett, P.H.; Teller, E.: *Adsorption of gases in multimolecular layers.* J.Am.Chem.Soc., 60, 309-319 (1938).

142. Bruno, T.J.: *Separation and instrumentation.* Prentice Hall: London, 1990.
143. Buchowski, H.: *Relation between partition coefficients and properties of solvents.* Nature, 194, 674-5 (1962).
144. Buckingham, E.: *On physically similar systems: Illustration of the use of dimensional equations.* Phys. Rev., Ser. 2, (4) 345 (1914).
145. Bugaevskii, A.A.: *Tablitsy konstant vazhnejshikh ravnovesii imeyushchikh znacheniye v analiticheskoi khimii (Tables of the equilibrium constants most important for analytical chemistry).* Izd. Khark. Univ.: Kharkov, 1967.
146. Bugaevskii, A.A.; Dunai, B.A.: *Computation of equilibrium composition and buffer properties of solutions.* Zh.Anal. Khim., 26(2) 205-9 (1971).
147. Bugaevskii, A.A.; Kholin, Yu.V.: *Simulation of equilibria of complex formation and extraction from data on the distribution of radioelements between phases.* Radiokhimiya, 27, 594-7 (1985).
148. Bumbalek, V.; Horak, V.; Haman, J.: *Posibilities how to improve chloride technology of magnesia production.* In: Proc. XVth Int. Mineral Processing Congress, Cannes, June 1985, vol.2, ps57-569. GEDIM: St. Etienne, 1985.
149. Bunge, A.L.; Noble, R.D.: *A diffusion model for reversible consumption in emulsion liquid membranes.* J. Membr. Sci., 21(1) 55-71 (1984).
150. Burmeister, M.; Ulanovsky, L. *Pulsed-field gel electrophoresis. Protocols, methods and theories.* Humana Press: New York, 1992.
151. Burnette, W.N.; Neul, W.: *"Western blotting": electrophoretic transfer of proteins from sodium dodecylsulfate - pol yacr ylamide gels to unmodified nitrocellulose and radiographic detection with antibody.* Anal. Biochem. 112(2) 195-203 (1981).

152. Burns, M.A.; Graves, D.J. *Continuous affinity chromatography using a magnetically stabilized fluidized bed.* Biotechnology Progress 1(2) 95 (1985).
153. Bush, M.S.; Densen, P.M.: *Systematic multiple fractional extraction procedures - application to the separation of mixtures.* Anal.Chem., 20, 121-9 (1948).
154. Butenandt, A.F.J.; Westphal, N.: *The isolation and characterization of the corpus luteum hormone.* Ber., 67, 1440-42 (1934).
155. Butler, J.P.: *Hydrogen isotope separation by catalyzed exchange between hydrogen and liquid water.* Sep.Sci. Technol., 15(3) 371-96 (1980).
156. Cabicar, J.; Gosman, A.; Plicka, J.; Stamberg, K.: *Study of kinetics, equilibrium and isotope exchange in ion exchange systems. II. Kinetics of isotope exchange in uranium(VI)-238-uranium(IV)-233- cation exchanger systems.* J.Radioanal. Chem., 80(1-2), 71-80 (1983).
157. Cadogan, D.F.; Purnell, J.H. *Concurrent solution and adsorption phenomena in chromatography. III. Measurement of formation constants of H-bonded complexes in solution.* J.Phys.Chem. 73(11) 3849-54 (1969).
158. Cadogan, D.F.; Purnell, J.H.: *Concurrent solution and adsorption phenomena in chromatography. III. Measurement of formation constants of H-bonded complexes in solution.* J.Phys.Chem. 73(11) 3849-54 (1969).
159. Caldwell, K.D; Nguyen, T.S.; Murray, T.M.; Marcus, N.; Giddings, J.C.: *Observation of anomalous retention in steric field-flow fractionation.* Sep.Sci.Technol., 14(10) 935-46 (1979).
160. Caldwell, K.D.: *Field -flow fractionation in biomedical analysis.* In: Chemical separations, v.1, p.1. C.J. King, J.D. Navratil, Eds., Litarvan Lit.: Denver, 1986.
161. Caldwell, K.D.; Karaiskakis, G.; Giddings, J.C.: *Characterization of T4D virus by sedimentation field-flow fractionation.* J.Chromatogr., 215, 323-32 (1981).

162. Caletka, R.; Hausbeck, R.; Krivan, V.: *Distribution of niobium and tantalum on Dowex 1 and polyurethane foam in HF-H_2SO_4 and HF-HCl medium.* J.Radioanal. Nucl. Chem., Articles, 131(2) 343-352 (1989).
163. Canon, R.M.; Begovich, J.M.; Sisson, W.G.: *Pressurized continuos chromatography.* Separ. Sci. Technol., 15(3) 655-78 (1980).
164. Cantor, C.R.; Smith, C.L.: *Pulsed-field gel electrophoresis of large DNA molecules.* Nature, (London) 319, 701-702 (1986).
165. Cantor, C.R.; Smith, C.L.; Mathew, M.K.: *Pulsed-field electrophoresis of very large DNA molecules.* Annu.Rev. Biophys.Biophys.Chem. 17, 287-304 (1988).
166. Carey, G.F.; Finlanson, B.A.: *Orthogonal collocation on finite elements.* Chem. Eng. Sci., 30, 587 (1975).
167. Carle, G.F.; Frank, M.; Olson, M.V.: *Electrophoretic separation of large DNA molecules by periodic inversion of the electric field.* Science 232, 65-70 (1986).
168. Carman, P.C.: *Fluid flow through granular beds.* Trans. Inst. Chem. Eng. (London) 15, 150-166 (1937).
169. Carman, P.C.: *Fundamental principles of industrial filtration.* Trans. Inst. Chem. Eng., 16, 168-188 (1938).
170. Carman, P.C.: *Molecular distillation and sublimation.* Trans. Faraday Soc., 44, 529-38 (1948).
171. Carney, T.P.: *Laboratory fractional distillation.* Macmillan: New York, 1949.
172. Cawley, L.P.; Minard, B.J.; Tourellotte, W.W.; Ma, B.I.; Chelle, C.: *Immunofixation electrophoretic technique applied to identification of proteins in aerum and cerebrospinal liquid.* Clin. Chem., 22(8) 1262-8 (1976).
173. Cechova, S.; Macasek, F.: *Solvent extraction of niobium cations with products of nitrobenzene radiolysis.* J. Radioanal.Nucl.Chem., Articles, 149(2) 271-276 (1991).

174. Celeda, J.; Skramovsky, S.; Zilkova, J.: *The metachors of polyvalent and associated electrolytes in aqueous solutions.* Coll. Czechosl. Chem. Commun., 49, 1079-89 (1984).
175. Celeda, J.; Tuck, D.G.: *Densimetric study of indium(III) species in mineral acid solutions.* J.Inorg.Nucl.Chem., 36(2) 373-8 (1974).
176. Chalabreyse, M.C.; Neige, R.; Aubert, J.: *Kinetic measurements in liquid liquid exchange applied to isotopic separation.* Sep.Sci.Technol., 15(3) 557-66 (1980).
177. Chapman, S.; Cawling, T.G.: *The mathematical theory of non-uniform gases.* Univ. Press: Cambridge, 1939.
178. Charm, S.E.; Matteo, C.C.: *Scale-up of protein isolation.* Methods Enzymol., 22, 476-556 (1971).
179. Chattoraj, D.K.; Birdi, K.S.: *Adsorption and the Gibbs surface excess.* Plenum Press: New York 1984.
180. Chen, B.; Liang, C.; Yang, J.; Yaghi O.M.: *A microporous metal-organic framework for gas-chromatographic separation of alkanes.* Angew. Chem. Int. Ed., 118, 1390-1393 (2006).
181. Cheremisinoff, N.P.; Azbel, D.S.: *Liquid filtration.* Butterworth: London, 1983.
182. Chiarizia, R.; Horwitz, E.P.: *Hydrolytic and radiolytic degradation of octyl(phenyl) -N, N-diisobutylcarbamoylmethylphosphine oxide and related compounds.* Solv. Extr. Ion Exch., 4(4) 677 (1986).
183. Chiarizia, R.; Horwitz, E.P.: *Study of uranium removal from groundwater by supported liquid membranes.* Solvent Extr. Ion Exch., 8(1) 65-98 (1990).
184. Chilton, T.H.; Colburn, A.P.: *Distillation and absorption in packed columns. A convenient design and correlation method.* Ind.Eng.Chem., 27, 255 (1935).
185. Chizhevskaya, S.V.; Sinegribova, O.A.; Danilova, S.S.; Korovin, Yu.F.; Yagodin, Yu.A.: *Behaviour of silicic acid along the interface in the nitric acid-silicic acid- decane system.* Izv.Vyssh.Ucheb.Zaved., Khimiya, 20(5) 694-7 (1977).

186. Chlopin, V.; Nikitin, B.: *Beitrag zur Kenntniss der fraktionierten Krystallisation radioaktiver Stoffe, nebst dem Versuche einer Theorie dieses Vorganges ll.:* Z. Anorg. Allg. Chem., 166, 311-38 (1927).
187. Chlopin, V.; Nikitin, B.A.: *The question concerning the existence of a new kind of mixed crystals of the type barium sulphate and potassium permanganate. Using radioactive indicators.* Z.Phys.Chem. (A) 145, 137 (1929).
188. Choppin, G.R.; Silva, R.I.: *Separation of the lanthanides by ion exchange with alpha-hydroxyisobutyric acid.* J.lnorg. Nucl.Chem., 3, 153-4 (1956).
189. Choppin, G.R.: *Studies of the synergistic effects.* Sep.Sci. Technol., 16(6) 1113-26 (1981).
190. Choppin, G.R.; Khankhasaeev, M.Kh. (Eds.): *Chemical separation technologies and related methods of nuclear waste management.* Kluwer Academic Publishers: Dordrecht, 1999.
191. Churchill, R.V.: *Modern operations mathematics in engineering.* McGraw Hill: New York, 1950.
192. Clark, J.H.; Anderson, R.G.: *Silane purification via laser-induced chemistry.* Appl. Phys. Lett., 32, 46 (1978).
193. Clarke, A.N.; Wilson, D.J.: *Foam flotation: Theory and applications.* Marcel Dekker: New York 1983.
194. Clarke, F.W.: *On a new method of separating tin from arsenic, antimony, and molybdenum.* Chem. News, 21, 124 (1870).
195. Clausen, J.: *Electrokinetic separation methods.* Elsevier: Amsterdam, 1979.
196. Clausing, P.: *The formation of beams in molecular streaming.* Z. Physik, 66, 471-476 (1930).
197. Clearfield, A.: *The preparation of crystalline zirconium phosphate and some observations on its ion-exchange behavior.* J. Inorg. Nucl. Chem., 26(1) 117-29 (1964).
198. Clearfield, A. (Ed.): *Inorganic ion exchange materials.* CRC Press: Boca Raton, 1982.

199. Clement, R.E.: *Gas chromatography: Biochemical, biomedical, and clinical applications.* Wiley Interscience: New York, 1991.
200. Clementi, E.; Corongiu, G.: *Simulation of the solvent structure for macro molecules: Solvation model for B-DNA and sodium ion-B-DNA helix at 300 K.* Ann.N.Y.Acad.Sci., 367, 83-107 (1981).
201. Clusius, K.; Dickel, G.: *The separation tube. I. Principles of a new method of gas and isotope separation by thermal diffusion.* Z. physik. Chem., B44, 397 (1939).
202. Coates, J.I.; Glueckauf, E.: *Theory of chromatography. III. Experimental separation of two solutes and comparison with theory.* J.Chem.Soc. (London) 1308-14 (1947).
203. Cognet, M.C.; Renon, H.: *Influence of aqueous phase composition upon copper extraction by cationic extractants: a thermodynamic interpretation.* Hydrometallurgy, 2, 305-14 (1976).
204. Cohen, K.: *Packed fractionating columns and the concentration of isotopes.* J.Chem.Phys., 81, 588-97 (1940).
205. Cohen, K.: *Theory of isotope separations as applied to the large scale production of U-235.* McGraw-Hill: New York, 1951.
206. Colburn, A.P.: *Relation between mass transfer (absorption) and fluid friction.* Ind.Eng.Chem., 22, 967-70 (1930).
207. Colburn, A.P.: *Simplified calculations of diffusional processes. Number of transfer units or theoretical plates.* Ind.Eng.Chem., 33, 459-67 (1941).
208. Coleman, C.F.; Brown, K.B.; Moor, J.G.; Crouse, D.J.: *Solvent extraction with alkyl amines.* Ind.Eng.Chem., 50, 1756-62 (1958).
209. Collander, R.: *The distribution of organic compounds between ether and water.* Acta Chem.Scand., 3, 717-47 (1949).

210. Comninellis, C.; Doyle, M.; Winnick, J. (Eds.): *Energy and electrochemical processes for a cleaner environment.* Electrochemical Society Proceedings, 2011.
211. Connick, R.E.; McVey, W.H.: *The aqueous chemistry of zirconium.* J.Am.Chem.Soc., 71, 3182-91 (1949).
212. Consden, R.; Gordon, A.H.; Martin, A.J.P.: *Quantitative analysis of proteins: a partition chromatographic method using paper.* Biochem. J., 38, 224-232 (1944).
213. Conway, W.D.: *Countercurrent chromatography: Apparatus, theory and applications.* VCH: Weinheim 1990.
214. Cooney, C.L.; Wang, H.Y.; Wang, D.I.C.: *Computer-aided material balancing for prediction of fermentation parameters.* Biotechnol. Bioeng., 19(1) 55-67 (1977).
215. Corradini, DF. (ed.): *Handbook of HPLC.* 2nd. Ed. CRC: Boca Raton, 2011.
216. Coster, D.; Hevesy, G.: *On the new element hafnium.* Chem. Ind., 42, 258 (1923).
217. Coughanowr, D.R.: *Process systems and analysis and control.* McGraw-Hill: New York, 1964.
218. Cox, G.B. (Ed.): *Preparative enantioselective chromatography*, Blackwell, Oxford 2005.
219. Cox, M.; Elizalde, M.; Castresana, J.: *Interfacial properties and metal extraction chemistry of the organophosphorous acids.* In: Proceedinga Intern. Solv. Extr. Con/. ISEC'89, Denver, 26.9.-2.9.1989, p. 268-269. ISEC: Denver, 1983.
220. Craig, L.C.: *Identification of small amounts of organic compounds by distribution* studies. *II. Separation by countercurrent distribution.* J. Biol. Chem., 155, 519-34 (1944).
221. Craig, L.C.: *Partition chromatography and counter-current distribution.* Anal.Chem., 22, 1346-52 (1950).
222. Cranford, M.: *Air pollution theory.* McGraw-Hill: New York, 1976.
223. Crank, J.: *The mathematics of diffusion.* Clarendon Press: Oxford, 1956.

224. Cremer, E.; Flugge, S.: *Adsorption an Oberflaechen mit eingefrorenen thermischen Gleichgewigkeit der aktiven Stelle.* Z.phys.Chem.B41 453-65 (1938).
225. Cremer, E.; Muller, R.: *Separation and determination of small quantities of gases by chromatography.* Mikrochem. Acta, 36/37, 553-60 (1951).
226. Cremer, H.W.; Davies, T. (Eds.): *Chemical engineering practice (Vols. I - XII).* Butterworth: London, 1956.
227. Criss, C.M.; Cobble, J.W.: *The thermodynamic properties of high temperature aqueos solutions. IV. Entropies of the ions up to 200° C and the correspondance principle.* J.Am.Chem. Soc., 86, 5385-90 (1964).
228. Crosland-Taylor, P.J., *A device for counting small particles suspended in a fluid through a tube.* Nature 171, 37-38 (1953).
229. Cuatrecasas, P.; Wilchek, M.; Anfinsen, C.B.: *Selective enzyme purification by affinity chromatography.* Proc. Natl. Acad. Sci. USA, 61(2) 636-42 (1968).
230. Cunningham, B.B.; Werner, L.B.: *First isolation of plutonium.* J.Am.Chem.Soc., 71, 1521-8 (1949).
231. Curie, M.: *Sur une substance nouvelle radio-active, contenue dans la pechblend.* C.R.Hebd.Acad.Sci.Physico-Chemie, 127, 175 (1898).
232. Curie, P.: *Oeuvres.* Societe Francaise de Physique: Paris, 1908.
233. Cussler, E.L.: *Membranes which pump.* Am.Inst.Chem.Eng. Journal, 17(6) 1300-1003 (1971).
234. Cussler, E.L.; Evans, D.F.: *How to design liquid membrane separations.* Separ.Purif.M., 3(2) 399-421 (1974).
235. Cvengros, J.; Tkac, A.: *Continuous processes on wiped films. 2. Distilling capacity and separating efficiency of molecular evaporator with convex evaporating surface.* Ind. Eng. Chem.Process Des.Dev., 17(3) 246-51 (1978).
236. Dahlstrom, D.A.; Cornell, C.A.: *Thickening and clarification.* Chem. Eng., 78(4) 63-69 (1971).

237. Damköhler, G.: *The influence of flow, diffusion and heat transfer on the performance of reaction furnaces. I. General considerations on the transfer of chemical processes from small to large.* Z.Electrochem., 42(12) 846-62 (1936).
238. Damköhler, G.: *The influence of flow, diffusion and heat transfer on performance of reaction furnaces. II. Isothermal, constant-volume homogeneous reactions of the first order.* Z.Electrochem., 43, 1-8 (1937).
239. Damköhler G., Theile H.: Stofftrennung durch Adsorption aus einem Hilfsgasstrom. Die Chemie, 56, 353-4 (1943).
240. Danckwerts, P.V.: *Absorption by Simultaneous diffusion and chemical reaction into particles of various shapes and into falling drops.* Trans.Faraday Soc., 47, 1014-23 (1951).
241. Danckwerts, P.V.: *Significance of liquid film coefficients in gas absorption* Eng. Chem., 43, 1460 (1951).
242. Danckwerts, P.V.: *Unsteady-state diffusion or heat conduction with moving boundary.* Trans. Farad. Soc., 46, 701-12 (1950).
243. Danesi, P.R.; Chiarizia, R.; Muhammed, M.: *Mass transfer rate in liquid anion exchange processes. I. Kinetics of the two phase acid-base reaction in the system TLA-toluene-HCl-water.* J.Inorg.Nucl.Chem., 40, 1581 (1978).
244. Danesi, P.R.; Horwitz, E.P.; Vandergriff, G.F.; Chiarizia, R.: *Mass transfer rate through liquid membranes: interfacial chemical reactions and diffusion as simultaneous permeability controlling factors.* Separ. Sci. Technol., 16(2) 201-11 (1981).
245. Danesi, P.R.: *Separation of metal species by supported liquid membranes.* Sep.Sci. Technol., 19, 857 (1984).
246. Danesi, P.R.: *Solvent extraction and supported liquid membranes: differences, similarities and their relevance in the design of practical membrane systems.* In: Proceedings ISEC'88, vol. III, p. 7. Academy of Science USSR: Moscow 1988.

247. Danesi, P.R.; Reichley-Yinger, L.: *Origin and significance of the deviations from a pseudo first order rate law in the coupled transport of metal species through supported liquid membranes.* J. Memb. Sci., 29, 195-206 (1986).
248. Danielli, J.F.; Davson, H.: *Theory of permeability of thin films.* J. Cellular Compar. Physiol., 5, 495-508 (1934).
249. Danilov, V.I.; Pluzhnik, E.E.; Teverovskii, B.M.: *Formation of centers of crystallization in a supercooled liquid. I. Crystallization of piperine in an ultrasound field.* Zh. Eksper. Teor.Fiz., 9, 66 (1939).
250. Darcy, H.P.G.: *Les fontainer publiques de la ville de Dijon.* V. Dalmont: Paris, 1856.
251. Darken, L.S.: *Thermodynamics of ternary metallic solutions.* Trans. Met. Soc. AIME, 239(1) 90-6 (1967).
252. Davankov, V.A.; Rogozhin, S.V.: *Ligand chromatography as a novel method for the investigation of mixed complexes: Stereoselective effects in α-amino acid copper(II) complexes.* J.Chromatog. 60, 280-283 (1971).
253. Davankov, V.A.; Navratil, J.D.; Walton, H.F.: *Ligand exchange chromatography*, CRC Press, Boca Raton 1988.
254. Davies, C.W.; Hoyle, J.: *The dissociation constants of calcium hydroxide.* J.Chem.Soc., (1) 233-234 (1951).
255. Davies, G.A.; Ponter, A.B.; Craine, K.: *The diffusion of carbon dioxide in organic liquids.* Can. J. Chem. Eng., 45(6) 372-6 (1967).
256. Davies, J.T.: *Turbulence phenomena.* Academ. Press: New York, 1972.
257. Davies, O.L.: *Statistical methods in research and production with special reference to the chemical industry.* London, 1949.
258. Davis, B.J.: *Disk electrophoresis. II. Method and application to human serum proteins.* Ann.New York Acad.Sci., 121, 404-427 (1964).

259. Davis, W. Jr.; Mrochek, J.; Judkins, R.R.: *Thermodynamics of the two phase system: water - uranyl nitrate - TBP - Amsco 125-82.* J. Inorg. Nucl. Chem., 32, 1689 (1970).
260. Davydov, Yu.P.: *Sostoyaniye radionuklidov v rastvorakh (Speciation of radionuclides in solutions).* Nauka i tekhnika: Minsk, 1978.
261. De Vault, D.: *The theory of chromatography.* J.Am.Chem. Soc., 65, 532-40 (1943).
262. Dean, J.A.: *Chemical separation methods.* Van Nostrand: New York, 1969.
263. Deane, W.A.: *Settling problems.* Trans. Am. Electrochem. Soc., 37, 71-102 (1920).
264. Debye, P.: *Die van der Waalschen Kohäsionskrafte.* Nachr.K.Ges.Wiss.Gottingen, 55-73 (1920).
265. Debye, P.; Hückel, E.: *Bemerkungen zu einem Satze uber die kataphoretische Wanderungsgeschwindigkeit suspendierter Teilchen.* Phys. Z., 25, 49-52 (1924).
266. Debye, P.; Hückel, E.: Zur *Theorie der Elektrolyte. I. Gefrierpunktserniedrigung* und *verwandte Erscheinnungen.* Phys. Z., 24, 185-206 (1923).
267. Debye, P.; Hückel, E.: *Zur theorie des Electrolyte. II. Das Grenzgesetz für die elektrische Leitfahigkeit.* Phys. Z., 24, 305-25 (1923).
268. DeGroot, S.R.: Thermodynamics of irreversible processes. North Holland: Amsterdam, 1961.
269. Denkova, A.G.; Terpstra, B.E.; Steinbach, O.M.; ten Dam J.; Wolterbeek, H. Th.: *Adsorption of molybdenum on mesoporous aluminum oxides for potential application in nuclear medicine.* Sep.Sci. Technol. 48 (9) 1331-1338 (2013).
270. DePoorter, G.L.; Rofer-DePoorter, C.K. *The effect of IR laser radiation on the $UO_2(NO_3)$-2-tributylphosphate - nitric acid solvent extraction system.* J.Inorg.Nucl.Chem.39 2061 (1977).

271. Determann, H.: *Gel chromatography, gel filtration, gel permeation, molecular sieves.* 2nd Ed. Springer Verlag: Berlin, 1969.
272. Deutsch, D.H.: *Can the second law of thermodynamics be circumvented?* An.Lab., 13(5) 54-65 (1981).
273. Deutsch, J.M.: *Dynamics of pulsed-field electrophoresis.* Phys.Rev.Lett., 59(11) 1255-8 (1987).
274. Devyatykh, G.G.; Churbanov, M.F.: *Current status of the problems of preparing substances of high purity.* Zh.Vsessoyuz.Khim.Obsch., 29(6) 606-14 (1984).
275. Devyatykh, G.G.; Vlasov, S.M.: *Dependence of the relative vapour pressure of a microcomponent on the interaction of its molecules in dilute nonelectrolyte solution.* Russ. J.Phys. Chem., 40(10) 1356-1359 (1966).
276. DeWachter, R.; Furs, W.: *Preparative two-dimensional polyacrylamide gel electrophoresis of phosphorous-32 labeled RN A.:* Anal. Bioch., 49(1) 184-97 (1972).
277. Deyl, Z. (Ed.): *Separation methods*. Elsevier Science: New York, 2011.
278. Deyl, Z.; Janak, J.; Schwarz, V.: *Analytical separations. Elsevier's electronic bibliography*. Elsevier Science: New York, 1998.
279. Ding, Xiaoyun; Sz-Chin S.L; Lapsley, M.I.; Li, Sixing; Guo, Xiang; Chan, Chung Yu; Chiang, Iao; Wang, Lin; J. Philip McCoy, J.P.; Huang, T.J.: *Standing surface acoustic wave (SSAW) based multichannel cell sorting.* Lab.Chip., 12, 4228-4231 (2012).
280. Ditkin, V.A.; Prudnikov, A.P.: *Spravochnik po operatsionnomu ischisleniyu (Handbook on operational calculus)*. Vysshaya Shkola: Moscow, 1965.
281. Dittrich, M.: *Ueber Filtrieren* und *Veraschen von schlemigen Niederschlagen.* Ber.Dtsch.Chem. Ges., 37, 1840 (1904).
282. Dixon, M.; Webb, E.C.: *Enzyme fractionation by salting-out: a theoretical note.* Advan. Protein Chem., 16, 197-219 (1961).

283. Dobre, T.G.; Marcano, J.G.S.: *Chemical engineering. Modelling, simulation and similitude.* Wiley-VCH Verlag: Weinheim, 2007.
284. Doerner, N.A.; Hoskins, W.M.: *Co-precipitation of radium and barium sulfates.* J.Am.Chem.Soc., 47(3) 662 (1925).
285. Dolezalek, F.: *Zur Theorie der binaeren Gemische und konzentrierter Loesungen.* Z.Phys.Chem., 64, 727-747 (1908).
286. Dolezalek, F.: *Zur Theorie der binaere Gemische und konzentrierte Loesungen. II.:* Z.Phys.Chem., 71, 191-213 (1910).
287. Dolezalek, F.; Schulze, A.: *The theory of binary mixtures and concentrated solutions. IV. The mixture: ether-chloroform.* Z.Phys.Chem., 83, 45-78 (1913).
288. Donnan, F.G.: *Theorie der Membrangleichgewichte und Membranpotentiale bei Vorhandeasein von nicht dialysierenden Elektrolyten. Ein Beitrag zur physikalisch chemischen Physiologie.* Z. Elektrochem., 17, 572-81 (1911).
289. Donohue, T. *Photochemical separation of europium from lanthanide mixtures in in aqueous solution.* J.Chem.Phys. 67, 5402 (1977).
290. Dorfner, K. (ed.): *Ion exchangers.* de Gruyter: Berlin 1991.
291. Dorr, J.V.N.; Franklin, P.L.: *Solid liquid separations.* In: *Colloid Chemistry,* vol. VI., p. 78P. Alexander, J., Ed., Reinhold Publ.Co: New York, 1946.
292. Douglas, J.M.: *Conceptual design of chemical processes,* McGraw-Hill, New York, 1988.
293. Dubinin, M.; Yavich, S.: *Theoretical basis for the calculation of recuperative adsorbers (especially for the selective adsorption of a mixture of gases and vapor).* J.Applied Chem.(USSR), 9, 1191-203 (1936).
294. Duhem, P.: *Sur les vapeurs emises par un melange de substances volatiles.* Compt. rend., 102, 1449 (1886).
295. Durrum, E.: *A microelectrophoretic and microionophoretic technique.* J.Am.Chem.Soc., 72, 2943-8 (1950).

296. Duursma, E.K.; Hoede, C.: *Theoretical, experimental and field studies concerning molecular diffusion of radioisotopes in sediments and suspended solid particles of the sea Part A: Theories and mathematical calculations.* Netherland J. Sea Res., 3(3) 423-57 (1967).
297. Dyrssen, D.; Jagner, D.; Wengelin, F.: *Computer calculation of ionic equilibria and titration procedures.* Almqvist and Wiksell: Stockholm, 1968.
298. Dyrssen, D.; Sillen, L.G.: *Two-parameter equations for a complex-formation sytem and their application to the two-phase distribution of metal complexes.* Acta Chem.Scand., 7, 663 (1953).
299. Eckert, C.A.; Newman, B.A.; Nicolaides, G.L.: *Measurement and application of limiting activity coefficients.* Am.Inst. Chem.Eng. Journal, 27, 33-40 (1981).
300. Efremov, A.A.; Blyum, G.Z.; Grinberg, E.E.: *Methods for the complete purification of liquid chemicals for new industrial areas.* Zh.Vsesssoyuz.Khim.Obsch., 29(6) 637-45 (1984).
301. Eidinoff, M.L.: *Cathodic protium-tritium separation factor.* J.Am.Chem.Soc., 69, 977 (1947).
302. Einstein, A.: *Eine neue Bestimmung der Molekuldimension.* Ann.Physik, 19, 289-306 (1905).
303. El-Garhy, M.; El-Bayoumy, S.; El-Alfy, S.: *Technetium-99m generator using saline solution.* Radiochim. Acta., 7, 163 (1967).
304. Emmett, P.H.; Brunauer, S.: *The use of low-temperature van der Waals absorption isotherm in determination of the surface area of iron synthetic ammonia catalysts.* J.Am. Chem.Soc., 59, 1553-69 (1937).
305. Erenferst-Afannasjewa, T.A.: *Dimensional analysis viewed from standpoint of the theory of similitude.* Phil. Mag., 1(1) 16-21 (1925).
306. Ergun, S.: *Fluid flow through packed columns.* Chem. Eng. Progr., 48, 89 (1952).

307. Esin, O. A.: *The theory of over-voltage and mutual discharge of ions.* Zh.Fiz.Khim., 6(6) 795-801 (1935).
308. Esser C.: *Historical and useful methods of preselection and preparative scale sorting.* In: *Cell Separation Methods and Applications.* D.Recktenwald, A.Radbruch (Eds.), p 1–14. Marcel Dekker: New York, 1998.
309. Eucken, A.; Knick, H.: *Automatic procedure for the microanalytical separation of low-boiling hydrocarbons by desorption.* Brennstoff. Chem., 17, 241 (1936).
310. Everaerts, F.M.; Beckers, J.L.; Verheggen, T.P.E.M.: *Isotachophoresis. Theory, instrumentation and applications.* Elsevier: Amsterdam, 1976.
311. Everaerts, F.M.; Geurts, M.; Mikkers, F.E.P.; Verheggen, T.P.E.M.: *Analytical isotachophoresis.* J.Chromatogr. (LIB) 119(6) 129-55 (1976).
312. Everearts, F.M.; Verheggen, T.P.E.M.; Mikkers, E.P.: *Determination of substances at low concentrations in complex mixtures by ITP with column coupling.* J.Chromatogr., 169, 21-38 (1979).
313. Ewell, R.H.; Harrison, J.M.; Berg, L.: *Azeotropic distillation.* Ind.Eng.Chem., 36, 871-5 (1944).
314. Eyring, H.: *Viscosity, plasticity and diffusion as examples of absolute reaction rates.* J.Chem. Phys., 4, 283-91 (1936).
315. Fair, J.R.; Humphrey, J.L.: *Liquid-liquid extraction: possible alternative to distillation.* Solvent Extr. Ion Exch., 2(3) 323 (1984).
316. Fallon, A.; Booth, R.F.G.; Bell, L.D.: *Application of HPLC in biochemistry: Laboratory techniques in biochemistry and molecular biology.* Elsevier: Amsterdam 1987.
317. Fan, Zh.; Zhang, X.F.; Su, X.S.: *Supported liquid membranes for copper extraction.* In: *Proceedings Int. Solvent Extraction Conference ISEC '88, vol. III, p. 82.* Academy of Science USSR: Moscow, 1988.

318. Fanali, S.; Haddad, P.R.; Poole, C.; Schoenmakers, P.; Lloyd, D.K.: *Liquid chromato-graphy: Fundamentals and instrumentation*, Newnes (Elsevier): Burlington, 2013.
319. Farron-Furstenthal, F.; Lightholder, J.R.: *The purification of nuclear protein kinase by affinity chromatography.* FEBS Letters, 84, 313 (1977).
320. Fay, H.; Quets, J.M.: *Density separation of solids in ferrofluids with magnetic grids.* Sep.Sci.Technol., 15(3) 339-69 (1980).
321. Federman, A.: *On the general methods of integration of partial derivative equation of the second order (*in Russian*).* Izv.Sankt-Peterb.Politekhn.Inst., 16 (1) 5-12 (1911).
322. Feibus, H.; Muchunas, P.J.: *Application of separations technology for the control of pollutants in advanced coal combustion systems. In: Chemical separations, vol. 2, p. 131.* King, C.J., Navratil, J.D., Litarvan Lit.: Denver, 1986.
323. Fell, C.J.D.: *New developments in water and effluent treatment.* In: Chemical separations, vol. 2, p. 61. King, C.J., Navratil, J.D., Eds., Litarvan Lit.: Denver, 1986.
324. Fendler, J.H.: *Membrane mimetic chemistry.* J. Wiley: New York, 1982.
325. Fendler, J.H.: *Membrane* mimetic separations. *In: Ordered media in chemical separations, p. 83-104.* W.L. Hinze, D.W. Armstrong, Eds., Am. Chem. Society: Washington, 1987.
326. Fenske, M.R.; Quiggle, D.; Tongberg, C.O.: *Composition of straight-run Pennsylvania gasoline. I. Design of fractionating equipment.* Ind.Eng.Chem., 24, 408-18 (1932).
327. Fenske, M.R.; Tongberg, S.O.; Quiggle, D.: *Packing materials for fractionating columns.* Ind.Eng.Chem., 26, 1169-77 (1934).
328. Ferse, A.: *On individual thermodynamic activity coefficients of some types of ions in concentrated solutions of electrolytes.* Z.Chemie (Leipzig) 18, 8-15 (1978).
329. Fialkov, B.Ya.; Zhitomirskii, A.N.; Tarasenko, Yu.A.: *Fizicheskaya khimiya nevodnykh rastvorov (Physical*

chemistry of non-aqueous solutions). Khimiya: Moskva, 1973.
330. Fick, A.: *On liquid diffusion.* Phil.Mag., 10 (4) 30 (1855).
331. Fick, A.: *Ueber Diffusion.* Ann.Phys.Chem., 40, 59 (1855).
332. Finch, J.A.; Dobby, G.S.: *Column flotation.* Pergamon Press: London, 1990.
333. Fleck, R.D.Jr.; Kirwan, D.J.; Hall, K.R.: *Mixed-resistance diffusion kinetics in fixed- bed adsorption under constant pattern conditions.* Ind.Eng.Chem., Fundam., 12, 95 (1973).
334. Fleetwood, J.G.: *Steady-state distribution - a new technique.* Chem. Process Eng., 43, 106-8 (1962).
335. Flett, D.S.: *Ion exchange membranes.* Ellis Horwood: Chichester, 1983.
336. Florey, H.W.; Chain, E.: *Penicillin: demonstration of its values as a chemotherapeutic agent. Preliminary report of a new method.* Med. Record., 158, 217-19 (1945).
337. Florkova, A.K.: *Razdelenie azeotropnykh smesej. (Separation of azeotropic mixtures).* VLADOS: Moscow, 2010.
338. Flory, P.J.: *Thermodynamics of high-polymer solutions.* J.Chem.Phys., 10, 51-61 (1942).
339. Fomin, V.V.: *Some problems in determining the composition and stability of compounds in solutions with constant ionic strength.* Radiokhimiya, 9(6) 652-64 (1967).
340. Fomin, V.V.; Maiorova, E.P.: *Determination of the stability constants for the $Th(NO_3)_4$ ions by extraction with tributylphosphate.* Zh.Neorg.Khim., 1, 1703-12 (1956).
341. Forciniti, D.: *Industrial bioseparations: Principles and practice.* Blackwell Publishers: Oxford, 2007.
342. Fredenslund, A.; Gruehling, J.; Michelsen, M.L.; Rassmussen, P.; Prausnitz, J.M.: *Computerized design of multicomponent distillation columns using the UNIFAC group contribution method for calculation of activity coefficients.* Ind.Eng.Chem., Process Des.Dev., 16(4) 450-62 (1977).

343. Fredenslund, A.; Iones, R.L.; Prausnitz, J.M.: *Group-contribution estimation of activity coefficients in non-ideal liquid mixtures.* Am.Inst.Chem.Eng. Journal, 21(6) 1086-99 (1975).
344. Freifelder, D.; Better, M.: *Dialysis of small samples in agarose gels.* Anal. Biochem., 123(1) 83-5 (1982).
345. Freiser, H.: *Relevance of solubility parameter in ion association extraction systems.* Anal.Chem., 41(10) 1354-5 (1969).
346. Freundlich, H.: *Colloid and capillary chemistry.* Dutton: New York, 1922.
347. Fried, B.; Sherma J. (eds): *Practical thin-layer chromatography: a multidisciplinary approach.* CRC: Boca Raton, 1996.
348. Fried, B.; Sherma, J.: *Thin-layer chromatography.* 4th eds., Marcel Dekker: New York, 1999.
349. Friedman, M.; Noma, As.: *Histamine analysis on a single column amino acid analyzer.* J.Chromatogr., 219(2) 343-8 (1981).
350. Fritz, W.; Merk, W.; Schleunder, E.V.: *Competitive adsorption of two dissolved organics onto activated carbon.* Chem. Eng. Sci., 36, 731 (1981).
351. Frolkova, A. K.: *Razdelenie azeotropnykh smesei: Fiziko-khimicheskie osnovy i tekhnologicheskie priemy (Separation of azeotropic mixtures: Physicochemical principles and techniques).* Vlados: Moscow, 2010.
352. Frolov, Yu.G.; Sergievskii, V.V.: *Relation between extraction constants and empiric parameters of polarity of solvents I. Correlation at physical distribution.* Radiokhimiya, 13(4) 530-4 (1971).
353. Frolov, Yu.G.: *Theory of mixed isoactive electrolyte solutions.* Russ.Chem.Rev., 50, 232-247 (1981).
354. Frolov, Yu.G.; Denisov, D.A.: *Calculation of complex formation constants from the activity coefficients of binary solutions.* Russ. J.Phys.Chem., 52(5) 750-751 (1978).

355. Frolov, Yu.G.; Sergievskii, V.V.: *Influence of diluent on extraction equilibrium. In: Khimiya protsessov ekstraktsii,* p. 97. Yu.A. Zolotov, B.Ya. Spivakov, Eds., Nauka: Moskva, 1972.
356. Frolov, Yu.G.; Sergievskii, V.V.; Zuev, A.P.: *The influence of hydration of ions on extraction equilibrium* Atom.Energiya, 35, 109 (1973).
357. Fronaeus, S.: *A new principle for the determination of complex equilibria and the determination of complexity constants.* Acta Chem.Scand., 4, 72-87 (1950).
358. Fronaeus, S.: *Use of cation exchangers for the quantitative investigation of complex systems.* Acta Chem.Scand., 5, 859-71 (1951).
359. Fuchs, N.A.: *The mechanics of aerosols.* McMillan: New York, 1964.
360. Fuller, E.J.: *Multiple extraction by slurry systems. In: Separation and purification methods. Vol.* I., p. 253. E. Perry, C.J. Van Oss, Eds., Marcel Dekker: New York, 1973.
361. Fulwyler, M.J.: *Electronic separation of biological cells by volume.* Science, 150(3698) 910-911 (1965).
362. Furry, W.H.; Iones, R.C.; Onsager, L.: *The theory of isotope separation by thermal diffusion.* Phys. Rev., 55, 1083 (1939).
363. Gaikar, V.G.; Sharma, M.M.: *Extractive separations with hydrotrops.* Solvent Extr.Ion Exch., 4(4) 839 (1986).
364. Gaile, A.A.; Somov, V.Ye.: *Protsessy razdelenia i ochistki produktov pererabotki nefti i gaza. (Processes of separation and purification of the products of oil and gas reprocessing).* Khimizdat: Sankt Peterburg, 2012.
365. Gaizer, F.: *Computer evaluation of complex equilibria.* Coord. Chem. Rev., 27, 195 (1979).
366. Gan, L.M.; Chew, C.H.; Wong, M.K.; Koh, L.L.; Ng, K.H.: *Continuous absorption of toluene vapor by micellar solutions and the resulting microemulsions* J. Dispersion Sci. Technol., 8(4) 385-40 (1987).

367. Ganetsos, G.; Barker, P.E.: *Preparative and production scale chromatography.* CRC Press: Boca Raton, 1992.
368. Gangrsky, Yu.P.; Hradecny, C.; Zemlyanoy, S.G.; Zuzaan, P.; Yermolayev, I.M.; Markov, B.N.; Mishinsky, G.V.; Slovak, J.; Tethal, T.; Stekl, I.: *Laser light-induced drift of radioactive isotopes ^{22}Na and ^{24}Na.* JETP, 79(3) 399-403 (1994).
369. Gapon, E.N.; Gapon, T.B.: *Chromatographic analysis of precipitates.* Dokl.Akad.Nauk.SSSR, 60(3) 401 (1948).
370. Garrels, R.M.; Christ, C.L.: *Solution, minerals and equilibria.* Harper and Row: New York, 1965.
371. Gassmann, E.; Kuo, J.E.; Zare, R.N.: *Electrokinetic separation of chiral compounds.* Science 230, 81 (1985).
372. Gates, S.C.; Becker, J.: *Laboratory automation using the IBM PC.* Prentice Hall: New York, 1989.
373. Gaudin, A.M.: *Principles of mineral dressing.* McGraw Hill, New York, 1939.
374. Geankoplis, C.: *Transport processes and separation process principles (includes unit operations).* 4th ed. Prentice Hall Press: Upper Saddle River, 2003.
375. Geiseler, D.; Ritter, M.: *Optimization of assay conditions for free ligand determination by immunoassay.* Anal.Chem., 54(12) 2062 (1982).
376. Gemski, M.J.: *Single step purification of monoclonal antibody from Murine Ascites a tissue culture fluids by anion exchange HPLC.* Bio.Techniques, 3, 378 (1985).
377. Gerritson, T. (Ed.): *Modern separation methods of macromolecules and particles. (Vol 2. of Progress in separations and purification).* Wiley-Interscience: New York, 1969.
378. Gevod, V.S.; Ksenzhek, O.S.; Reshetnyak, I.L.: *Artificial membrane structures and prospects of their practical applications.* Biol.Membr., 5(12) 1237- 1269 (1988).
379. Ghiorso, A.: *Isotopes of element 102 with masses 251 to 258.* Phys.Rev.Lett., 18(11) 401-4 (1967).

380. Ghiorso, A.; Choppin, G.R.; Harvey, B.G.; Thompson, S.G.; Seaborg, G.S.: *New element mendelevium, atomic number 101.* Phys. Rev., 98, 1518 (1955).
381. Gibbs, J.W.: *On the equilibrium of heterogeneous substances.* T. Connecticut Acad. Sci, 3, 108 (1874).
382. Giddings, J.C.; Eyring, H.: *A molecular-dynamic theory of chromatography.* J.Phys.Chem. 59, 416 (1955).
383. Giddings, J.C.: *Plate height contributions in gas chromatography.* Anal.Chem., 33, 962 (1961).
384. Giddings, J.C.; Robinson, R.A.: *Failure of the eddy diffusion concept of gas chromatography.* Anal.Chem., 34, 885 (1962).
385. Giddings, J.C.: *Generalized nonequilibrium theory of plate height in largescale gas chromatography.* J. Gas Chrom., 1(4) 38-42 (1963).
386. Giddings, J.C.: *Dynamics of chromatography.* Marcel Dekker: New York, 1965.
387. Giddings, J.C. *Maximum number of components resolvable by gel filtration and other chromatographic methods.* Anal. Chem. 39 1927 (1967).
388. Giddings, J.C.; Kucera, E.; Russel, C.P.; Myers, M.N.: *Statistical theory for the equilibrium distribution of rigid molecules in inert porous network I. Exclusion chromatography* J.Phys.Chem., 72(13) 4379-408 (1968).
389. Giddings, J.C.; Myers, H.N.; McLaren, L.; Keller, R.A.: *High-pressure gas chromatography of nonvolatile species. Compressed gas is used to cause migration of interactable solutes.* Science, 162, 67-73 (1968).
390. Giddings, J.C.: *Generation of variance, theoretical plates, resolution, and peak capacity in electrophoresis and sedimentation.* Separ.Sci., 4(3) 181-9 (1969).
391. Giddings, J.C.; Dahlgren, K.: *Resolution and peak capacity in equilibrium-gradient methods of separation.* Separ.Sci. 6 345 (1971).

392. Giddings, J.C.: *The conceptual basis of field-flow fractionation.* J.Chem.Ed., 50, 667-9 (1973).
393. Giddings, J.C.; Yang, F.J.F.; Myers, M.N.: *Sedimentation field-flow fractionation.* Anal.Chem., 46(13) 1917-24 (1974).
394. Giddings, J.C.; Myers, M.N.; Caldwell, K.D.: *Analysis of biological macromolecules and particles by field-flow fractionation.* Methods Bioch.Anal., 26, 79-136 (1980).
395. Giddings, J.C.: *Two-dimensional separations: Concept and promise.* Anal.Chem., 56(12) 1259 (1984).
396. Giddings, J.C.: *Unified separation science.* J. Wiley Interscience: New York, 1991.
397. Gil-Av, E.; Feibush B.; Charles-Sigler, R.: *Separation of enantiomers by gas-liquid chromatography with an optically active stationary phase.* Tetrahedr.Lett. 1009-1015 (1966).
398. Gindin, L.M.: *General principles of ion-exchange extraction and its applications.* Lzv.Sib.Otd.Akad.Nauk SSSR, 6, 36-47 (1967).
399. Glajch, J.L.; Kirkland, J.J.; Minor, J.M.; Squire, K.M.: *Optimization of solvent strength and selectivity for reversed-phase liquid chromatography using an interactive mixture-design statistical technique.* J.Chromatogr. 199, 57-59 (1980).
400. Glazov, V.M.; Pavlova, L.M.: *Chemical thermodynamics and phase equilibria* (in Russian) Metalurgiya: Moscow, 1988.
401. Glueckauf, E.: *Theory of chromatography.* Proc. Roy. Soc. (London) A186, 35-57 (1946).
402. Glueckauf, E.; McKay, H.A.C.; Mathieson, A.R.: *The partition of uranyl nitrate between water and organic solvents. II. The partition data and their interpretation.* Trans.Faraday Soc., 47, 437-49 (1951).
403. Glueckauf, E.: *Theory of chromatography. IX. Theoretical plate concept in column separations.* Trans. Faraday Soc., 51, 34-44 (1955).

404. Glueckauf, E.: *Theory of chromatography. X. Formulae for diffusion into spheres and their application to chromatography.* Trans. Faraday Soc., 51, 1540- 51 (1955).
405. Gmehling, J.; Rasmunssen, P.; Fredenslund, A.: *Vapor-liquid equilibria by UNIFAC group contribution revision and extension. 2.* Ind.Eng.Chem. Process Des. Dev., 21, 118-27 (1982).
406. Gmehling, J: *Azeotropic data.* VCH: Weinheim, 1994.
407. Golay, M.J.E.: *Vapor-phase chromatography and the telegrapher's equation.* Anal.Chem., 29 928-32 (1957).
408. Golay, M.J.E.: *Performance index for gas chromatographic columns.* Nature, 180, 435-6 (1957).
409. Golden, K.E.; Hatton, T.A.: *Protein extraction using reverse micelles.* Biotech.Prog., 1, 69-74 (1985).
410. Goldschmidt, V.M.: *Krystallbau und chemische Zusamensetzung.* Chem. Ber., 60, 1263-1296 (1927).
411. Golovnya, R.V.; Misharina, T.A.: *The thermodynamic interpretation of polarity and selectivity of sorbents in gas chromatography.* Usp.Khim., 49, 171-91 (1980).
412. Gooch, F.A.: *On a new method for the separation and subsequent treatment of precipitates in chemical analysis.* Proc. Am. Acad. Arts and Sci., 13, 342 (1878).
413. Gooding, K.M.; Regnier, F.E.: *HPLC of biological macromolecules: Methods and applications.* Marcel Dekker: New York, 1990.
414. Gordon A.H.; Martin, A.J.P.; Synge, R.L.M.: *The amino acid composition of tyrocidin.* Biochem. J., 37, 313 (1943).
415. Gordon, A.; Keil, B.; Sebesta, K.: *Electrophoresis of proteins in agar jelly.* Nature, 164, 498-9 (1949).
416. Gordon, A.H.; Martin, A.J.P.; Synge, R.L.M.: *Partition chromatography in the study of protein constituents.* Biochem. J., 37, 79-86 (1943).
417. Gordon, S.; McBride, B.J.: *Computer program for calculation of complex chemical equilibrium composition*

and applications. NASA Reference Publication 1311. NASA Lewis Research Centre: Cleveland,1994.
418. Gorshkov, V.I.: *Counter-current ion exchange.* Zh.Vses. Khim.Obshch., 28(1) 63 (1983).
419. Gorshkov, V.I.; Murav'ev, D.N.; Saurin, A.D.; Kovalenko, Yu.A.; Medvedev, G.A.; Ferapontov, N.B.: *The dependence of the equilibrium separation factor for sodium chloride and glutamic acid in ion exchange on the of the solution by the method of output curves.* Russ. J.Phys.Chem., 53(7) 954-956 (1979).
420. Gortler, H.: *Dimensionanalyse. Eine Theorie der physikalischen Dimensionen mit Anwendungen.* Springer Verlag: Berlin, 1975.
421. Gosling, J.P.; Basso, L.V.: *Immunoassay. Laboratory analysis and clinical application.* Butterworth-Heinemann: Boston, 1994.
422. Gosman, A.; Linkonnen, S.; Passiniemi, P.: *Adsorption and diffusion at low electrolyte concentrations.* J.Phys.Chem., 90, 6051-3 (1986).
423. Grabar, P.; Burtin, P.: *Immuno-electrophoretic analysis.* Elsevier: New York, 1964.
424. Grabar, P.; Williams, C.A.: *Method permitting the simultaneous study of electrophoretic and immunochemical properties of a mixture of proteins. Application to blood serum.* Biochim.Biophys. Acta., 10, 193-4 (1953).
425. Grace, H.P.: *Resistance and compressibility of filter cakes.* Chem.Eng.Progr., 49, 303-427 (1953).
426. Graham, T.: *On the molecular mobility of gases.* Trans.Roy. Soc. London, 153, 385 (1863).
427. Graham, T.: *The Bakerian lecture - On osmotic force.* Phil. Trans.Roy.Soc. (London) 144, 177 (1854).
428. Gran, G.: *Determination of the equivalence point in potentiometric titrations.* Analyst, 77, 661-71 (1952).

429. Grandison, A.S.; Lewis, M.J.: *Separation processes in the food and biotechnology industries. Principles and applications.* Woodhead Publishing: Cambridge, 1996.
430. Grassmann, W.; Hannig, K.: *Separation of mixtures on filter paper by migration in an electrical field.* Z. Physiol. Chem., 292, 132-50 (1953).
431. Grebenyuk, V.D.: *Elektrodializ (Electrodialysis).* Naukova Dumka: Kiev, 1976.
432. Greco, F.A.: *On the 2nd law of thermodynamics. A summary of the replies to Deutsch's article.* An.Lab., 16(1) 38-47 (1983).
433. Gregg, S.J., Sing, K.S.W.: *Adsorption, surface area and porosity.* Acad. Press: London, 1967.
434. Gregor, H.P.; *Gibbs-Donnan equilibria in ion-exchange resin systems.* J.Am.Chem.Soc. 73, 642 (1951).
435. Griffin, W.C., *Calculation of HLB values of non-ionic surfactants,* J.Soc.Cosmetic Chemists 5 (4) 249-56 (1954).
436. Grimm, H.G.: *Mixed crystals.* Z. Electrochem., 30, 467-473 (1924).
437. Grossman, P.D.; Colburn, J.C.: *Capillary electrophoresis: Theory and practice.* Academic Press: San Diego, 1992.
438. Grushka, E.: *Characterization of exponentially modified Gaussian peaks in chromatography.* Anal.Chem., 44(11) 1733-8 (1972).
439. Grushka, E.; Caldwell, K.D.; Myers, M.N.; Giddings, J.C.: *Field-flow fractionation.* Separ.Purif.M., 2(1) 127-51 (1973).
440. Grushka, E.: *New developments in separation methods.* E. Gruska, Ed., Marcel Dekker: New York, 1976.
441. Grushka, E.; Colin, H.; Guichon, G.: *Retention behavior of alkylbenzenes as a function of temperature and mobile phase composition in reversed-phase chromatography.* J.Chromatogr., 248, 325-39 (1982).
442. Gu, T.: *Mathematical modeling and scale-up of liquid chromatography.* Springer, Berlin1995.

443. Gu, Zh.; Navratil, J.D.; Cheng, H.: *Rare earth recovery using pseudo liquid membranes.* In: Proceed.Intern.Solvent Extraction Conf. ISEC'90, 18-21 July, Tokyo, 1990.
444. Gu, Zh.M.: *Electrostatic pseudoliquid membrane separation technology.* J. Chem. Ind. Eng. (China) 5(1) 44-55 (1990).
445. Guerritore, D.; Bellelli, L.: *Interactions between nucleic acid and a cationic detergent.* Nature, 184, 1638 (1959).
446. Guggenheim, E.A.: *Mixtures. The theory of the equilibrium properties of some simple cases of mixtures, solutions and alloys.* Clarendon Press: Oxford, 1952.
447. Guggenheim, E.A.: *Statistical thermodynamics of mixtures with zero energies of mixing.* Proc.Roy.Soc. (London) A 183, 203-212 (1944).
448. Guillaumont, R.; Bouissieres, G.: *Chimie a l'echelle des indicateura et a l'echelle de l'atome (Chemistry on the tracer scale and the atom scale).* Bull. Soc.Chim.Fr., 12, 4555-9 (1972).
449. Guiochon, G.; Golshan-Shirazi, S.; Felinger, A.: *Fundamentals of preparative and nonlinear chromatography.* 2nd. Ed., Academic Press: New York, 2006.
450. Guldberg, B.L., Waage, P.: *Études sur les affinités chimiques.* Christiania University: Oslo, 1867.
451. Gupta, B.L.; Moreton, R.B., Oschman, J.L., Wall, B.J. (Eds.) *Transport of ions and water in animals.* Acad. Press: London, 1977.
452. Haase, R.; Muenster, A.: *The theory of the infinitely diluted solutions* Z.phys.Chem., 194, 253-277 (1950).
453. Haase, R.; Schonert, H.: *Solid-liquid equilibrium.* Pergamon Press: Oxford, 1969.
454. Habashi, F.: *Principles of extractive metallurgy. Vol. 1. General principles. Vol. 2. Hydrometallurgy.* Gordon and Breaeh Science Publishers: New York, 1969.
455. Hagen, G.: *Ueber die Bewegung des Wassers in eigen cylindrischen Roehren.* Annal.Phys. (Poggendorffs Ann.) 46, 423 (1839).

456. Haglund, H.: *Isoelectric focusing in pH gradients. Technique for fractionation and characterization of ampholytes.* Methods Biochem. Anal., 19, 1-104 (1971).
457. Haglund, H.: *Isotachophoresis - a principle for analytical and preparative separation of substances such as proteins, peptides, nucleotides, weak acids, metals.* Sci.Tools, 17(1), 2-13 (1970).
458. Hahn, O.: *Gesetzmassigkeit bei der fällung und Adsorption kleiner Substanzmengen und ihre Beziehung zur radioaktiven Fällungsregel.* Ber.Dtsch.Chem.Ges., 59, 2014-2025 (1926).
459. Hais, I.M.; Macek, K. (Eds.): *Paper chromatography: A comprehensive treatise.* 9rd ed. Academic Press: New York, CSAV: Prague, 1963.
460. Hala, E.; Pick, J.; Fried, V.; Vilim, O.: *Vapor- liquid equilibrium.* Pergamon Press: Oxford, 1967.
461. Hala, J.: *The solvent extraction of hafnium(V) and zirconium(IV) by N-benzoyl N-phenylhydroxylamine and 2-thenoyltrifluoracetone from strongly acidic solutions.* J.Inorg. Nucl. Chem., 29(1) 187-98 (1967).
462. Hala, S.; Kures, M.; Popl, M.: *Comprehensive analytical chemistry. XIII. Analysis of complex hydrocarbon mixtures. A. Separation methods.* G. Svehla (Ed.) Elsevier: Amsterdam, 1981.
463. Halicioglu, T.; Sinanoglu, O.: *Solvent effects on cis trans azobenzene isomerization: a detailed application of a theory of solvent effects on molecular association.* Ann. N. Y. Acad. Sci., 158(1) 308-17 (1969).
464. Hamel, J.-F.; Sikdar, S.K.; Hunter, J.B. (eds.): *Downstream processing and bioseparation.* American Chemical Society: Washington, 1989.
465. Hammett, L.P.; Deyrup, A.J.: *A series of simple basic indicators. I.The acidity functions of mixtures of sulfuric and perchloric acids with water.* J.Am.Chem. Soc., 54, 2721-39 (1932).

466. Hanai, T.: *Quantitative in silico chromatography: Computational modelling of molecular interactions.* Royal Soc.Chemistry: London, 2014.
467. Hanauer, M.; Pierrat, S.; Zins, I.; Lotz A.; Sönnichsen C.: *Separation of nanoparticles by gel electrophoresis according to size and shape.* Nano Lett., 7 (9) 2881–2885 (2007).
468. Hand, D.B.: *Dineric distribution.* J.Phys.Chem., 34, 1961-2000 (1930).
469. Hanna, G.J.; Noble, R.D.: *Measurement of liquid-liquid interfacial kinetics.* Chem. Rev., 85, 583 (1985).
470. Hannig, K.: *Carrier-free continuous electrophoresis and its use.* Z. Anal.Chem., 181, 244-54 (1961).
471. Hansch, C.; Muir, R.M.; Fujita, T.; Maloney, P.P.; Geiger, F. Streich, M.: *The correlation of biological activity of plant growth regulators and chloromycetin derivatives with Hammett's constant and partition coefficients.* J.Am.Chem. Soc., 84, 2817-24 (1963).
472. Hanson, C.: *Solvent extraction - an economically competitive process.* Chem. Eng., 86 (10) 83-87 (1979).
473. Hanson, C.: *Solvent extraction. Theory, experiment, and commercial operations.* Chem. Eng., 75(18) 76-98 (1968).
474. Hanson, C.; Hughes, M.A.; Whewell, R.J.: *The rate of masa transfer to single drops. The copper hydroxyoxime systems.* J. Appl. Chem. Biotech nol., 28(6) 426-34 (1978).
475. Harned, H.S.; Hamer, W.J.: *The thermodynamics of aqueous sulfuric acid solutions from electromotive-force measurements.* J.Am.Chem.Soc., 57, 27-33 (1935).
476. Hartland, S.: *Counter-current extraction.* Pergamon Press: Oxford, 1970.
477. Hartland, S.: *Profile of the draining film between a rigid sphere and deformable fluid-liquid interface.* Chem. Eng. Sci., 24(6) 987-95 (1969).
478. Harwell, J.H.; Scamehorn, J.F.; *Surfactant based separation processes.* Marcel Dekker: New York, 1989.

479. Hassas, B.V.; Karakas, F.; Celik, M.S.: *Ultrafine coal dewatering: Relationship between hydrophilic lipophilic balance (HLB) of surfactants and coal rank.* Int.J.Mineral Process. 133, 97-104 (2014).
480. Hasselbach, K. A. *Die Berechnung der Wasserstoffzahl des Blutes aus der freien und gebundenen Kohlensaeure desselben, und die Sauerstoffbindung des Blutes als Funktion der Wasserstoffzahl.* Biochemische Zeitschrift 78, 112–144 (1917).
481. Hatta, S.: *Study of adsorption rate of gases with liquids. l. Adsorption of carbon dioxide with alkali.* J. Soc. Chem. Ind., Japan, 31, 210B (1928).
482. Hatton, T.A.: *Extraction of proteins and amino acids using reversed micelles.* In: *Ordered media in chemical separations*, p. 170-183. W.L. Hinze, D.W. Armstrong, Eds., ACS: Washington, 1987.
483. Hatton, T.A.; Lightfoot, E.N.: *Dispersion, mass transfer and chemical reaction in multiphase contactors. II. Numerical examples.* Am.Inst.Chem.Eng. Journal, 30(2) 243-9 (1984).
484. Hatton, T.A.; Lightfoot, E.N.: *Dispersion, mass transfer and chemical reaction in multiphase contactors. I. Theoretical development.* Am.Inst.Chem.Eng. Journal, 30(2) 235-43 (1984).
485. Hearn, M.S.W.: *High performance liquid chromatography and its application to protein chemistry.* Adv.Chromatog., 20, 1-82 (1982).
486. Hearn, M.S.W.: *High-performance liquid chromatography of peptides.* HighPerform. Liq. Chromatogr. 3, 87-155 (1983).
487. Hearn, M.S.W.: *Reversed-phase high-performance liquid chromatography.* Methods Enzymol., 104, 190-212 (1984).
488. Hearn, M.S.W.; Grego, B.: *High-performance liquid chromatography of amino acids, peptides and proteins. LV. Studies on the origin of band broadening of polypeptide., and proteins separated.* J.Chromatogr.1 296, 61-82 (1984).

489. Heckmann, K.; Stroeble, Ch.; Bauer, S.: *Hyperfiltration through cross-linked monolayers.* Thin Solid Films, 99, 265-69 (1983).
490. Hedstrom, B.O.A.: *The hydrolysis of metal ions. VI. The hydrolysis of the iron(II) ion Fe^{++}.* Arkiv Kemi, 5, 457-68 (1953).
491. Heidelberger, M.; Kendall, F.E.: *Quantitative theory of the precipitation reaction. II. A study of an azoprotein-antibody system.* J. Exptl. Med., 62, 467 (1935).
492. Heitler, W.; London, F.: *Wechselwirkung neutraler Atome und homopolare Bindung nach der Quantenmechanik.* Z.Phys., 44, 455-72 (1927).
493. Helfferich, F.: *Ionenaustauscher. Band I. Grundlagen, Struktur, Herstellung, Teorie.* Verlag Chemie: Weinheim, 1959.
494. Helfferich, F.; Klein, G.: *Multicomponent chromatography: Theory of interference.* Marcel Dekker: New York, 1970.
495. Helfferich, F.G.: *Ion-exchange kinetics. V. Ion exchange accompanied by reactions.* J.Phys.Chem., 69(4) 1178-87 (1965).
496. Helfferich, F.G.: *Ion-exchange kinetics. A nonlinear diffusion problem.* J. Chem. Phys., 28, 418-24 (1958).
497. Helfferich, F.G.: *Travel of molecules and disturbances in chromatographic columns. A paradox and its resolution.* J. Chem. Educ., 41(8) 410-13 (1964).
498. Helmholtz, H.L.F.: *Studien ueber elektrische Grenzschichten.* Ann. Physik, 7, 337 (1879).
499. Henderson, G.M., Rule, H.G.: *A new method of resolving a racemic compound.* J.Chem.Soc. 1568-1573 (1939).
500. Henderson, L.M.; Kracek, F.C.: *The fractional precipitation of barium and radium chromates.* J.Am.Chem.Soc., 49(3) 738 (1927).
501. Hendry, J.E.; Hughes, R.R.: *Generating separation process flowsheets.* Chem.Eng.Progr. 68(6), 71 (1972).

502. Henley, E.J.; Rosen, E.M.: *Material and energy balance computations.* J. Wiley, New York, 1969.
503. Henley, E.J.; Seader, J.D.: *Equilibrium-stage separation operations in chemical engineering.* J. Wiley: New York, 1981.
504. Henry, W.: *Appendix to Mr.William Henry's paper: On the quantity of gases adsorbed by water, at different temperatures, and under different pressures.* Phil.Trans., 29, 274 (1803).
505. Herold, K.E.; Rasooly, A. (Eds.): *Lab on a chip technology: Biomolecular separation and analysis,* Vol.2., Caister Academic Press: Norfolk, 2009.
506. Herr, W.: *On the isolation of the radioisotopes of Zn, Ga, In, V, Mo, Pd, Os and Pt in a practically carrier-free state by the (n,γ) recoil processes from metallic complexes of phthalocyanines.* Z. Naturf., 7b, 201 (1952).
507. Heusch, R.: *Eine experimentelle Methode zur Bestimmung des HLB-Wertes von Tensiden.* Kolloid Ztschr.u.Ztschr.f.Polymere, 236, 31-8 (1970).
508. Higbie, R.: *The rate of absorption of a pure gas into a still liquid during short periods of exposure.* Trans. Amer. Inst. Chem. Eng., 31, 365 (1935).
509. Hildebrand, J.H.: *Solubility XII. Regular solutions.* J.Am. Chem.Soc., 51, 66-80 (1929).
510. Hildebrand, J.H.; Wood, S.E.: *The derivation of equations for regular solutions.* J.Chem.Phys., 1(12) 817-22 (1933).
511. Hildebrand, J.H.; Scott, R.L.: *Solubility of nonelectrolytes.* Reinhold: New York, 1950.
512. Hille, B.: *Ionic selectivity, saturation and block in sodium channels. A four-barrier model.* J. Gen. Physiol., 66, 535-60 (1975).
513. Hille, B.: *Ionic channels of nerves: question for theoretical chemists.* Biosystems, 8, 195-9 (1977).
514. Himmelblau, D.M.: *Applied nonlinear programming.* McGraw-Hill: New York, 1972.

515. Hinton, J.F.; Amis, E.S.: *Solvation numbers of ions.* Chem. Rev., 71(6) 627-74 (1971).
516. Hinze, W.J.: *Organized surfactant assemblies in separation science.* In: Ordered media in chemical separations, p. 2-82, ACS Symp. Ser. 942, W.J. Hinze, D.W.Armstrong (Eds.) ACS: Washington, 1987.
517. Hinze, W.L.; Armstrong, D.W. (Eds.): *Ordered media in chemical separations. ACS Symposium Serie1 2.* American Chemical Society: Washington, 1987.
518. Hirata, M.; Ohe, S.; Nagahama, K.: *Computer-aided data book of vaporliquid equilibria.* Elsevier: Amsterdam, 1976.
519. Hirschfelder, J.O.; Curtiss, C.F.; Bird, R.B.: *Molecular theory of gases and liquids.* J.Wiley: New York, 1954.
520. Hirschler, A.E.; Amon, S.: *Adsorption - a tool in the preparation of highpurity saturated hydrocarbons.* Ind. Eng.Chem., 39, 1585-96 (1947).
521. Hlavacek, V.; Vaclavek, V.; Kubicek, M.: *Balance and simulation computing of complicated processes of chemical technology (in Czech).* Academia: Prague, 1979.
522. Hobza, P.; Zahradnik, R.: *Weak intermolecular interactions in chemistry and biology.* Acaaemia: Prague 1980.
523. Hoffmann, U.; Hofmann, H.: *Einfuehrung in die Optimierung* mit *Anwendungs beispielen aus dem Chemie- Ingenieur-Wessen.* Verlag Chemie: Weinheim, 1971.
524. Hogfeldt, F.: *Stability constants of metal-ion complexes. A. Inorganic ligands.* Pergamon Press: London, 1980.
525. Hood, L.E.; Smith, L.M.: *Automated synthesis and sequence analysis of biological macromolecules.* Anal.Chem., 60(6) 384A-390A (1988).
526. Horsley, L.H.: *Azeotropic data.* Am. Chem. Soc.: Washington, 1952.
527. Horvath, C.; Melander, W.; Molnar, I.: *Solvophobic interactions in liquid chromatography with nonpolar stationary phase.* J.Chromatogr., 125(1) 129-56 (1976).

528. Horvath, C.; Melander, W.: *Liquid chromatography with hydrocarbonaceous bonded phases; theory and practice of reversed phase chromatography.* J.Chrom.Sci., 15(9) 393-404 (1977).
529. Horvath, C.; Melander, W.; Molnar, I.: *Liquid chromatography of inorganic substances with nonpolar stationary phase.* Anal.Chem., 49(1) 142-54 (1977).
530. Horvath, C.; Nahum, A.; Freuz, J.H.: *High-performance displacement chromatography.* J.Chromatogr., 218, 365-93 (1981).
531. Horvath, C.; Ettre, L.S. *Chromatography in biotechnology.* Am.Chem.Soc.: New York 1993.
532. Horwitz, E.P.; Kalina, D.G.; Diamond, H.; Vandegrift, G.F.; Schulz, W.W.: *The TRUEX process - a process for the extraction of the transuranium elements from nitric acid wastes utilizing modified PUREX Solvent.* Solvent Extr.Ion. Exch., 3(1-2) 75-109 (1985).
533. Howard, G.S.; Martin, A.J.P. *Separation of the C12-C18 fatty acids by reversed phase partition chromatography.* Biochem.J. 46, 532 (1950).
534. Howe, E.D.: *Fundamentals of water desalination.* Marcel Dekker: New York, 1974.
535. Huang, H-J.; Ramaswamy, S.; Tschirner, U.W.; Ramarao, B.V.: *A review of separation technologies in current and future biorefineries.* Sep.Pur.Technol. 62(1) 1-21 (2008).
536. Huber, J.F.K.; Hulsman, J.A.R.J.: *A study of liquid chromatography in columns.* Anal. Chim. Acta., 38(1-2) 305-13 (1967).
537. Huggins, M.L.: *Some properties of solutions of long-chain compounds.* J.Phys.Chem., 46, 151-8 (1942).
538. Huggins, M.L.: *Thermodynamic properties of liquids, including solutions. 2. Polymer solutions considered as ditonic systems.* Polymer, 12(6) 389-99 (1970).

539. Hughes, M.A.; Rod, V.: *On the use of the constant interface stirred cell for kinetic studies.* Hydrometallurgy, 12(2) 267-70 (1984).
540. Hunter, T.G.; Nash, A.W.: *Applications of physicochemical principles to the design of liquid-liquid contact equipment. I. General theory.* J.Soc.Chem.Ind. London, 51, 285-97 (1932).
541. Huntley, H.E.: *Dimensional analysis.* Dover Pub.: New York, 1967.
542. Hustedt, H.; Kroner, K.H.; Kula, M.-R.: *Continuous enzyme workup by cross-current extraction.* Chem. Ind., 37(8) 527-9 (1985).
543. Hutta, M.; Moskalova, M.; Zemberyova, M.; Foltin, M.: *Off-line combination of RPHPLC and Trace Mercury Analyser for organomercurial speciation in environmental samples.* J.Radioanal.Nucl.Chem. 208, 403-415 (1996).
544. Hwang, S.T.; Yuen, K.H.; Thorman, J.M.: *Gas separation by a continuous membrane column.* Sep.Sci.Technol., 15, 1069-90 (1980).
545. Hyun, S.H.; Danner, R.P.: *Determination of gas adsorption equilibriums by the concentration-pulse technique.* Am.Inst. Chem.Eng. Symp. Ser., 78(219) 19-28 (1982).
546. Ilani, A.: *K-Na discrimination by porous filters saturated with organic solvents as expressed by diffusion potentials.* J. Gen. Physiol., 46, 836-50 (1963).
547. Inczedy, J.: *Some remarks on the quantitative expression of the selectivity of analytical procedure.* Talanta, 29(7) 595-9 (1982).
548. Ingri, N.; Sillen, L.G.: *High-speed computers as a supplement to graphical methods. IV. An ALGOL version of LETAGROP VRID.*: Acta Chem.Scand., 18, 1085-98 (1964).
549. Ingri, N.; Kakolowicz, W.; Sillen, L.G.; Warnqvist, B.: *High speed computers as a supplement to graphical methods. V. HALTAFALL, a general program for calculating the composition of equilibrium mixtures.* Talanta, 14(11) 1261-86 (1967).

550. Iofa, B.Z.; Nesmeyanov, An.N.; Kireev, G.I.: *Study of iodine extraction kinetics by means of the isotope exchange* (In Russian). Teoret.Osnovy Khim.Tekhnologii, 4, 429-432 (1970).
551. Irving, H.; Pierce, T.B.: *Observations on Job's method of continuous variations and its extension to two-phase systems.* J.Chem.Soc., 2565-74 (1959).
552. Issaq, H.J. (Ed.): *A century of separation science.* Marcel Dekker: New York, 2002.
553. Ishii, D.: *Introduction to micro-scale high-performance liquid chromatography.* VCH: Weinheim, 1988.
554. Ito, Y.; Conway, W.D.: *Development of countercurrent chromatography (CCC).* Anal.Chem., 56(4) 534 (1984).
555. Izatt, R.M.; Dearden, D.V.; Brown, P.R.; Bradshaw, J.S.; Lamb, J.D.; Christensen, J.J.: *Cation fluxes from binary Ag^+-M^+ mixtures in a watertrichlormethane-water liquid membrane system containing a series of macrocyclic ligand carriers.* J.Am.Chem.Soc., 105(7) 1785-90 (1983).
556. Izmailov, A.V.; Maltsev, P.M.; Mitskewich, Yu.G.: *Mathematical modelling of steady states in countercurrent extraction processes.* Teor. Osn. Khim. Tekhnol., 9(5) 651-6 (1975).
557. Izmailov, N.A.; Shraiber, M.S.: *A drop-chromatographic method of analysis and its utilization in pharmacy. I.:* Farmaciya, 3, 1 (1938).
558. Jain, R.; Schulz, J.S.: *A numerical technique for solving carrier-mediated transport problems.* J. Membr. Sci., 11(1) 79-106 (1982).
559. Jakoby, W.B. (Ed.): *Methods in Enzymology. Vol. XXII. Enzyme purification and related techniques.* Jakoby, W.B., Ed., Acad. Press: New York, 1971.
560. James, A.T.; Martin, A.J.P.: *Gas-liquid partition chromatography. A technique for the analysis of volatile minerals.* Analyst, 77, 915-32 (1952).

561. James, A.T.; Martin, A.J.P.: *Gas-liquid partition chromatography: the separation and microestimation of volatile fatty acids from formic to dodecanoic acid.* Biochem. J., 50, 679-90 (1952).
562. James, A.T.; Martin, A.J.P.; Smith, G.H.: *Gas-liquid partition chromatography. Separation and microestimation of ammonia and the methylamines.* Biochem. J., 52, 238-42 (1952).
563. Janak, J.: *The chromatographic semimicroanalysis of gases.* Coll. Czechosl. Chem. Commun., 18, 798 (1953).
564. Janata, J.; Huber, R.J. (Eds.): *Solid state chemical sensors.* Academ. Press: New York, 1985.
565. Janca, J.: *Recent developments in particle size separation.* Newnes (Elsevier): Burlington, 2014.
566. Jandera, P.; Churacek, J.: *Gradient elution in liquid chromatography. Theory and practice.* Elsevier: Amsterdam, 1985.
567. Janecke, E.: *Ueber die Anwendung der thermischen Analyse auf Dreistoff-systeme.* Z.Phys. Chem., 59, 697-702 (1907).
568. Jansen, B.C.P.; Donath, W.F.: *Antineuritische Vitamine.* Chem. Weekblad, 23, 201-3 (1926).
569. Janson, J-C.; Ryden, L. (Ede.): *Protein purification: Principles, high resolution methods, and applications.* VCH: Weinheim, 1989.
570. Jantzen, E.: *Das fraktionierte Destillieren und das fraktionierte Verteilung.* Dechema Monogr., 5 (48), (1932).
571. Jedinakova, V.; Celeda, J.: *Aquo and halo complexes of Zn^{2+} ion in isomolar series perchlorate - halide.* Coll.Czechosl. Chem.Commun., 41, 2829-37 (1975).
572. Jeelani, S.A.K.; Hartland, S.: *Prediction of steady date dispersion height from batch settling data.* Am. Inst. Chem. Eng. Journal, 31(5) 711-20 (1985).
573. Jenkins, I.L.: *Ion exchange in the atomic energy industry with particular reference to actinide and fissiion product separation.* Solvent Extr. Ion Exch., 2(1) 1-27 (1984).

574. Jenkins, I.L.; McKay, H.A.C.: *The partition of uranyl nitrate between water and organic solvent. VI. Salting-out by a second nitrate.* Trans. Faraday Soc., 50, 107-119 (1954).
575. Jenkins, J.D.: *The effect of various factors upon the velocity of crystallization of substances from solution.* J.Am.Chem. Soc., 47, 903-22 (1925).
576. Jensen Skytte, B.: *Solvent extraction of metal chelates. I. Application of a titration procedures to the study of the extraction of metal chelates.* Acta Chem.Scand., 13, 1347 (1959).
577. Jensen, R.J.; Sullivan, J.A.; Finch, F.S.: *Laser isotope separation.* Sep.Sci.Technol., 15(3) 509-32 (1980).
578. Jitoh, F.; Imura, H.; Suzuki, N.: *Substoichiometric separation for inorganic arsenic and methylated arsenic and its application to samples of marine organisms.* Anal. Chim. Acta, 228, 85 (1990).
579. Johansson, G., Kopperschlaeger, G.; Albertsson, P.A.: *Affinity partitioning of phosphofructokinase from baker's yeast using polymer-bound Cibacron Blue F3G-A.* Eur.J.Biochem., 131(3) 585-94 (1983).
580. Johansson, G.: *Affinity partitioning.* Methods Enzymol., 104, 356-64 (1984).
581. Johansson, G.: *Aqueous two-phase systems in protein purification.* J.Biotechnol., 3, 11-18 (1985).
582. Johansson, G.; Andersson, M.: *Parameters determining affinity partitioning of yeast enzymes using polymer-bound triazine dye ligands.* J.Chromatogr.Biomed., 303, 39-51 (1984).
583. Johnson, E.L.; Stevenson, R.: *Basic liquid chromatography.* Varian Associates: Palo Alto, 1978.
584. Johnson, J.S.; Minturn, R.E.; Wadia, P.H.: *Hyperfiltration. XXI. Dynamically formed hydrous zirconium oxide - polyacrylate membranes.* J. Electroanal.Chem., 37, 267-81 (1972).

585. Jokl, V.: *Complex compounds in solution using paper electrophoresis. II. Electrophoretic mobilities and stabilities of mononuclear complexes.* J.Chromatogr., 14, 71-8 (1964).
586. Jones, A.G.: *Crystallization process systems.* Butterworth-Heinemann: Oxford, 2002.
587. Jones, A.S. *Isolation of bacterial nucleic acids using cetyltrimetylammonium bromide (Cetavlon).* Biochim. Biophys.Acta 10 607-12 (1953).
588. Jones, R.C.; Furry, W.H.: *The separation of isotopes by thermal diffusion.* Rev. Mod. Phys., 18(2) 151 (1946).
589. Jost, W.: *Diffusion on solids, liquids and gases.* Acad. Press: New York, 1960.
590. Jovin, T.M.: *Multiphasic zone electrophoresis I. Steady state moving boundary systems formed by different electrolyte combinations.* Biochemistry, 12(5) 871-9 (1973).
591. Jovin, T.M.: *Multiphasic zone electrophoresis. III. Further analysis and new forms of discontinuous buffer systems.* Biochemistry, 12(5) 890-8 (1973).
592. Kafarov, V.V.: *A general form of equation for* mass *transfer processes.* Zh.Prikl.Khim., 33, 1495-9 (1960).
593. Kafarov, V.V.: *Osnovy massoperedachi (Fundamentala of mass transfer).* Izd.Vysshaya Shkola: Moskva, 1962.
594. Kafarov, V.V.; Schelgov, V.N.; Dorokhov, I.N.: *The modelling of complex chemico-technological processes based on logical algebra methods.* Dokl.Akad.Nauk SSSR, 231, 1415-18 (1976).
595. Kafarov, V.V.; Shestopalov, V.V.; Dorokhov, I.N.; Zheleznova, G.L.: *Direct method of determining longitudinal* mixing *of a packed bed.* Dokl. AN SSSR, 174(4) 897-99 (1967).
596. Kaiser, R.: *Gas phase chromatography.* Vol. 1-3. Butterworths: London, 1963.
597. Kalinitchev, A.I.: *Nonlinear theory of multicomponent sorption dynamics and chromatography*, Uspekhi Khim, 65(2) 103-124 (1996).

598. Kaminski, Marie; Ouchterlony, O.: *Qualitative and quantitative immunochemical studies of the proteins of egg white.* Bull. Soc. Chim. Biol., 33, 758-70 (1951).
599. Kammermayer, K.; Hagenbaumer, D.H.: *Membrane separations in the liquid phase.* Am.Inst.Chem.Eng. Journal, 1, 215 (1955).
600. Kammermeyer, K.: *Silicone rubber as a selective barrier-gas and vapor transfer.* Ind.Eng.Chem., Ind. E., 49, 1685-6 (1957).
601. Kammermeyer, K.: *Technical gas permeation proceses.* Chem. Ing. Tech., 48(8) 672-5 (1976).
602. Kaniansky, D.; Havasi, P.: *Instrumentation for capillary isotachophoresis.* Trends Anal.Chem., 2(9) 197-202 (1983).
603. Karger, B.L.; Snyder, L.R.; Horvath, C.: *An introduction to separation science.* J.Wiley: New York, 1973.
604. Karlish, S.J.D.; Yates, D.W.; Glynn, I.H.: *Elementary steps of the sodium potassium ion-dependent ATPase mechanism, studied with formycin nucleotides.* Biochim. Biophys. Acta, 525(1) 230-251 (1978).
605. Karlova, E.K.; Karlov, N.V.; Kuzmin, G.I.; Laskorin, B.I.; Prokhorov, A.M.; Stupin, N.I.; Shurmel', L.B.: *Smeshchenie khimicheskogo ravnovesiya v rastvorakh pri rezonansnom vozdeistvii IK izlucheniya.* Pisma Zh.Eksp.Teor.Fiz. 22, 459 (1975).
606. Karr, A.E.: *Performance of a reciprocating-plate extraction column.* Am.Inst.Chem.Eng. Journal, 5, 446-52 (1959).
607. Karr, A.E.: *Design, scale-up, and application of the reciprocating plate extraction column.* Sep.Sci.Technol., 15(4) 877-905 (1980).
608. Kasatkin, A.G.: *Osnovnye protsessy i apparaty khimicheskoy promyslennosti (Fundamental processes and apparates in chemical industry).* Khimiya.: Moskva 1971.
609. Kataoka, T.; Nishiki, T.; Kimura, S.: *Phenol permeation through liquid surfactant membrane - permeation model and effective diffusivity.* J.Membr.Sci., 41, 197 (1989).

610. Katchalsky, A.; Curran, P.F.: *Non-equilibrium thermodynamics in biophysics.* Harvard Univ. Press: Cambridge, 1967.
611. Kedem, O.; Katchalsky, A.: *Permeability of composite membranes. Part Electric current, volume flow and flow of solute through membranes.* Trans. Faraday Soc., 59, 1918 (1963).
612. Kedem, O.; Katchalsky, A.: *Thermodynamic analysis of the permeability of biological membranes to non-electrolytes.* Biochim. Biophys. Acta., 27, 229-46 (1958).
613. Keey, R.B.: *Drying: Principles and practice.* Pergamon Press: New York, 1972.
614. Keil, B.; Herout, V.; Hudlicky, M.; Ernest, I.; Protiva, M.; Gut, J.; Komers, R.; Moravek, J.: *Laboratorni technika organicke chemie (Laboratory technique in organic chemistry)* 2nd Ed., Nakl.CSAV: Praha 1963.
615. Kelly, R.M.: *General processing considerations.*, In: *Handbook of separation technology.* R.W.Rousseau, Ed., Wiley-Interscience: New York 1987.
616. Kertes, A.S.: *Solubility and activity of high-molecular amine hydrochlorides in organic solvents.* J.Inorg. Nucl. Chem., 27(1). 209-17 (1965).
617. Kertes, A.S.; Beck, A.: *Metallic nitrates in paper chromatography. IV systems containing tributylphosphate and nitric acid.* J.Chromatogr., 3, 195-6 (1960).
618. Kertes, A.S.; Grauer, F.: *Effect of chain length on heats of mixing in tri-n alkyl-amine-benzene systems.* J.Phys.Chem., 77, 3107 (1973).
619. Kertes, A.S.; Gutmann, H.: *Surfactants on organic solvents: the physical chemistry of aggregation and micellization.* Surface Colloid Sci., 8, 193-295 (1975).
620. Kertes, A.S.; King, C.J.: *Extraction chemistry of fermentation product carboxylic acids.* Biotech. Bioeng., 28, 269-82 (1986).

621. Kertes, A.S.; Markovits, G.: *Activity coefficients, aggregation, and thermodynamics of tridodecylammonium salts in nonpolar solvents.* J.Phys.Chem., 72(12) 4202-10 (1968).
622. Keulemans, A.I.M.; Kwantes, A.; Zaal, P.: *The selectivity of the stationary liquid in vapor -phase chromatography.* Anal. Chim. Acta, 13, 357-72 (1955).
623. Khalafalla, S.E.; Reimers, G.W.: *Separating nonferrous metals in incinerator residues using magnetic fluids.* Sep. Sci., 8(2) 161-78 (1973).
624. Khoury, F.M.: *Multistage separation processes.* 4th Ed. CRC Press: Boca Raton, 2014.
625. Khurana, Tarun K.: *On-chip isotachophoresis assays for high sensitivity electrophoretic preconcentration, separation, and indirect detection.* BiblioBazaar: Charleston, 2011.
626. Khym, J.H.: *Analytical ion exchange procedures in chemistry and biology.* Prentice Hall: New York, 1974.
627. Kielland, J.: *Individual activity coefficients of ions in aqueous solutions.* J.Am.Chem.Soc., 59, 1675-78 (1937).
628. Kim, S.N.; Kammermayer, K.: *Actual concentration profiles in membrane permeation.* Separ.Sci., 5 (6) 679-97 (1970).
629. King, C.J.: *Separation* processes. 2nd Ed. McGraw Hill: New York, 1980.
630. King, C.J.: *The additivity of individual phase resistances in mass transfer operations.* Am.Inst.Chem.Eng. Journal, 10(5) 671-7 (1964).
631. King, C.J.: *Turbulent liquid-phase mass transfer as a free gas-liquid interface.* Ind.Eng.Chem. Fundam., 5(1) 1-8 (1966).
632. Kipling, J.J.: *Adsorption from solution of non-electrolytes.* Acad. Press: London, 1965.
633. Kireev, V.A.: *Changes of the free energy in the formation of mixtures and solutions of liquids.* Zh. Fiz. Khim., 16, 124 (1942).

634. Kirkwood, J.G.; Scatchard, G.: *Das Verhalten von Zwitterionen und von mehrwertigen Ionen mit weit entfernten Ladungen in Elektrolytloesungen.* Phys. Ztschr., 33, 297-300 (1932).
635. Kirpichev, M.V.: *Teoriya podobiya (Similitude theory).* Izd. AN SSSR: Moskva, 1953.
636. Kirpichev, V.L.: *O podobii pri uprugikh yavleniyakh (On the similitude of elasticity phenomena).* Zh.Russ.Fiz.Khim. Obsch., Phys., 6 (9) 152 (1874).
637. Kiselev, A.V.; Yashin, Ya.I.: *Gas adsorption chromatography.* Plenum Press: New York, 1969.
638. Kislik, V.S.: *Solvent extraction: Classical and novel approaches.* Elsevier: Amsterdam 2012.
639. Kister, H.Z.: *Distillation design.* McGraw-Hill: New York, 1992.
640. Klas, J.: Some *problems of quantitative precipitation from solutions. I. Precipitation surface and optimal precipitation yield.* Coll.Czechosl.Chem.Commun., 35, 684-88 (1970).
641. Klas, J.; Tolgyessy, J.; Lesny, J.: *The sub- and super-equivalence method of isotope dilution analysis. V. Principle of the universal isotope dilution method.* Radiochem. Radioanal.Lett., 31(3) 171-9 (1977).
642. Klein, E.: *Evaluation of hemodialyzers and dialysis membranes. III. Water and dialysate.* Clin.Nephrol., 9, 131 (1978).
643. Klotz, I.M.: *In: The proteins, Vol. 1B, p. 748.* H. Neurath, K. Bailey, Eds., Academic Press: New York 1953.
644. Knox, J.H.; Grant, I.H.: *Miniaturization in pressure and electroendoosmotically driven liquid chromatography: some theoretical considerations.* Chromatographia, 24, 135-43 (1987).
645. Knox, J.H.: *Gas chromatography.* J. Wiley: New York, 1962.
646. Knox, J.H.: *The speed of analysis by gas chromatography.* J.Chem.Soc., 433-41 (1961).

647. Knudsen, M.: *Die Gesetze der Molecularstroemung und der inneren Reibungstroemung der Gase durch Roehren.* Ann. Phys., 28, 75-130 (1909).
648. Knyazev, I.N.; Kudryavtsev, Yu.A.; Kuzmina, N.P.; Letokhov, Vs.; Sarkisyan, A.A.: *Laser isotope separation of carbon by multiple IR photon and subsequent UV excitation of triftuoroiodomethane molecules.* Appl. Phys., 17(4) 427-9 (1978).
649. Kohlrausch, F.: *Ueber Koncentrations-Verschiebungen durch Electrolyse im inneren von Lōsungen und Lōsungsgemischen.* Ann. Phys. Leipzig, 62, 208 (1897).
650. Kohn, J.: *Small-scale membrane -filter electrophoresis and immunoelectrophoresis.* Clin. Chim. Acta, 3, 450-4 (1958).
651. Kohn, J.: *A microelectrophoretic method.* Nature, 181, 839-40 (1958).
652. Kolff, W.J.; Berk, H.S.J.: *Technique and chemical results of dialysis in vivo. Treatment with the artificial kidney.* Arch. Neerland Physiol., 28, 166-190 (1946).
653. Kolthoff, I.M.: *Adsorption on ionic lattices.* J.Phys.Chem. 40 (8) 1027-1040 (1936).
654. Kolthoff, I.M.; Sandell, E.B.: *A quantitative expression for the extractability of metals in the form of dithizonates from aqueous solutions. The equilibrium constant of zinc dithizonate.* J.Am.Chem.Soc., 63, 1906 (1941).
655. Komar, N.P.: *Khimicheskaya metrologiya. Geterogennye ionnye ravnovesiya (Chemical metrology. Heterogeneous ionic equilibria).* Vysshaya Shkola: Kharkov 1983.
656. Komarov, E., V.: *Molecular theory of solutions and extraction of metals and acids with associated reagents.* Radiokhimiya, 12, 312-18 (1970).
657. Komarov, E.V.; Komarov, V.N.: *Distribution of metal ions in the systems with associated extractants.* Radiokhimiya, 12(2), 291-7 (1970).

658. Komsta, L.; Waksmundzka-Hajnos, M.; Sherma J.: *Thin layer chromatography in drug analysis.* CRC Press: Boca Raton, 2014.
659. Konstantinov, B.P.; Oshurkova, O.V.: *Apparatus for the analysis of electrolyte solutions by ion mobilities.* Sov.Phys.-Techn.Phys., 11, 693 (1966).
660. Konstantinov, B.P.; Oshurkova, O.V.: *Rapid microanalysis of chemical elements by the moving boundary method.* Dokl. Akad.Nauk SSSR, 148(5) 1110-13 (1963).
661. Kopunec, R.; Kovalancik, J.: *Separation of cerium and europium by extraction with tributyl phosphate and di(2-ethylhexyl)phosphoric acid in alkane from nitrate solutions.* J.Radioanal Nucl.Chem., 129(2) 295-303 (1989).
662. Korpusov, G.V.; Yeskevich, I.V.; Patrusheva, Ye.N.; Yerchenkov, V.V.; Alekseeva, L.P.: *Regularities in extraction distribution behaviour of rare earth elements in nitrate solutions.* In: *Ekstraktsiya. Teoria, primeneniye, apparatura.* vol. 2, p. 117. Zefirova, A.P., Senyavina., M.M., Eds., Gosatomizdat: Moscow, 1962.
663. Koryta, J.; Skalicky, M.: *Electrolysis at the interface of two immiscible electrolyte solutions and extraction kinetics* J.Radioanal.Nucl.Chem., Articles, 129(2) 279-288 (1989).
664. Kotyk, A.; Janacek, K.: *Membrane transport - An interdisciplinary approach.* Plenum Press: New York, 1977.
665. Kourim, V.; Rais, J.; Million, B.: *Exchange properties of complex cyanides. Ion exchange of cesium on ferrocyanides.* J.Inorg.Nucl.Chem., 26(6) 1111-15 (1964).
666. Kovats, E.; Simon, W.; Heilbronner, E.: *Program-controlled gas chromatography for preparative separation of organic compounds. 11.* Helv. Chim. Acta., 41, 275-88 (1958).
667. Kozeny, J.: *Ueber kapillare Leitung des Wassers in Boden (Aufstieg, Versickerung, und Anwendung auf die Bewässerung).* Sitzber.Akad.Wiss.Wien, Math.Naturw., II a, 136 (1927).

668. Kraus, K.A.; Carlson, T.A.; Johnson, J.S.: *Cation-exchange properties of zirconium(IV)-tungsten(VI) precipitates.* Nature, 177, 1128-9 (1956).
669. Krejci, M.; Pajurek, J.; Komers, R. (Eds.): *Výpočty a veličiny v sorpční kolonové chromatografii (Calculations and parameters in column chromatography)*, SNTL: Praha, 1990.
670. Krell, E.: *Distillation analysis of mixtures of synthetic fatty acids.* Chem. Techn., 4, 200-7 (1952).
671. Krell, E.: *Handbuch der Laboratoriums Destillation.* VEB Deutscher Verlag der Wissenschaften: Berlin, 1958.
672. Kremser, A.: *Theoretical analysis of absorption columns.* Natl.Petroleum News, 22(21) 43-49 (1930).
673. Krichevskii, I.P.: *Thermodynamics of critical infinitely diluted solutions (in Russian)* Khimiya: Moscow, 1975.
674. Kroll, K.: *Trockner und Trocknungverfahren.* Springer: Berlin, 1959.
675. Krtil, J.: *Exchange properties of ammonium salts of 12-heteropolyacids. IV. Cs exchange on ammonium phosphotungstate and phosphomolybdate.* J. Inorg. Nucl. Chem., 24, 1139-44 (1962).
676. Krylov, V.S.: *Theoretical problems of transfer processes in the systems with intensive mass exchange.* Usp. K him., 49, 118-146 (1980).
677. Kubin, M.: *Beitrag zur Theorie der Chromatographie.* Coll. Czechosl.Chem.Commun., 30, 1104-16 (1965).
678. Kubo, R.: *Thermodynamics. An advanced course with problems and solutions.* North Holland: Amsterdam, 1968.
679. Kuca, L.; Hogfeldt, E.: *Extraction with long chain tertiary amines. X. The mechanism of extraction of trivalent iron by trilaurylammonium bromide.* Acta Chem.Scand., 22(1) 183-92 (1968).
680. Kucera, E.: *Theory of chromatography. Linear nonequilibrium elution chromatography.* J.Chromatogr., 19(2) 237-48 (1965).

681. Kuhn, H.: *Functionalized monolayer assembly manipulation.* Thin Solid Films, 99, 1-16 (1983).
682. Kuhn, R.; Lederer, E.: *The separation of carotene into its components. I. The growth vitamine.* Ber., 64, 1349-57 (1931).
683. Kula, M-R.; Kroner, K.H.; Hustedt, H.: *Purification of enzymes by liquid liquid extraction.* Adv. Biochem. Eng., 24, 73-118 (1982).
684. Kulprathipanja, S.: *Reactive separation process.* New York: Taylor & Francis, 2002.
685. Kutter, J.P.; Fintschenko, Y. (Eds.): *Separation methods in microanalytical systems.* CRC Press, Boca Raton, 2005.
686. Kuzmin, N.M.; Zolotov, Yu.A.: *Kontsentrirovanie sledov elementov (Preconcentration of traces of elements).* Nauka: Moscow, 1998.
687. Kyrs, M.: *The method of concentration dependent distribution in the quantitative use of radioisotopes.* Anal. Chim. Acta., 33, 245 (1965).
688. Kyrs, M.; Kadlecova, L.: *Separation of alkali metals by extraction chromatography with dipicrylamine and nitrobenzene.* J. RadioAnal.Chem., 1(2) 103 12 (1968).
689. Lacey, R.L.: *Electromembrane processes.* Wiley Interscience: New York, 1972.
690. Lacey, R.L.; Loeb, S.: *Industrial processes with membranes.* Wiley-Interscience: New York, 1972.
691. Ladisch, M.R. et al. (eds.): *Protein purification. From molecular mechanism to large-scale processes.* American Chemical Society: Washington, 1991.
692. Ladisch, M.R.: *Bioseparations engineering: Principles, practice, and economics.* Wiley-Interscience, 2001.
693. Lalegerie, P.; Bailly, M.: *Dynamic aspects of the protein ligand bond. Problem of specificity in biology.* Ann. Biol. Clio. (Paris) 39(5) 259-71 (1981).
694. Landers, J.P. (Ed.) *Handbook of capillary electrophoresis.* CRC Press: Boca Raton, 1994.

695. Lang, J.: *Ueber das Gleigewicht nach Einwirkung einerseits von Salzsaure auf Antimontrisulfid und anderseits von Schwefelwasserstof auf Salzsaure Antimontrichloridloesung.* Ber., 18, 2714-2724 (1885).
696. Langmuir, I.: *The constitution and fundamental properties of solids and liquids.* J.Am.Chem.Soc., 38, 2221 (1916).
697. Langmuir, I.: *Overturning and anchoring of monolayers.* Science, 87, 493 (1938).
698. Langmuir, I.: *The distribution and orientation of molecules.* In: 3rd. *Colloid Symposium Monograph.* The Chem. Catalogue Co.Inc.: New York, 1925.
699. Lapidus, L.; Amundson, N.R.: T*he effects of longitudinal diffusion in ion exchange and chromatographic columns.* J.Phys.Chem., 56, 484-8 (1952).
700. Lathe, G.H.; Ruthven, C.R.: *Separation of substances and estimation of their relative molecular sizes by the use of columns of starch in water.* Biochem. J., 62, 665-74 (1956).
701. Latterelt J. J., Walton H. F., *Separation of amines by ligand exchange.* II, Anal. Chim. Acta, 32, 101 (1965).
702. Lederer, M.: *The periodic table for chromatographers.* J. Wiley: New York, 1992.
703. Lee, K.H.; Evans, D.F.; Cussler, E.L.: *Selective copper recovery with two types of liquid membrane.* Am.Inst. Chem.Eng. Journal, 24, 860 (1978).
704. Legget, D.J. (Ed.): *Computational methods for the determination of formation constants.* Plenum Press: New York, 1985.
705. Leibnitz, E.; Struppe, H.G. (eds.): *Handbuch der Gaschromatographie.* 3rd Ed. Akademische Verlagsgesellschaft: Leipzig 1984.
706. Leja, J.: *Surface chemistry of froth flotation.* Plenum Publishing: New York, 1982.
707. Lemlich, R. (Ed.): *Absorptive bubble separation techniques.* Academic Press: New York, 1972.

708. Lennard-Jones, J.E.: *Cohesion.* Proc. Phys. Soc., 43, 461-82 (1931).
709. Letokhov, V.S.: *Non -linear selective processes in atoms and molecules (in Russian).* Nauka: Moscow, 1983.
710. Lev, A.A.: *Modelirovaniye ionnoi selektivnosti kletochnykh membran (Modelling of ionic selectivity of cell membranes).* Nauka: Leningrad, 1976.
711. Levene, S.D.; Zimm, B.H.: *Separations of open-circular DNA using pulse-field electrophoresis.* Proc.Natl.Acad.Sci.USA, 84(12) 4054-7 (1987).
712. Levich, V.G.: *Fiziko-khimicheskaya gidrodinamika (Physico-chemical hydrodynamics).* Mosk.Gos.Izdat.Tekhn.Lit.: Moscow, 1954.
713. Levin, V.I.: *Quantitative interpretation of experimental data on TBP extraction from chloride solutions. In: Khimiya protsesov ekstraktsii, p. 151.* Zolotov, Yu.A., Spivakov, B.Ya., Eds., Nauka: Moskva, 1972.
714. Levin, V.I.; Kozlova, M.D.; Sevastyanova, A.S.: *Extraction with tributylphosphate from chloride solutions. III. Extraction of cobalt.* Radiokhimiya, 14(1) 48-54 (1972).
715. Levine, L.: *Specifities of prostaglandins B1, F1α, and F2α antigen-antibody reactions.* J. Biol. Chem., 246(22) 6782-5 (1971).
716. Levine, L.; Morgan, R.A.; Lewis, R.A.; Austen, K.F.: *Radioimmunoassay of the leukotrienes of slow reacting substance of anaphylaxis.* Proc.Natl.Acad.Sci. USA, 78(12) 7692-6 (1981).
717. Lewin, S.: *The solubility product principle: an introduction to its uses and limitations.* J. Wiley Interscience: New York, 1960.
718. Lewis, D.C.; Shibamoto, T.: *Analysis of toxic anthraquinones and related compounds with a fused silica capillary column.* J. High Resolut. Chrom., 8(6) 280-2 (1985).
719. Lewis, G.N.: *The law of physico-chemical change.* Zeit.phys. Chem. 38, 25 (1901).

720. Lewis, G.N.: *Outlines of a new system of thermodynamic chemistry.* Proc. Am.Acad. Arts and Sciences 43 (7) 257-294 (1907).
721. Lewis, G.N., Randall, M., *The activity coefficients of strong electrolytes.* J.Am.Chem.Soc. 43, 1112 (1921).
722. Lewis, G.N.; Randall, M.: *Thermodynamics.* McCraw-Hill: New York, 1923.
723. Lewis, G.N.; Cornish, R.E.: *Separation of the isotopic forms of water by fractional distillation.* J.Am.Chem.Soc., 55, 2616-17 (1933).
724. Lewis, G.N.; MacDonald, R.T.: *Concentration of ^2H isotope.* J.Chem.Phys., 1, 341-4 (1933).
725. Lewis, J.B.: *The mechanism of mass transfer of solutes across liquid-liquid interfaces. I. The determination of individual transfer coefficients for binary systems.* Chem. Eng. Sci., 3(6) 248-59 (1954).
726. Lewis, J.B.: *The mechanism of mass transfer of solutes across liquid-liquid interfaces. II. The transfer of organic solutes between solvent and aqueous phases.* Chem. Eng. Sci., 3(6) 260-78 (1954).
727. Lewis, J.B.: *The mechanism of mass transfer of solutes across liquid-liquid interface. III. The transfer of uranyl nitrate between solvent and aqueous phase.* Chem. Eng. Sci., 8, 825 (1958).
728. Lewis, J.B.; Pratt, H.R.C.: *Oscillating droplets.* Nature, 171, 1155-6 (1953).
729. Lewis, W.K.; Whitman, W.G.: *Principles of gas absorption.* Ind.Eng.Chem., 16(12) 1215-20 (1924).
730. Li, N.N.: *Permeation through liquid surfactant membranes.* Am. Inst. Chem.Eng. Journal, 17(2) 459-63 (1971).
731. Li, N.N.: *Separation of hydrocarbons by liquid membrane permeation.* Ind. Eng.Chem., Process Des. Dev., 10(2) 215-21 (1971).
732. Li, N.N.; Long, R.B.: *Permeation through plastic films.* Am. Inst. Chem.Eng. Journal, 15{1) 73-80 (1969).

733. Lichtenberg, D.; Barenholz, Y.: *Liposomes: preparation, characterization and preservation.* Methods Biochem. Anal., 33, 337-462 (1988).
734. Lieser, K.H.: *Radiochemical measurement of heterogeneous exchange of labeled ions on the surface of alkaline earth carbonates.* Radiochim. Acta., 3(1/2) 93-96 (1964).
735. Lieser, K.H.; Guetlich, P.; Rosenbaum, I.: *Rate of heterogenous isotope exchange on the surface of ionic crystals.* Radiochim. Acta., 4(4) 216-22 (1965).
736. Lightfoot, E.N., Jr.: *Low-order approximations for membrane blood oxygenators.* Am.Inst.Chem.Eng. Journal, 14(4) 699-70 (1968).
737. Lightfoot, E.N.: *Transport phenomena and living systems. Biomedical aspects of momentum and mass transfer.* J. Wiley Interscience, New York, 1974.
738. Lightfoot, E.N.; Noble, P.S.; Chiang, As.; Ugulini, T.A.: *Characterization of an improved electropalarization chromatographic system using homogeneous proteins.* Sep. Sci.Technol., 16(6) 619-56 (1981).
739. Liteanu, C.; Gocan, S.: *Gradient liquid chromatography.* E. Horwood: Chichester, 1974.
740. Littlewood, A.B.: *Gas chromatography. Principles, techniques, and applications* 2nd. ed. Academic Press: New York, 1970.
741. Liu, Y.A. *Process synthesis: Some simple and practical developments, In: Recent developments in chemical processes and design.*, Y.A.Liu, H.A.McGee, W.R.Experly (Eds.) Wiley-Interscience: New York, 1987.
742. Loffler, F.: *Problems and recent advances in aerosol filtration.* Sep.Sci.Technol., 15(3) 297-305 (1980).
743. London, F.: *Properties and applications of molecular forces.* Z. Physik. Chem., 11, 221-51 (1930).
744. London, H.: *Separation of isotopes.* G. Newnes: London, 1961.

745. Long, R.B.: *Separation processes in waste minimization.* CRC Press: Boca Raton, 1995.
746. Lonsdale, H.K.: *The growth of membrane technology.* J.Membr. Sci., 10(2-3) 81-181 (1982).
747. Lonsdale, H.K.: *Transport properties of cellulose acetate osmotic membranes* J.Appl. Polym. Sci., 9, 1341-62 (1965).
748. Lough, W.J.; Wainer, I.W.: *High performance chromatography. Fundamental principles and practice.* Chapman and Hall: London, 1995.
749. Lowe, C.R.; Dean, P.D.G.: *Affinity chromatography.* J. Wiley: London, 1974.
750. Lozinskii, V.I.; Rogozhin, S.V.: *Chemospecific (covalent) chromatography of biopolymers.* Usp. Khim., 49, 879 (1980).
751. Lu, S.; Pugh, R.J.; Forssberg, E.: *Interfacial separation of particles.* Elsevier: Amsterdam, 2005.
752. Lumetta, G.J.; Nash, K.L.; Clark, S.B. (Eds.): *Separations for the nuclear fuel cycle in the 21st century.* Am.Chem.Society, Washington DC, 2006.
753. Lumpkin, O.J.; Dejardin, P.; Zimm, B.H.: *Theory of gel electrophoresis of DNA.* Biopolymers 24(8) 1573-83 (1985).
754. Lykov, M.V.: *Sushka v khimicheskoi promyshlennosti (The drying in chemical industry).* Khimiya: Moskva, 1970.
755. Macasek, F.: *Simplifications in the description of solvent extraction equilibria of complex compounds.* Chem.Zvesti (Chem. Papers) 28, 3-16 (1974).
756. Macasek, F.; Mikulaj, V.; Kopunec, R.; Rajec, P.: *On the behaviour of the daughter species resulting from chelates by beta decay in separation processes.* J.Radioanal.Chem., 30, 15-328 (1976).
757. Macasek, F.; Matel, L.; Kyrs, M.: *Radiolysis of the bis(1,2-dicarbollyl) cobalt(III) ion in nitrobenzene-bromoform mixtures. III. Extraction properties of the dibromo derivative*

synthesized by radiation. Radiochem.Radioanal.Letters, 35, 247-254 (1978).

758. Macasek, F.: *Deviations from thermodynamic ideality at slope analysis of distribution of microamounts of metal complexes.* Acta F.R.N. Univ.Comen. Chimia, 27, 79-91 (1979).

759. Macasek, F.: *Corresponding extraction systems and conditions of separation of elements.* Sov.Anal.Chem., 46, 3038-87 (1980).

760. Macasek, F.; Vanco, D.: *Description of liquid-liquid extraction equilibria in exchange extraction of chelates. Part 3. Calculation of pH-pA diagrams and enrichment factors in pH-pA coordinates.* Anal.Chim.Acta, 132, 175-185 (1981).

761. Macasek, F.; Cech, R.: *Macrokinetics of radiolysis in the systems with liquid liquid partition of substrates. I. A general approach to mathematical models of simulated solvent extraction systems.* Radiat.Phys.Chem., 23, 473-479 (1984).

762. Macasek, F.: *Macrokinetics of radiolysis in the systems with liquid-liquid partition of substrates. II. Radiation yields in two-liquid systems.* Radiat.Phys.Chem., 23, 481-484 (1984).

763. Macasek, F.; Rajec, P.; Kopunec, R.; Mikulaj, V.: *Membrane extraction in preconcentration of some uranium fission products.* Solvent Extr. Ion Exch., 2, 227-252 (1984).

764. Macasek, F.: *Membrane extraction instead of solvent extraction. What does it give?* J.Radioanal.Nucl.Chem., Articles, 129(2) 233-244 (1989).

765. Macasek, F.; Klas, J.: *A generalized separation reaction scheme for assessment of the radioanalytical method of concentration dependent distribution.* J.Radioanal.Nucl. Chem., Articles 172(2) 231-238 (1993).

766. Macasek, F.; Gerhart, P.; Malovikova, A.: *Membraneless dialysis of strontium in aqueous liquid- liquid milk - pectin*

system. J.Radioanal.Nucl.Chem., Letters 186(2) 99-111 (1994).
767. Macek, K.; Deyl, Z.; Smrz, M.: *Two-dimensional thin-layer chromatography of Dns-amino acids on reversed-phase silica gel.* J. Chromatog., 193(3) 421-6 (1980).
768. Macek, K.; Hais, I.M.: *Stationary phase in paper and thin-layer chromatography.* Elsevier: Amsterdam, 1965.
769. Madelung, E.: *Das elektrische Feld in Systemen von regelmassig angeordneten Punktladungen.* Physik. Ztschr., 19, 524-33 (1918).
770. Magee, E.M.: *Course of a reaction in a chromatographic column.* Ind.Eng.Chem.Fundamentals 2, 32-36 (1963).
771. Magnussen, T.; Rasmussen, P.; Fredenslund, A.: *UNIFAC parameters table for prediction of liquid-liquid equilibria.* Ind.Eng.Chem. Process Des. Dev., 20, 331-9 (1981).
772. Majer, J.; Trinh Van Quy; Valaskova, I.: *Studies on complex-forming properties of N-(phosphomethyl)-iminodiacetic acid and glycine-N, N-bis(methylenephosphonic) acid by electrophoresis.* Chem. Zvesti (Chem. Papers) 34(5) 637-44 (1980).
773. Maksimovic, Z.B.; Reichardt, C.; Spizic, A.: *Determination of empirical parameters of solvent polarity in binary mixtures by solvatochromic N-hydroxyphenylpyridinium betaine.* Z. Anal.Chem., 270(2) 100-4 (1974).
774. Mancini, G.; Carbonare, A.O.; Heremans, J.S.: *Immunochemical quantitation of antigens by single radial immunodiffusion.* Immunochemistry, 2(3) 235-54 (1965).
775. Manning, G.S.: *Correlation of solute permeability and reflection coefficient for rigid membranes with high solvent content.* J.Phys.Chem., 76, 393 (1972).
776. Mansour, A.R.: *Comparison of equilibrium and non-equilibrium modes in the simulation of multicomponent sorption proces.* In: Chemical separations, vol. 1., p.217. King, C.J., Navratil, J.D., Eds., Litarvan: Denver, 1986.

777. Marcus, Y.; Coryell, C.D.: *The anion exchange of metal complexes. I.Theory.* Bull. Res. Council Israel, Sec. A, 8, 1-16 (1959).
778. Marcus, Y.; Kertes, A.S.: *Ion exchange and solvent extraction of metal complexes.* Wiley Interscience: London, 1969.
779. Marcus, Y.: *Development and publication of solvent extraction methods.* Talanta, 23, 203-209 (1976).
780. Marcus, Y.: *Introduction to liquid state chemistry.* Wiley&Sons: London, 1977.
781. Marcus, Y.; Asher, L.E.: *Extraction of alkali halides from their aqueous solutions by crown ethers.* J.Phys.Chem., 82(11) 1246-54 (1978).
782. Marcus, Y.: *Ionic radii in aqueous solutions.* J. Solution Chem., 12(4) 271-5 (1983).
783. Margules, M.: *Ueber die Zusammensetzung der gesaettigten Daempfe von Mischungen.* Sitzber.Akad.Wiss.Wien, Math. Naturw., 104, 1243 (1895).
784. Marhol, M.: *Ion exchangers containing phosphorus in their functional group. Sorption of cations from nitric acid solutions and from acetate medium.* J. Appl. Chem., 16(6) 191-6 (1966).
785. Marinsky, J.A.: *Ion binding in charged polymers.* Coord. Chem. Rev., 19(2) 125-71 (1976).
786. Marinsky, J.A.; Glendenin, L.E.; Coryell, C.D.: *The chemical identification of radioisotopes of neodymium and element 61.* J.Am. Chem. Soc., 69, 2781-5 (1947).
787. Marr, R.; Kopp, A.: *Liquid membrane technology - a survey of phenomena, mechanisms and models.* Int.Chem.Eng., 22(1) 44-60 (1982).
788. Martin, A.J.P.; Everaerts, F.M.: *Displacement electrophoresis.* Proc. Roy. Soc. London, A 316, 493-514 (1970).
789. Martin, A.J.P.; Synge, R.L.M.: *A new form of chromatogram employing two liquid phase1. I. A theory of chromatography. II. Application to the microdetermination of the higher*

monoamino acids in proteins. Biochem. J., 35, 1358-68 (1941).

790. Martin, A.J.P.; Synge, R.L.M.: *Separation of the higher monoaminoacids by counter-current liquid-liquid extraction: the aminoacids composition of wool.* Biochem.J., 35, 91-121 (1941).

791. Martire, D.E.; Riedl, P.: *Thermodynamic study of hydrogen bonding by means of gas-liquid chromatography.* J.Phys. Chem. 72, 3478-88 (1968).

792. Masseyeff, R.F.; Albert, W.H.; Staines, N.A. (eds.): *Methods in immunological analysis,* Vol.1. J.Wiley: New York, 1992.

793. Masson, D.O.: *Geloeste Molekularvolumina in Beziehung zur Solvatation und Ionization.* Phil.Mag., 8(7) 218-35 (1929).

794. Matsuura, T.: *Synthetic membranes and membrane separation processes.* CRC Press: Boca Raton, 1994.

795. Matsuura, T.; Sasaki, T.: *Separation of recoiling chromium(II) species from neutron irradiated hexaaquochromium(III) ion adsorbed on ion-exchanger.* Radiochim. Acta., 8, 33 (1967).

796. Maurer, H.R.: *Disc electrophoresis and related techniques of polyacrylamide gel electrophoresis.* 2nd ed., Walter de Gruyter: Berlin, 1971.

797. McCabe, W.L.; Smith, J.C.: *Unit operations of chemical engineering.* 3rd. ed. McGraw Hill: New York, 1976.

798. McCabe, W.L.; Thiele, E.W.: *Graphical design of fractionating columns.* Ind.Eng.Chem., 17, 605-11 (1925).

799. McCoy, B.J.: *Counteracting chromatographic electrophoresis - a multiple-field fractionation process.* In: Chemical Separations, v.1, p.169-184. C.J. King, J.D. Navratil, Eds., Litarvan: Denver, 1986.

800. McCoy, B.J.: *Modeling and optimizing large-scale chromatographic separations.* In: *Chemical separation,* vol. 1, p. 113. King, J.B., Navratil, J.D., Eds., Litarvan: Denver, 1986.

801. McDowell, W.J.: *Extraction of alkaline earths from sodium nitrate solutions by bis(2-ethylhexyl)phosphate in benzene: Mechanism and equilibria.* J. Inorg. Nucl. Chem., 28(4) 1083-9 (1966).
802. McGregor, W.C.: *Membrane separations in biotechnology.* Marcel Dekker: New York, 1986.
803. Mcinnes, D.A.: *Activities of the ions of strong electrolytes.* J.Am.Chem.Soc., 41, 1086 (1919).
804. McKay, H.A.C.: *Kinetics of exchange reactions.* Nature, 142, 997-8 (1938).
805. McKay, H.A.C.: *Kinetics of some exchange reactions of the type* RI + I* = RI* + I *in alcoholic solution.* J.Am.Chem.Soc., 65, 702 (1943).
806. McKay, H.A.C.: *Activities and activity coefficients in ternary mixtures.* Trans. Faraday Soc., 41, 237-42 (1953).
807. McKay, H.A.C.: *TBP - meeting point of science and technology.* In: Solvent extraction chemiatry, p.185. Dyrssen, D., Liljenzin, J.O., Rydberg, J., Eds., North Holland: Amsterdam 1967.
808. McKay, H.A.C.: *Tri-n-butylphosphate as an extracting agent for the nitrates of the actinide elements.* In: Proc.Int.Conf. on the Peaceful Uses of Atomic Eneryy, vol. 1, p. *311*, United Nations: New York 1956.
809. Melikhov, I.V.; Berdonosov, S.S.: *On classification of co-precipitation phenomena.* Radiokhimiya, 16(1) 3 (1974).
810. Melikhov, I.V.; Merkulova, M.S.: *Sokristallizatsiya (Co-crystalization).* Khimiya.: Moskva 1975.
811. Melikhov, I.V.: *Crystalization of salts from supersaturated solutions. Kinetic systems.* Teor.Osn.Khim.Tekh., 13(4) 530-7 (1979).
812. Melikhov, I.V.; Berliner, L.B.: *Effect of fluctuations on the kinetics of crystalization.* Dokl.Akad.Nauk SSSR, 245(5) 1159-63 (1979).
813. Melikhov, I.V.; Berliner, L.B.: *Simulation of batch crystalization.* Chem.Eng. Sci., 36(6) 1021-34 (1981).

814. Meloan, C.E.: *Chemical separations. Principles, techniques and experiments.* Wiley-Interscience, New York, 1999.
815. Meloun, M.; Havel, J.; Hogfeldt, E.: *Computation of solution equilibria: A guide to methods in potentiometry, extraction and spectrophotometry.* Ellis Horwood: Chichester, 1988.
816. Menestrina, G.; Pasquali, F.: *Reconstitution of the complement channel into lipid vesicles and planar bilayers starting from the fluid phase complex.* Bioscience Rep., 5, 129-136 (1985).
817. Meselson, M.S.; Stahl, F.W.; Vinograd, J.R.: *Equilibrium sedimentation of macromolecules in density gradients.* Proc. Nat. Acad. Sci. USA, 43, 581-8 (1957).
818. Meselson, M.S.; Stahl, F.W.; Vinograd, J.R.: *The replication of deoxyribonucleic acid in Escherichia coli.* Proc.Natl.Acad. Sci. USA, 44, 671-82 (1958).
819. Metzsch, F.A.: *Automatic apparatus for fractional countercurrent distribution.* Chem.-Ing.Technik, 25(2) 66-72 (1953).
820. Meyer, K.H.; Strauss, W.: *Permeability of membranes (VI). The passage of the electric current through selective membranes.* Helv. Chim. Acta., 23, 795 (1940).
821. Michaels, A.S.: *Polyelectrolyte complexes.* Ind.Eng.Chem., 57(10) 32-40 (1965).
822. Michaels, A.S.: *Separation techniques for the CPI (chemical process industry).* Chem.Eng.Progr., 64, 31 (1968).
823. Michaels, A.S.: *Simplified method of interpreting* kinetic *data in fixed-bed ion exchange.* Ind.Eng.Chem., 44, 1922-30 (1952).
824. Michal, J.: *Inorganic chromatographic analysis.* Van Nostrand, Reinhold: New York, 1974.
825. Mie, G.: *Zur kinetische Theorie der einatomiger Koerper.* Ann. Physik, 11, 657 (1903).
826. Miertus, S.; Miertusova, I.: *Theoretical and experimental study of the sorption processes of gases on NaY zeolites.* J.Chromatogr., 286, 31-36 (1984).

827. Mikes, O. (Ed.): *Laboratory handbook of chromatographic and allied methods.* Ellis Horwood: Chichester, 1979.
828. Mikes, O.: *Proteins. X XXVIII. Descending paper electrophoresis of protein hydrolyzates and peptides.* Collect. Czechosl. Chem. Commun., 22, 831-50 (1957).
829. Mikes, O.; Strop, P.; Hostomska, Z.: *Ion-exchange derivatives of spheron. V. Sulphate and sulpho-derivatives.* J.Chromatogr.-Biomed., 301, 93-105 (1984).
830. Mikheev, N.B.; El-Garhy, M.; Moustafa, Z.: *Generator for production of ^{99m}Tc from irradiated molybdenum.* Atompraxis, 10, 263 (1964).
831. Mikulaj, V.; Rajec, P.; Faberova, V.: *Chemical stabilization of daughter ^{99m}Tc after beta decay in chelate systems with molybdenyl(^{99}Mo).* Chem. Papers, 28(1) 37-46 (1974).
832. Mikulaj, V.; Macasek, F.; Steinerova, M.: *Chelate extraction in repeating separations of ^{99m}Tc from parent ^{99}Mo using N-benzoyl-N-phenylhydroxylamine.* Radiochem.Radioanal. Letters, 29, 199-206 (1977).
833. Mikulski, J.; Stronski, I.: *Radiochemical separation of some metals by partition chromatography with reversed phases on Teflon in the systems trioctylamine electrolyte.* J.Chromatogr., 17(1) 197-200 (1965).
834. Miller, J.M.: *Separation methods in chemical analysis.* J. Wiley: New York, 1975.
835. Miltenyi, S.; Muller, W.; Weichel, W.; Radbruch, A.: *High gradient magnetic cell separation with MACS.* Cytometry, 11, 231-238 (1990).
836. Milton, T.; Hearn, M.T.W.; Anspach, B.: In: *Separation processes in biotechnology*, p.17, J. Asenjo (Ed.), Marcel Dekker: New York, 1990.
837. Minczewski, J.; Chwastowska, J.; Dybzcynski, R.: *Separation and preconcentration methods in inorganic trace analysis.* J. Wiley: New York, 1982.
838. Misak, N.Z.: *Langmuir isotherm and its application in ion-exchange reactions.* React.Polym 21, 53-64 (1993).

839. Misak, N.Z.: *Some aspects of the application of adsorption isotherms to ion exchange reactions.* Reactive Polymers 43, 153-164 (2000).
840. Misek T.: *Breakup of drops by a rotating disk.* Coll.Czechosl. Chem.Commun., 28, 426-35 (1963).
841. Misek, T.: *Hydrodynamic behavior of pulsed liquid-liquid extractors.* Coll.Czechosl.Chem.Commun., 29(8) 1755-66 (1964).
842. Mitchell, P.: *Coupling of phosphorylation to electron and hydrogen transfer by a chemiosmotic type of mechanism.* Nature, 191, 144-8 (1961).
843. Mitchell, P.: *Stoichiometry of proton translocation through the respiratory chain and adenosinetriphosphatase systems of rat liver mitochondria.* Nature, 208, 147- 51 (1965).
844. Moldoveanu, S.; David, V.: *Essentials in modern HPLC separations.* Elsevier Science, New York, 2012.
845. Moody, G.J.; Thomas, J.D.R.: *Chromatographic separation and extraction with foamed plastics and rubbers.* Marcel Dekker: New York, 1982.
846. Moore, G.E.; Kraus, K.A.: *Anion exchange studies. IV. Cobalt and nickel in hydrochloracid solutions.* J.Am.Chem.Soc., 74, 843-4 (1952).
847. Moore, R.L.: *The* mechanism *of extraction of uranium by tributyl phosphate.* AECD-3196 Report, 1951.
848. Morris, C.J.O.R.: *Molecular-sieve chromatography and electrophoresis in polyacrylamide gels.* Biochem. J., 124, 517-28 (1971).
849. Morris, C.J.O.R.; Morris, P.: *Separation methods in biochemistry.* 2nd ed. Pitman, London, 1976.
850. Moser, L.; Maxymovicz, W.: *Erfahrungen* uber *die Verwendbarkeit der Glasfiltertiegel in der Gewichtaanalyse.* Chem.Ztg., 48, 693 (1924).
851. Moskvin, L.N.; Tserkovnitskaya, L.G: *Metody razdeleniya i kontsentrirovaniya v analiticheskoy khimii (Separation*

and *preconcentration techniques in analytical chemistry).* Khimiya: Moscow, 1991.
852. Mukerjee, P.; Mysels, K.J.: *Critical micelle concentrations of aqueous surfactants systems.* NBS: Washington, 1971.
853. Mulder, M.: *Basic principles of membrane technology.* Kluwer: Amsterdam, 1991.
854. Mullin, J.W.: *Crystallization. 2nd. Ed.* Chemical Rubber Co: Cleveland, 1972.
855. Murphree, E.V.: *Evaluation of rectification columns.* Ind. Eng.Chem., 17, 747-50 (1925).
856. Muscatello, A.C.; Navratil, J.D.; Killion, M.E.: *The extraction of americium(III) by mixtures of tributylphosphate with tributyl-N, N-diethylcarbamoyl phosphonate and methyl phosphonate (DBDECP and DBDECPM).* Solvent Extr.Ion Exch., 1(1) 127-39 (1983).
857. Myasoedov, B.F.; Milyukova, M.S.; Malikov, D.A.: *Extraction of berkelium(IV) by neutral organophosphorus compounds and high molecular weight amines.* Solvent Extr. Ion Exch., 2(1) 61-77 (1984).
858. Myasoedov, B.F.; Vinogradov, A.P.: *Analiticheskaya khimiya transplutonievykh elementov (Analytical chemistry of transplutonium elemnts).* GEOKHI: Moscow 1972.
859. Myerson, A.S.: *Handbook of industrial crystallization.* 2nd ed. Butterworth-Heinemann: Woburn, 2001.
860. Nardone, M.S.: *Direct digital control systems. Application. Commissioning.* Kluwer Academic Publisher: Dordrecht, 1999.
861. Nash, K.L.; Lumetta, G.J. (Eds.): *Advanced separation techniques for nuclear fuel reprocessing and radioactive waste treatment.* Woodhead Publishing: London, 2011.
862. Naushad, M.; Khan, M.R.: *Ultra Performance Liquid Chromatography Mass Spectrometry: Evaluation and Applications in Food Analysis.* CRC Press: Boca Raton, 2014.

863. Navarenko, V.A.; Antonovich, V.P.; Nevskaya, Ye.M.: *Gidroliz ionov metallov v razbavlennykh rastvorakh (Hydrolysis of metal ions in dilute solutions)*. Atomizdat: Moscow 1979.
864. Navratil, J.D.; Murgia, E.; Walton, H.F.: *Ligand exchange chromatography of amino sugars.* Anal.Chem., 47(1) 122-5 (1975).
865. Navratil, J.D.; Martella, L.L.: *Extraction behavior of americium and plutonium with mixed solvent extractants.* Sep.Sci.Technol., 16(9) 1147-55 (1981).
866. Navratil, J.D.; Murphy, A.; Sun, D.: *Mixed solvent extraction-annular chromatographic systems for f-element separation and purification.* In: Proceedings Intern. Solv. Extr. Conf. ISEC'90, 18-21 July, Tokyo. Tokyo, 1990.
867. Navratil, O.; Hala, J.; Kopunec, R.; Leseticky, L.; Macasek, F.; Mikulaj, V.: *Nuclear chemistry.* E. Horwood: Chichester, 1992.
868. Neplenbroek, A.M.; Bargeman, D.; Smolders, C.A.: *Supported liquid membrane degradation by emulsion formation.* In: Proceedings ISEC '88, vol. III, p. 61. Academy of Science USSR: Moscow, 1988.
869. Nernst, W.: *Theorie der Reaktionsgeschwindigkeit in heterogenen Systemen.* Z.Phys.Chem., 47(1) 52 (1904).
870. Nernst, W.: *Verteilung eines Stoffes zwischen zwei Loesungsmitteln und zwischen Loesungsmittel und Dampfraum.* Z.phys.Chem., 8, 110 (1891).
871. Nernst, W.: *Zur kinetischen Theorie der Einatomigen Köorper.* Z.phys.Chem., 2, 613 (1888).
872. Nernst, W. *Die elektromotorische Wirksamkeit der Jonen.* Z.phys.Chem. 4(2), 129-181 (1889).
873. Nernst, W.; Riesenfeld, E.H.: *Ueber elektrolytische Erscheinungen an der Grenzfläche zweier Loesungsmittel.* Nachr.k.Ges.Wiss.Gottingen, 2(8) 54-61 (1901).
874. Newton, I.S.: *Philosophiæ Naturalis. Principia Mathematica.* S.Pepys: London, 1686.

875. Nesterov, A.Ye.: *Inverse gas chromatography of polymers* (in Russian). Naukova Dumka: Kiev, 1988.
876. Nguyen, Anh V.; George, P.; Jameson, G.J.: *Demonstration of a minimum in the recovery of nanoparticles by flotation: Theory and experiment.* Chem.Eng.Sci. 61(8) 2494-2509 (2006).
877. Nikolaev, A.V.; Yakovlev, I.I.; Dyadin, Yu.A.: *Phase equilibria in binary and ternary systems. In: Solvent extraction chemistry, p. 919.* Dyrssen, D., Liljenzin, P.O., Rydberg, J., Eds., North Holland: Amsterdam, 1967.
878. Nikolaev, N.I.: *Diffuziya v membranakh (Diffusion in membranes).* Khimiya: Moscow, 1980.
879. Noble, R.D.: *Slope factors in facilitated transport through membranes.* Ind.Eng.Chem., Fundam., 22(1) 139-44 (1983).
880. Noble, R.D. (Ed.): *Liquid membranes: Theory and applications.* Am.Chem.Society, 1987.
881. Noble, R.D.; Stern, A.S. (Eds.): *Membrane separations technology: Principles and applications.* Elsevier Science, 1995.
882. Noble, R.D.; Terry, P.A.: *Principles of chemical separations with environmental applications.* Cambridge University Press, 2010.
883. Noel, D.F.; Mellon, C.E.: *Some empirical correlations in solvent extraction.* Sep. Sci., 7(1) 75 (1972).
884. Nollet, L.M.L. Toldra F. (eds.): *Food analysis by HPLC.* 3rd ed. CRC: Boca Raton, 2013.
885. Northrop, J.H.: *Crystalline pepsin I. Isolation.* J. Gen. Physiol., 13, 739-60 (1930).
886. Novak, J.: *Quantitative analysis by gas chromatography.* Marcel Dekker: New York, 1975.
887. Novak, J.P.; Vonka, P.; Suska, J.; Matous, J.; Pick, J.: *Applicability of the three-constant Wilson equation to correlation of strongly nonideal systems. I.* Coll.Czechosl. Chem. Commun., 39, 3593 (1974).

888. Noyes, A.A.; Whitney, W.R.: *Ueber die Aufloesungsgeschwindigkeit von festen Stoffen und ihren eigenen Loesungen.* Z.Phys.Chem., 23, 689 (1897).
889. Noyes, R.M.: *Thermodynamics of ion hydrations as a mesure of effective dielectric properties of water.* J.Am.Chem.Soc., 84, 513-17 (1962).
890. Null, H.R.: *Energy economy in separation processes.* Chem. Eng.Prog., 76(8) 42-9 (1980).
891. Null, H.R.: *Heat pumps in distillation.* Chem.Eng.Prog., 72(7) 58-64 (1976).
892. Null, H.R.: *Phase equilibrium in process design.* J. Wiley-Interscience: New York, 1970.
893. Nusselt, W.: *Heat transmission, diffusion and evaporation.* Z.angew.Math.u.Mechan., 10(2) 105-121 (1930).
894. Nusselt, W.: *Der Warmeaustausch am Berusehlungskuhler.* Z.Ver.Dtsch.Ing., 67, 206-10 (1923).
895. Nylander, C.: *Chemical and biological sensors.* J.Phys. E, 18, 736-50 (1985).
896. Nyvlt, J.: *Solid-liquid phase equilibria.* Elsevier: Amsterdam, 1977.
897. Ochkin, A.V.: *Thermodynamic description of the influence of alcohols on the activity of ammonium salts.* Russ. J.Phys. Chem., 54(7) 1058-1059 (1980).
898. O'Farell, P.H.: *High-resolution, two-dimensional electrophoresis of proteins.* J.Biol. Chem., 250, 4007 (1975).
899. O'Farell, P.O.: *Separation techniques based on the opposition of two counteracting force a to produce a dynamic equilibrium.* Science (Washington) 227, 1586-9 (1985).
900. Olander, D.R.: *Simultaneous mass transfer and equilibrium chemical reaction.* Am.Inst.Chem.Eng. Journal, 6, 233-9 (1960).
901. Oliver, E.D.: *Diffusional separation processes: theory, design and evaluation.* J.Wiley: New York, 1966.

902. Olshanova, K.M.; Kopylova, V.D.; Morozova, N.M.: *Osadochnaya khromatografiya (Precipitation chromatography)*. Izdat.Akad.Nauk SSSR: Moskva, 1963.
903. Onsager, L.: *Reciprocal relations in irreversible processes.* Phys.Rev., 37, 405-26 (1931).
904. Onsager, L.: *Reciprocal relations in irreversible processes II.* Phys.Rev., 38, 2265 (1931).
905. Onsager, L.; Fuoss, R.M.: *Irreversible processes in electrolytes. Diffusion, conductance, and viscous flow in arbitrary mixtures of strong electrolytes.* J.Phys.Chem., 36, 2689-2778 (1932).
906. Onsager, L.: *Electric moments of molecules in liquids.* J.Am.Chem.Soc., 58, 1486-93 (1936).
907. Orlicek, A.F.; Hackl, A.E.; Kindermann, P.E.: *Filtration.* Dechema: Frankurt/Main 1964.
908. Ouchterlony, O.: *Antigen-antibody reactions in gels.* Acta Path. Microbiol., 26, 507-15 (1949).
909. Oudin, J.: *Method of immunochemical analysis by specific precipitation in gelled medium.* Compt. rend., 222, 115 (1946).
910. Palagyi, S.: *Theoretical efficiency of pulsed polyurethane foam column separations.* Solvent Extr. Ion Exch., 3, 517 (1985).
911. Palagyi, S.; Braun, T.: *Separations and preconcentration of trace elements and inorganic species on solid polyurethane foam sorbents.* In: *Preconcentration techniques for trace element.* Z.B. Alfassi, Ch.M. Wai, Eds., CRC Press: Boca Raton, 1992.
912. Palkin, V.A.; Gadel'shin, V.M.; Aleksandrov, O.E.; Seleznev, V.D.: *Multicomponent separation potential. Generalization of the Dirac theory.* J.Eng.Phys.Thermophys., 87(3) 515 (2014).
913. Panchenkov, G.M.; Moiseev, V.D.; Makarov, A.V.: *Chemical method for the separation of boron isotopes.* Zh.fiz.khim., 31, 1851-59 (1957).

914. Paneth, F.: *Ueber* eine *Methode zur Bestimmung der Oberflaecheadsorbierenden Pulver.* Z. Electrochem., 28, 113-15 (1922).
915. Papadoyannis, I.N.: *HPLC in clinical chemistry.* Marcel Dekker: New York, 1989.
916. Parker, A.J.: *Solvation of ion-enthalpies, entropies and free energy of transfer.* Electrochim. Acta, 21, 671-9 (1976).
917. Pasteur, L.: *Sur les relations qui peuvent exister entre la forme crystalline, la composition chimique et le sens de la polarisation rotatoire.* Ann. Chim. Phys., 24, 442 (1848).
918. Paulhamus, J.A.: *Airborne contamination.* In: Ultrapurity, p. 55. Zief, M., Speights, R., Eds., Marcel Dekker: New York, 1972.
919. Pedersen, C.J.: *Cyclic polyethers and their complexes with metal salts.* J.Am.Chem.Soc., 89, 2495-6 (1967).
920. Peppard, D.F.: *Fractional extraction of the lanthanides as their di-alkylortho-phosphates.* J. Inorg. Nucl. Chem., 4, 334-43 (1957).
921. Perez de Los Rios, A.; Fernandez F.J.H.: *Ionic liquids in separation technology.* Elsevier Science: Amsterdam 2014.
922. Perlmutter-Hayman, B.: *Cooperative binding to macromolecules. A formal approach.* Acc.Chem.Res. 19, 90-96 (1986).
923. Pertoff, H.; Laurent, T.C.; Laas, T.; Kagedal, L.: *Density gradients prepared from colloidal silica particles coated by polyvinylpyrrolidone (Percoll).* Anal. Biochem., 88, 271 (1978).
924. Peter, S.: *Thermodynamics of multicomponent systems as a base for physico chemical separation processes.* Int. Chem. Eng., 19, 410-19 (1979).
925. Petlyiuk, F.B.; Platonov, V.M.; Slavinskii, D.M.: *Thermodynamically optimal method for separating multicomponent mixtures.* Int.Chem.Eng., 5, 555 (1965).
926. Pfann, W.G.: *Principles of zone-melting.* J.Metals T., 4, 747 (1952).

927. Pfann, W.G.; Olsen, K.M.: *Purification and prevention of segregation in single crystals of germanium.* Phys. Rev., 89, 322-3 (1953).
928. Pichler, H.; Schulz, H.: *Continuous separation of gases by a new process of countercurrent distribution.* Brennstoff-Chem., 39, 148-53 (1958).
929. Pigford, R.L.: *Hydrodynamic stobility of a fluidized bed.* Ind. Engn.Chem.Fundam., 4(1) 81-7 (1965).
930. Pigford, R.L.; Baker, B.; Blum, D.E.: *Equilibrium theory of the parametric pump.* In. Eng. Chem., Fund., 8(1) 144-9 (1969).
931. Pigford, R.L.; Sliger, G.: *Rate of diffusion controlled reaction between a gas and a porous solid state. Reaction of sulfur dioxide with calcium chloride.* Ind.Eng.Chem., Process Des. Dev., 12(1) 85-91 (1973).
932. Piljac, G.; Piljac, V.: *Genetic engineering. Liquid chromatography.* TIZ: Zrinski Calcovec, 1986.
933. Piljac, V.; Piljac, G.: *Genetic engineering. Centrifugation and electrophoresis.* TIZ: Zrinski Calcovec, 1986.
934. Plaksin, I.N.; Bessonov, S.V.: *Influence of various gases on the wetting of metals and sulfidic minerals.* Dokl. AN SSSR, 61(5) 865-8 (1948).
935. Planck, M.: *Ueber die Erregung von Electricitaet und Waerme in Electrolyten.* Ann. Physik, 39, 161 (1890).
936. Planck, M.: *Ueber die kanonische Zustandsgleichung einatomiger Gase.* Sitzungsber.Kgl.Pr. Akad. Wiss. Berlin, 633-47 (1908).
937. Planck, M.: *Ueber die Potenzialdifferenz zwischen zwei verduenten Loesungen binaerer Electrolyte.* Ann. Physik, 40, 561 (1890).
938. Poole, C.: *Essence of chromatography.* Elsevier Science: Amsterdam, 2003.
939. Poole, C. (Ed.): *Handbook of methods and instrumentation in separation science.* Academic Press: London, 2009.

940. Porath, J.; Flodin, P.: *Gel filtration: a method for desalting and group separation.* Nature, 183, 1657-9 (1959).
941. Porath, J; Garlsson, J.; Olsson, I.; Belfrage, G.: *Metal chelate affinity chromatography, a new approach to protein fractionation.* Nature 258, 598 (1975).
942. Porath, J.: *IMAC - immobilized metal ion affinity based chromatography.* Trends Anal. Chem, 7, 254–259 (1988).
943. Poulik, M.D.: *Starch electrophoresis in a discontinuos system of buffers.* Nature, 180, 1477-8 (1957).
944. Pourbaix, M.J.N.: *Atlas of electrochemical equilibria in aqueous solutions at 25 °C.* Pergamon Press: London, 1966.
945. Powell, M.J.D.: *An efficient method for finding the maximum of a function of several variables without calculating derivatives.* Computer J. 7 (1964).
946. Powell, R.E.; Rosevare, W.E.; Eyring, H.: *Diffusion, thermal conductivity and viscous flow of liquids.* Ind.Eng.Chem., 33, 430-5 (1941).
947. Prandtl, L.: *Bemerkung über den Wärmeübergang im Röhr.* Z. Physik, 29, 487 (1928).
948. Prandtl, L.: *Eine Beziehung zwischen Waermeaustausch und Stroemungs-widerstand der Flüssigkeiten.* Z. Physik, 11, 1072 (1910).
949. Pratt, H.R.C.: *Countercurrent separation processes.* Elsevier: Amsterdam, 1967.
950. Pratt, H.R.C.: *Simplified analytical design method for differential extractors with backmixing. 1. Linear equilibrium relations.* Ind.Eng.Chem., Process Des. Dev., 14(1) 74-80 (1975).
951. Prausnitz, J.M.: *Calculations of phase equilibria for separation operations.* Trans.I. Chem. E., 59, 3-16 (1981).
952. Prausnitz, J.M.: *Molecular thermodynamics of fluid phase equilibrium.* Prentice-Hall, Englewood Cliffs: New York, 1969.
953. Preetz, W.; Pfeifer, H.L.: *Countercurrent ionophoresis. II. Experimental studies.* Talanta, 14(2) 143-53 (1967).

954. Pretlow, T.G.; Pretlow, T.P.; Cheret, A.M. (Eds.) *Cell separation*. Academic Press: London, 1987.
955. Price, C.A.: *Centrifugation in density gradients*. Academic Press: New York, 1982.
956. Prigogine, I.: *Introduction to thermodynamics of irreversible processes*. J. Wiley: New York, 1967.
957. Prigogine, I.: *The fluctuation of chemical equilibrium*. Physica, 16, 134 (1950).
958. Prokhorov, A.; Bokhan, P.; Bukhanov, V.; Zakrevskii. D.; Kazaryan, M.; Fateev, N.: *Opticheskoe i lazerno-khimicheskoe razdelenie izotopov v atomarnykh parakh (Optical and laser-chemical separation of isotopes in atomic vapors)*. LitRes: Moscow, 2015.
959. Prusik, Z.: *Free-flow electromigration separations*. J. Chroma.tog., 87, 73 (1973).
960. Pshezhetskii, S.Ya.; Rubinshtein, R.N.: *On formal theory of multiple (complicated) reactions*. Zh.Fiz.Khim., 21(6) 659 (1947).
961. Purnell, J.H.: *Comparison of efficiency and separating power of packed and capillary gas chromatographic columns*. Nature, 184/826, 2009 (1959).
962. Purnell, J.H.: *Gas chromatography*. J. Wiley: New York, 1962.
963. Qiu, Lu-Fu; Kang, Xi-Hui; Wang, Tong-Sheng.: *A study on photochemical separation of rare earths: The separation of europium from an industrial concentrate material of samarium, europium, and gadolinium*. Sep.Sci.Technol. 26, 199-221 (1991).
964. Rabiller, C.: *Stéréochimie et chiralité en chimie organique*. De Boeck&Larcler: Paris, 1999.
965. Rachinskii, V.V.; Zhukova, L.A.: *Formulae for calculations of the solubilities of electrolytes*. Russ. J.Phys.Chem., 52(6) 913-914 (1978).
966. Rachinskii, V.V.: *Theory of ion-exchange sorption filters*. Zh.prikl.khim., 27, 831-42 (1954).

967. Rachinskii, V.V.; Lurie, A.A.: *The precipitation isotherm.* Dok.I. Akad. Nauk SSSR, 152(6) 1365 (1963).
968. Rais, J.: *Individual extraction constants of univalent ions in water-nitrobenzene systems.* Coll. Czechosl. Chem. Comm., 36(1) 3253-62 (1971).
969. Rajec, P.; Macasek, F.; Belan, J.: *Membrane extraction of pertechnetate with quaternary ammonium salts. Comparison with solvent extraction.* J.Radioanal.Nucl.Chem.Articles, 101 (1) 71-76 (1986).
970. Rajec, P.; Mikulaj, V.: *Preparation of dioxo(8-hydroxyquinolinato)technetium(V) chelate by extraction method.* Radioch. Radioanal. Lett., 17, 375-80 (1974).
971. Ralston, A.: *A first course in numerical analysis.* McGraw Hill: New York, 1965.
972. Ramachandra, R.S.: *Surface chemistry of froth flotation. Vol.I. Fundamentals.* 2nd Ed. Springer Science: New York, 2004.
973. Ramaswamy, S.; Huang, H-J.; Ramarao, B.V.: *Separation and purification technologies in biorefineries.* J.Wiley&Sons: New York, 2013.
974. Ramey, D.W.; Petek, M.; Taylor, R.D.; Fisher, P.W.; Kobisk, E.H.; Ramey, J.; Sampson, C.A.: *Hydrogen isotope separation by bipolar electrolysis with countercurrent electrolyte flow.* Sep.Sci.Technol., 15(3) 405-21 (1980).
975. Raoult, F.M.: *Ueber die Dampfdrucke Aetherischer Loesungen.* Z.Phys.Chem., 2, 352 (1886).
976. Ratner, A.P.: *On the theory of the distribution of electrolytes between a solid crystalline and a liquid phase.* J. Chem. Phys., 1(11) 789 (1933).
977. Raub, E.; Mueller, K.: *Fundamentals of metal deposition.* Elsevier: Amsterdam, 1968.
978. Rautenbach, R.; Albrecht, R.: *Separation of organic binary mixtures by pervaporation.* J. Memb. Sci., 7(2). 203-23 (1980).

979. Rayleigh, D.: *On the questions of the stability of the flow of fluids.* Philos.Mag.and J.Sci.Sers., 34, 59 (1892).
980. Rayleigh, D.: *The principle of similitude.* Nature, 95, 66-68 (1915).
981. Rayleigh, J.W. (Strutt, J.W.): *On the distillation of binary mixtures.* Phil.Mag. 6th Series, 4 (23). 521 (1902).
982. Raymond, S.: *Acrylamide gel as a supporting medium for zone electrophoresis.* Science, 130, 711 (1959).
983. Raymond, S.; Nakamichi, M.: *Electrophoresis in synthetic gels. 1. Relation of gel structure to resolution.* Anal. Biochem., 3, 23-30 (1962).
984. Recktenwald, D.: *Cell separation methods and applications.* CRC Press: Boca Raton, 1997.
985. Reid, R.C.; Sherwood, T.K.: *The properties of gases and liquids. Their estimation and correlation.* McGraw Hill: New York, 1958.
986. Reinhardt, H.; Rydberg, J.: *Solvent extraction studies by the AKUFVE method. II. A new centrifuge for absolute phase separation.* Acta Chem. Scand. 23, 2773-2780 (1969).
987. Renon, H.; Prausnitz, J.M.: *Liquid-liquid and vapor-liquid equilibria for binary and ternary systems with dibutylketone, dimethylsulfoxide, n-hexane, and hexene-1.* Ind.Eng.Chem., Process Des. Dev., 7(2). 220 (1968).
988. Reschke, M.; Schugerl, K.: *Reactive extraction of penicillin. I. Stability of penicillin G in the presence of carriers and relationship for distribution coefficients and degrees of extraction.* Chem. Eng. J., 28, Bl -B9 (1984).
989. Reynolds, O.: *On the destruction of sound by fog and the inertness of a heterogeneous fluids.* Proc.Manchester Lit. Phil. Soc., 8, (1874).
990. Reynolds, O.: *On the theory of lubrication and its application to Mr.Beauchamp tower's experiments, including an experimental determination of the viscosity of olive oil.* Phil.T.R.Soc. London A, 177, 157 (1886).

991. Rhee, H.; Rutherford, A.; Amundson, N.R.: *First order partial differential equations. Vol. II. Theory and application of hyperbolic systems of quasilinear equations.* Prentice Hall: New York, 1989.
992. Ricci, J.E.: *The phase rule and heterogeneous equilibria.* Van Nostrand: New York, 1951.
993. Rice, S.A.: *Comprehensive chemical kinetics. Vol. 25. Diffusion-limited reactions.* C.H. Bamford, Ed., Elsevier: Amsterdam, 1985.
994. Rice, S.A.; Nagasawa, M.: *Polyelectrolyte solutions.* Acad. Press: London, 1961.
995. Rickwood, D.(Ed.): *Biological separations in iodinated density gradient media.* IRL Press: Oxford, 1976.
996. Rickwood, D.; Ford, T.; Graham, J.: *Nycodenz: a new noniodinated gradient medium.* Anal. Biochem., 123(1) 23-31 (1982).
997. Rickwood, D.: *Iodinated density gradient media: A practical approach.* IRL Press: Oxford, 1983.
998. Rieman, W. III, Walton, H.F.: *Ion exchange in analytical chemistry.* Pergamon Press: Oxford, 1970.
999. Rietema, K.: *Efficiency in separating mixtures of two constituents.* Chem. Eng. Sci., 7, 89-96 (1957).
1000. Rietema, K.; Verver, C.G.: *Cyclones in industry.* Elsevier: Amsterdam, 1961.
1001. Righetti, P.G.: *Isolectric focusing.* Elsevier: Amsterdam, 1983.
1002. Rigler, R.: *Fluorescence relaxation spectroscopy in the analysis of macromolecular structure and motion.* Q. Rev. Biophys., 9(1) 1-129 (1976).
1003. Rilbe, H.: *Historical and theoretical aspects of isoelectric focusing.* Ann. N. Y. Acad. Sci., 209, 11 (1973).
1004. Ringbom, A.: *The analyst and the inconstant constants.* J. Chem. Educ., 35, 282-288 (1958).

1005. Ritcey, G.M.: *Crud in solvent extraction processing - a review of causes and treatment.* Hydrometallurgy, 5(2-3) 97-107 (1980).
1006. Ritcey, G.M.; Ashbrook, A.W.: *Solvent extraction principles and applications to process metallurgy.* Elsevier: New York, 1979.
1007. Ritcey, J.M.: *Hydrometallurgy - its development and future.* In: *Chemical separations*, vol.2, p.257-28. King, C.J., Navratil, J.D., Eds., Litarvan: Denver, 1986.
1008. Rizvy, S. (Ed.): *Separation, extraction and concentration processes in the food, beverage and nutraceutical industries.* Elsevier, New York, 2010.
1009. Robel, H.; Vogel, P.: *Verfahrenstechnische Berechnungsamethoden. Teil 3. Mechanische Trennen in fluider Phase.* VCH Verlags: Weinheim, 1985.
1010. Roberts, S.M.: *Dynamic programming in chemical engineering and process control.* Academic Press: New York, 1964
1011. Robinson, C.S.; Gilliland, E.R.: *The elements of fractional distillation.* 4th Ed. McGrawHill: New York, 1950.
1012. Robinson, P.J.; Holbrook, K.A.: *Unimolecular reactions.* Wiley Interscience: New York, 1972.
1013. Rod, V.; Strnadova, L.; Hancil, V.; Sir, Z.: *Kinetics of metal extraction by chelate formation. Part II. Extraction of copper(II) by hydroxyoximes.* Chem. Eng.J., 21(3) 187-93 (1981).
1014. Roesch, F.; Reimann, T.; Buklanov, G.V.; Milanov, M.; Khalkin, V.A.; Dreyer, R.: *Electromigration of carrier-free radionuclides. 13. Ion mobilities and hydrolysis of ^{241}Am-Am(III) in aqueous inert electrolytes.* J.Radioanal Nucl. Chem., Articles, 134(1) 109-28 (1989).
1015. Roginskii, S.Z.: *Kinetic fundamentals of radioactive indicator techniques.* Izv.AN SSSR, Otd. Khim. Nauk, (5) 601-15 (1940).

1016. Rolia, E.: *Theory and practice of precipitation and coprecipitation.* Dept. Energy, Mines and Resources: Ottawa, 1974.
1017. Romankov, P.G.: *Massoobmennye protsessy khimicheskoi tekhnologii. (Mass transfer processes of chemical technology).* Khimizdat: Sankt Peterburg, 2011.
1018. Rony, P.R.: *Extent of separation. Unification of the field of chemical separation.* Chem.Eng.Prog., Syropser., 68(120), 89-104 (1972).
1019. Roozeboom, H.W.B.: *Over de hydraten van zwaveligzuur, chloor, broom en chloorwaterstof.* D.Donner: Leyden, 1884.
1020. Rose, A.: *Batch fractionation. Calculation of theoretical plates required for separation of two normal liquids.* Ind. Chem. Eng., 33, 594-7 (1941).
1021. Rosen, J.B.: *General numerical solution for solid diffusion in fixed beds.* Ind.Eng.Chem., 46, 1590-94 (1954).
1022. Rosen, J.B.: *Kinetics of a fixed-bed system for solid diffusion into spherical particles.* J. Chem. Phys., 20, 387-94 (1952).
1023. Rosensweig, R.E.: *Buoyancy and stable levitation of a magnetic body immersed in magnetizable fluid.* Nature, 210, 613-4 (1966).
1024. Rossotti, F.J.; Rossotti, H.: *The determination of stability constants.* McGraw Hill: New York, 1961.
1025. Rothe, J.W.: *Trennung des Eisen von anderen Elementen nach einem neuen Verfahren.* Stahl u. Eisen, 12, 1052 (1892).
1026. Rousseau, R.W. (Ed.) *Handbook of separation process technology.* American Chemical Society: Washington, 1987.
1027. Rozen, A.M.: *Continuous mass transfer in a rectification column.* Proc. Acad.Sci. USSR, Sect. Khim. Tekhnol, 107, 27-31 (1956).
1028. Rozen, A.M.: *Thermodynamic consideration of component separation in three component systems.* Dokl. Akad. Nauk SSSR, 81, 863-6 (1951).

1029. Rozen, A.M.: *Teoriya razdeleniya izotopov v kolonakh (Theory of isotopes separation in columns)*. Atomizdat: Moskva, 1960.
1030. Rozen, A.M.; Nikolotova, Z.I.: *Dependence of the extracting capacity of organic compounds on their structure and the electronegativity of the substituent.* Russ.J.Inorg. Chem., 9(7) 933-944 (1964).
1031. Rozen, A.M.: *Problems in the physical chemistry of solvent extraction.* In: Solvent extraction chemistry, p. 195. Dyrssen, D.; Liljenzin, P.O., Rydberg, J., Eds., North Holand: Amsterdam, 1967.
1032. Rozen, A.M.; Ionin, M.V.: *Physico-chemical interpretation of limiting activity coefficients and extraction constants of electrolytes.* Radiokhimiya, 13(2) 187-9 (1971).
1033. Rozen, A.M.; Yurkin, B.G.; Fedoseev, D.A.: *Influence of solvent on extraction.* In: Khimiya protsesov ekstraktsii, p. 88. Zolotov, Yu.A., Spivakov, B.Ya., Eds., Nauka: Moskva, 1972.
1034. Rozen, A.M.; Krylov, V.S.: *Theory of scaling up and hydrodynamic modelling of industrial mass transfer equipment.* Chem. Eng. J., 7, 85 (1974).
1035. Rozenkevich, M.B.: *Termodinamika i kinetika razdelenia izotopov (Thermodynamics and kinetics of isotope separation)*. RKhTU: Moscow, 2011.
1036. Rudd, D.F.; Powers, G.J.; Siirola, J.J.: *Process synthesis.* Prentice Hall: New York 1973.
1037. Rushton, A.; Ward, A.S.; Holdich, R.G.: *Solid-liquid filtration and separation technology.* 2nd ed. Wiley-VCH: Weinheim, 2000.
1038. Rushton, J.H.: *The use of pilot plant mixing data.* Chem. Eng. Progr., 47, 485-8 (1951).
1039. Ruthven, D.M.: *Principles of adsorption and adsorption processes.* J. Wiley: New York, 1984.

1040. Ruthven, D.M.; Ching, C.B.: *Counter-current and simulated counter-current adsorption separation processes.* Chem. Eng. Sci., 44(5) 1011-38 (1989).

1041. Rutten, P.W.M.: *Diffusion in liquids.* Delft University Press: Delft, 1992.

1042. Ruzicka, J.; Stary, J.: *A new principle of activation analysis separation. I. Theory of substoichiometric determinations.* Talanta, 10, 287 (1963).

1043. Ruzicka, J.; Stary, J.: *Substoichiometry in radiochemical analysis.* Pergamon Press: Oxford, 1968.

1044. Ruzicka, J.; Hansen, E.H.: *Flow injection analysis.* 2nd ed., J.Wiley&Sons: New York, 1988.

1045. Rydberg, J.: *Complex formation between thorium and acetylacetone.* Acta Chem.Scand., 4, 1503-22 (1950).

1046. Rydberg, J.: *The extraction of metal complexes. XII A. The formation of composite, mononuclear complexes. A. Theoretical.* Arkiv Kemi, 8, 101-12 (1955).

1047. Rydberg, J.; Skarnemark, G.: *Rapid multistage solvent extraction using SISAK technique.* In: Proc. Intern. Solvent Extr.Conf. ISEC '86, Munich 11-16.9.86, vol. 3, p. 21. Dechema: Frankfurt am Main, 1986.

1048. Rydberg, J.; Musikas, C.; Choppin, G.R.: *Principles and practices of solvent extraction.* Marcel Dekker: New York 1992.

1049. Rydberg, J.: *Solvent extraction principles and practice. Revised and expanded.* CRC Press: Boca Raton, 2004.

1050. Sada, E.; Katoh, S.; Terashima, M.; Takada, Y.: *Carrier-mediated transport across phospholipid-composite membranes containing valinomycin.* Am.Inst.Chem.Eng. Journal, 31, 311 (1985).

1051. Sada, E.; Terashima, M.: *Simulation of biomembrane function and mass transfer manipulation.* Hyomen, 27(4) 313 (1989).

1052. Said, A.S.: *Theory and mathematics of chromatography.* Huthig: Heidelberg, 1981.

1053. Saito, M.: *Enrichment reliability of solid polymer electrolysis for tritium water analysis.* J.Radioanal.Nucl.Chem., 173, 407-410 (2008).
1054. Sakodinskii, K.I.; Volkov, S.A.: *Preparativnaya gazovaya khromatografiya (Preparative gas chromatography).* Khimiya: Moscow, 1972.
1055. Samoilov, O.Ya.: *Structure of dilute aqueous solutions of electrolytes and hydration ions.* Zh. Neorg. Khim., 1, 1202-9 (1956).
1056. Samuelson, O.: *Fractionation of sulfite waste liquor.* Svensk. Papperstidn.,46, 583 (1943).
1057. Samuelson, O.: *Ion exchange separations in analytical chemistry.* J. Wiley: London, 1963.
1058. Samuelson, O: *Use of base-exchanging substances in analytical chemistry II.* Svensk. Kem. Tidskr., 51, 195 (1939).
1059. Sanchez, V.; Clifton, M.: *An empirical relationship for predicting the variation with concentration of diffusion coefficients in binary liquid mixtures.* Ind. Chem. Eng., Fundam., 16, 318 (1977).
1060. Sanger, F.; Air, G.M.; Barrell, B.G.; Brown, N.L.; Coulson, A.R.; Fiddes, J.C.: *Nucleotide sequence of bacteriophage OX114 DNA.* Nature, 265, 687-95 (1977).
1061. Sanger, F.; Coulson, A.R.; Hong, G.F.; Hill, D.F.; Peterssen, G.B.: *Nucleotide sequence of bacteriophage DNA.* J. Mol. Biol., 162(4) 729-73 (1982).
1062. Sato, T.: *Complex f ormed in the copper(II)-hydrochloric acid-tri-n-octylamine extraction system. J.Inorg.Nucl. Chem., 31(5) 1995-401 (1969).*
1063. Sattler, K.; Feindt, H.J.: *Thermal separation processes.* VCH: Weinheim, 1995.
1064. Saxen, U.: *Ueber die Reciprocitaet der Electrischen Endoosmose und der Stroemungsstroeme.* Ann. Phys. Chem., 47, 46 (1892).

1065. Scamehorn, J.F.; Harwell, J.H.: *Surfactant-based separation processes.* Marcel Dekker: New York, 1989.
1066. Scatchard, G.: *Equilibria in non-electrolyte solutions in relation to the vapor pressures and densities of the components.* Chem. Rev., 8, 321-333 (1931).
1067. Scatchard, G.: *The attraction of proteins for small molecules and ions.* Ann. N. Y. Acad. Sci., 51(4) 660-672 (1949).
1068. Scatchard, G.: *Molecular interactions in protein solutions.* Am. Scientist., 40, 61-83 (1952).
1069. Schell, W.J.: *Commercial applications for gas permeation membrane systems.* J. Membr. Sci., 22, 217 (1985).
1070. Schindler, H.; Nelson, L.: *Proteolipid of adenosinetriphosphatase from yeast mitochondria forms proton-selective channels in planar lipid bilayers.* Biochemistry, 21(23) 5787-94 (1982).
1071. Schissel, P.; Orth, R.A.: *Separation of ethanol-water mixtures by pervaporation through thin, composite membranes.* J. Membr. Sci., 17, 109-120 (1984).
1072. Schloegl, R.: *The theory of the diffusion potentials and ion transport in free solution and in charged membranes.* Z. Elektrochem., 58, 672-3 (1954).
1073. Schloegl, R.: *Theory of anomalous osmosis.* Z.Phys.Chem., 3, 73-102 (1955).
1074. Schlosser, S.; Kossaczky, E.: *Comparison of pertraction through liquid membranes and double liquid-liquid extraction.* J.Memb.Sci. 6(1) 83-105 (1980).
1075. Schmid, G.: *Electrochemistry of fine-pore capillary systems. VJ. Convection conductivity (theoretical consideration).* Z. Elektrochem., 56, 181-93 (1952).
1076. Schmidt, E.: *Verdunstung und Wärmeübergang.* Gesundheits-Ing., 52, 525 (1929).
1077. Scholander, P.F.: *Oxygen transport through hemoglobin solutions.* Science, 131, 585 (1960).
1078. Schrodt, V.N.; Saunders, A.M.: *Interactive image processing in research.* Comput. Chem. Eng., 5(4) 299-305 (1981).

1079. Schrodt, V.N.; Sommerfeld, Js.; Martin, O.R.; Chieu, H.H.: *Plant scale study of controlled cyclic distillation.* Chem. Eng. Sci., 22(5) 759-67 (1967).
1080. Schubert, J.: *The use of ion exchangers for the determination of physicalchemical properties of substances, particulary radiotracers, in solution.* J. Phys. Colloid Chem., 52(2) 340-50 (1948).
1081. Schugerl, K.: *Solvent extraction in biotechnology: Recovery of primary and secondary metabolites.* Springer: Berlin, 2010.
1082. Schulz, G.; Werner, N.: *Gas separation using the membrane fractionation technique.* Desalination, 61, 123-33 (1984).
1083. Schulz, J.S.; Goddard, J.D.; Suckdeo, S.R.: *Facilitated transport via carrier-mediated diffusion in membranes. Part I. Mechanistic aspect, experimental systems and characteristic regimes.* Am. Inst. Chem. Eng. Journal, 20, 417 (1974).
1084. Schulz, W.W.; Navratil, J.D.: *Bifunctional organophosphorous liquid-liquid extraction reagents: development and applications.* Sep.Sci.Technol., 19, 927- 41 (1984).
1085. Schumacher, E.: *Focusing ion exchange. IV. Theory of focusing effects.* Helv.Chim. Acta, 40, 2322-40 (1957).
1086. Schweitzer, P.A.: *Handbook of separation techniques for chemical engineers.* Mc-Graw Hill, New York, 1997.
1087. Scibona, G.; Fabiani, C.; Scuppa, B.: *Electrochemical behavior of Nafion-type membrane.* J. Memb. Sci., 16, 37-50 (1983).
1088. Scott, K.; Hughes, R.: *Industrial membrane separation technology.* Chapman&Hall: London, 1996.
1089. Scott, R.P.W.: *Contemporary liquid chromatography.* J. Wiley: New York, 1976.
1090. Seaborg, G.S.; McMillan, E.M.; Kennedy, J.W.; Wahl, A.: *Radioactive element 94 from deuterons on uranium* [Jan 1941 - not published]. Phys. Rev., 69, 366 (1946).

1091. Seader, J.D.; Henley, E.J.: *Separation process principles.* J. Wiley: New York, 1998.
1092. Sebba, F.: *Foams and biliquid foams - aphrons.* J. Wiley: New York, 1987.
1093. Sebba, F.: *Ion flotation.* Elsevier: New York, 1962.
1094. Sebesta, F.; Stary, J.: *A generator for preparation of carrier-free 224Ra.* J.Radioanal.Chem., 21, 151-5 (1974).
1095. Secor, R.M.; Buetler, J.A.: *Penetration theory for diffusion accompanied by an reversible chemical reaction with generalized kinetics.* Am. Inst. Chem. Eng. Journal, 13(2) 365-73 (1967).
1096. Seitz, W.R.: *Chemical sensors based on fiber optics.* Anal. Chem., 56, 16A-34A (1984).
1097. Sekine, T.; Hasegawa, Y.: *Solvent extraction chemistry: Fundamentals and applications.* Marcel Dekker: New York, 1977.
1098. Sekine, T.; Hasegawa, Y.; Ibara, N.: *Acid dissociation and two-phase distribution constants of eight beta-diketones in several solvent extraction systems.* J.Inorg.Nucl.Chem., 35, 3968-70 (1973).
1099. Serrato, R.M.: *Laser isotope separation and the future of nuclear proliferation.* Universal Publishers: Osborne Park, 2010.
1100. Setchenow, J.: *Ueber die Constitution der Salzloesungen auf Grund ihres Verhaltens zu Kohlensäure.* Z. Physik. Chem., 4, 117 (1889).
1101. Shakhparonov, M.I.: *Relation among solubility, activity coefficient and properties of solvent and solute.* Zh.Fiz.Khim., 26, 1103-10 (1951).
1102. Sharma, M.M.; Danckwerts, P.V.: *Chemical methods of measuring interfacial area and mass transfer coefficients in two-liquid systems.* Brit. Chem. Eng., 15(4) 522-28 (1970).
1103. Sheeler, P.: *Centrifugation in biology and medicine.* J. Wiley: Chichester, 1982.

1104. Shemyaldn, M.M.; Ovchinnikov, Yu.A.; Ivanov, Vs.: *Cyclodepsipeptides as chemical tools for studying* ionic *transport through membranes.* J. Membr.Biol., 1, 402 (1969).
1105. Sherwood, T.K.: *Mass transfer and friction in turbulent flow.* Trans. Am. Ind.Chem. Eng., 36, 817-40 (1940).
1106. Sherwood, T.K.; Brian, P.Ls.; Fisher, R.E.; Dresner, L.: *Salt concentration at phase boundaries in desalination by reverse osmosis.* Ind.Eng.Chem. Fundam., 4, 113 (1965).
1107. Sherwood, T.K.; Pigford, R.L.: *Absorption and extraction.* McGraw Hill: New York, 1952.
1108. Sherwood, T.K.; Pigford, R.L.; Wilke, C.R.: *Mass transfer.* McGraw Hill: New York, 1975.
1109. Shimiza, S.: *Technology of ferrite manufacture and heavy metal removal.* Hyomen, 15, 564 (1977).
1110. Shmidt, V.S.; Shesterikov, V.N.; Gertskin, M.G.; Novozhilov S.S.; Rubisov, V.N.; Scherbatykh, V.I.: *A mathematical model of distribution of tetravalent actinides between* nitrate *solutions and tri-n-octylamine.* Radiokhimiya, 17(1), 44-51 (1975).
1111. Schoenmakers, P.J.; Billet, H.A.H.; Degalan, L.: *The solubility parameter as a tool in understanding liquid chromatography.* Chromatographia 15, 205 (1982).
1112. Shoenmakers, P.J.: *Optimization of chromatographic selectivity.* Elsevier: Amsterdam, 1986.
1113. Shpigun, O.A.; Zolotov, Yu.A.: *Ionnaya khromatografiya (Ion chromatography).* Moscow State University: Moscow, 1993.
1114. Shvydko, N.S.; Ivanova, N.P.; Rushonik, S.I.: *Physico-chemical state and exchange of plutonium and americium in organism (in Russian).* Energoatomizdat: Moscow, 1987.
1115. Siddall, T.H.: *Trialkylphosphates and dialkyl phosphates in uranium and thorium extraction.* Ind.Eng.Chem., IH, 41-4 (1959).

1116. Siddall, T.H. III *Bidentate organophosphorus compounds as extractants. I. Extraction of cerium, promethium, and americium nitrates.* J. Inorg. Nucl. Chem., 25(5) 883-92 (1963).
1117. Sidhu, K.S., Tuch, B.E.: *Derivation of three clones from human embryonic stem cell lines by FACS sorting and their characterization.* Stem Cells Dev., 15(1) 61-9 (2006).
1118. Siekierski, S.; Fidelia, I.: *Separation of some rare earths by reversed-phase partition chromatography.* J.Chromatogr., 4, 60-64 (1960).
1119. Siekierski, S.; Kotlinskaya, B.: *Separation of zirconium-niobium mixtures by the method of reversed-phase partition chromatography.* Atom. Energ. (USSR) 7, 160 (1959).
1120. Siggia, S.: *Continuous analysis of chemical process systems.* J. Wiley: New York, 1959.
1121. Sillen, L.G.: *On equilibria in systems with polynuclear complex formation. V. Some useful differential expresions.* Acta Chem.Scand., 15, 1481-92 (1961).
1122. Sillen, L.G.: *Polynuclear complexes: criticism invited.* Acta Chem.Scand., 15, 1421-2 (1961).
1123. Sillen, L.G.; Martell, A.E.: *Stability constants of metal-ion complexes.* Chemical Society: London, 1971.
1124. Sillen, L.G.; Martell, A.E.; Bjerrum, J.: *Stability constants of metal-ion complexes. Vol.1.* Chem. Soc.: London, 1964.
1125. Silva, R.J.; Harris, J.; Nurmia, M.; Eskola, K.; Ghiorso, A.: *Chemical separation of rutherfordium.* Inorg. Nucl. Chem. Lett., 6(12) 871-7 (1970).
1126. Silva, R.J.; Sikkeland, T.; Nurmia, M.J.; Ghiorso, A.: *Tracer chemical studies of lawrencium.* Inorg. Nucl. Chem. Lett., 6(9) 733-9 (1970).
1127. Sinaiski, E.G.; Lapiga E.J.: *Separation of multiphase, multicomponent systems.* J.Wiley 2007.
1128. Sine, J-P.: *Séparation et analyse des biomolécules: Méthodes physico-chimiques.* Ellipses: Paris, 2003.

1129. Sips, R., *On the structure of catalyst surface II*, J.Chem.Phys. 18 (8) 1024-1026 (1960).
1130. Sirkar, Kamalesh K.: *Separation of molecules, macromolecules and particles. Principles, phenomena and processes*. Cambridge University Prass, 2014.
1131. Skey, W.: *On the production of some new metallic sulphocyanides, and the separation of certain bases from each other by the method therein employed*. Chem. News., 16, 201 (1867).
1132. Slattery, J.C.: *Interfacial transport phenomena*. Springer Verlag: Berlin, 1990.
1133. Small, H.: *Modern inorganic chromatography*. Anal.Chem., 55(2) 235 (1983).
1134. Small, H.; Langhorst, M.A.: *Hydrodynamic chromatography*. Anal. Chem., 54(8) 893-8 (1982).
1135. Smith, C.M.; Navratil, J.D.: *Removal and pre concentration of surfactants from waste water with open-pore polyurethane*. Sep.Sci.Technol., 14, 255 (1979).
1136. Smith, E.L.: *Conditions covering the extraction of a solution by an immiscible solvent*. J. Soc. Chem. Ind., 47, T159-60 (1928).
1137. Smith, L.C.: *Digital computer process control*. Intext: Toronto, 1972.
1138. Smith, R.M. *Supercritical fluid chromatography*. Royal Society of Chemistry: London 1993.
1139. Smithies, O.: *Zone electrophoresis in starch gels: group variations in the serum proteins of normal human adults*. Biochem.J., 61, 629-41 (1955).
1140. Smithies, O.; Poulik, M.D.: *Two-dimensional electrophoresis of serum proteins*. Nature, 177, 1033 (1956).
1141. Smolders, K.; Franken, A.C.M.: *Terminology for membrane distillation*. Desalination, 72, 249-62 1989.
1142. Snyder, L.R.: *Classification of the solvent properties of common liquids*. J.Chromatogr., 92(2) 223-30 (1974).

1143. Snyder, L.R.: *Mobile-phase effects in liquid-solid chromatography.* High-Perform. Liq. Chromatogr., 3, 157-223 (1983).
1144. Snyder, L.R.; Kirkland, J.J.: *Introduction to modern liquid chromatography.* 2nd Ed. Wiley Interscience: New York 1979.
1145. Snyder, L.R.; Poppe, H.: *Mechanism of solute retention in liquid-solid chromatography and the role of the mobile phase in affecting separation. Competition versus "sorption".* J. Chromatogr., 184(4) 363-413 (1980).
1146. Snyder, L.R.; Kirkland, J.J.; Glajch, J.L.: *Practical HPLC method development.* 2nd ed. Wiley Interscience: New York, 2012.
1147. Sojak, L.; Rijks, J.A.: *Capillary gas chromatography of alkylbenzenes. I. Some problems encountered with the precision of the retention indices of alkylbenzenes.* J.Chromatogr., 119, 505-21 (1976).
1148. Sojak, L.; Kraus, G.; Ostrovsky, I.; Kralovicova, E.; Krupcik, I.: *Highperformance gas chromatography with liquid crystal glass capillaries 1. Separation of hydrocarbon isomers on nematic mesophases.* J.Chromatogr., 206(3) 463-74 (1981).
1149. Solovkin, A.S.: *Thermodynamics of extraction of zirconium, present in a monomeric state, from* nitric *acid solutions by tri-n-butylphosphate.* Russ. J. Inorg. Chem., 15, 983-984 (1970).
1150. Solovkin, A.S.: *Vysalivaniye i kolichestvennoye opisaniye ekstraktsionnykh ravnovesii (The salting-out phenomena and quantitative description of extraction equilibria).* Atomizdat: Moskva 1969.
1151. Solovkin, A.S.; Zakharov, Yu.N.: *Use of the tributylphosphate extraction method for determining the mechanism and thermodynamic constant of the complexing of plutonium(IV) with acetic acid in aqueous solutions.* Radiokhimiya, 22(2) 225-30 (1980).

1152. Sotnikov, V.S.; Belanovskii, As.: *Adsorption of some metal ions from electrolyte solutions on the surfaces of Ge, Si and quartz.* Radiokhimiya, 8(2) 171-82 (1966).
1153. Sourirajan, S.: *Reverse osmosis.* Acad. Press: New York, 1970.
1154. Southern, E.M.: *Detection of specific sequences among DNA fragments separated by gel electrophoresis.* J.Mol.Biol., 98(3) 503-517 (1975).
1155. Sova, O.: *Autofocusing - a method for isoelectric focusing without* carrier *ampholytes.* J.Chromatogr., 320(1) 15-22 (1985).
1156. Spangenberg, B.; Poole, C.F.; Weins, Ch.: *Quantitative thin-layer chromatography. A practical survey.* Springer: Heidelberg, 2011.
1157. Spedding, F.H.; Powell, J.E.: *The separation of rare earths by ion exchange. VIII. Quantitative theory of the mechanism involved in elution by dilute citrate solutions.* J.Am.Chem.Soc., 76, 2550-7 (1954).
1158. Sperry, D.R.: *Effect of pressure on fundamental filtration equation when solids are non-rigid or deformable,* Ind.Eng.Chem.20, 892-5 (1928).
1159. Stahl, E.; Schroter, G.; Kraft, G.; Renz, R.: *Thin layer chromatography (the method, affecting factors, and a few examples of application).* Pharmazie, 11, 633-7 (1956).
1160. Stahl, E. (ed.) *Thin layer chromatography. A laboratory handbook.* Academic Press: New York, 1965.
1161. Stahl, F.: *Thin-layer chromatography.* Springer: Berlin, 1969.
1162. Stamberg, K.; Cabicar, J.; Havlicek, L.: *Ion-exchange kinetics in systems with non-linear equilibrium isotherms. Sorption of uranium(VI) on strong acid cation exchangers.* J.Chromatogr.201, 113 (1980).
1163. Stamberg, K.; Cabicar, J.: *"Separation work"- the quantity characterizing the sorption and liquid-liquid extraction*

separation processes. In: Proc. CHISA'84. Czechosl. Chem. Eng. Soc.: Prague, 1984.
1164. Starkenstein, E.: *Ueber Fermentwirkung und deren Beeinflussung durch Neutralsalze.* Biochem. Z., 24, 210 (1910).
1165. Stary, J.: *Bestimmung der Zusamennsetzung und der Stabilitatskonstanten von Metallkomplexen durch Extraktionsmethode.* Coll.Czechosl.Chem.Commun., 25, 2630 (1960).
1166. Stary, J.: *Systematic study of the solvent extraction of metal oxinates.* Anal.Chim. Acta, 28(2) 132-149 (1963).
1167. Stary, J.: *The solvent extraction of metal chelates.* Pergamon Press: Oxford, 1964.
1168. Stary, J.; Kratzer, K.: *The accumulation of toxic metals in algae.* Int. J.Environ. Anal.Chem., 12(1) 65-71 (1982).
1169. Stary, J.; Kratzer, K.; Prasilova, J.: *The cumulation of elements on algal cell walls* J.Radioanal Nucl.Chem., 101(1) 127-134 (1986).
1170. Stary, J.; Kyrs, M.; Marhol, M.: *Separacni metody v radiochemii (Separation methods in radiochemistry).* Academia: Praha, 1975.
1171. Steele, W.A.: *The interaction of gases with solid surfaces.* Pergamon Press: Oxford, 1974.
1172. Stein, W.D.: *The movement of molecules across cell membranes.* Acad. Press: New York, 1967.
1173. Steinfeld, J.I. (ed.) *Laser-induced chemical processes.* Plenum Press: New York, 1981.
1174. Stepanov, A.V.; Gedeonov, A.D.: *On the kinetic theory of the substoichiometric analysis of trace amounts of a metal.* J.Radioanal.Chem 30 (1) 197-203 (1976).
1175. Stepanov, N.I.: *The metrics of equilibrium chemical diagram.* Usp.Khim., 5, 75 (1936).
1176. Stepin, B.D.: *The entropy factors in the separation of potassium, rubidium and cesium by crystalization.* Zh.Neorg.Khim., 12(3) 720-4 (1967).

1177. Stern, O.: *Zur theorie der Elektrolytischen Doppelschicht.* Z. Elektrochem., 30, 508 (1924).
1178. Stevenson, D.; Wilson, I.D. (eds.) *Chiral separations.* Plenum Press: New York, 1988.
1179. Stokes, G.G.: *On the theories of the internal friction of fluids in motion and of the equilibrium and motion of elastic solids.* Trans. Cambridge Phil. Soc., 8, 287 (1845).
1180. Stokes, R.H.; Robinson, R.A.: *Ionic hydration and activity in electrolyte solution1.* J.Am.Chem.Soc., 70, 1870-8 (1948).
1181. Storey, S.H.; Zeggeren, F.: *Computation of chemical equilibrium composition.* Can. J. Chem. Eng., 48(5) 591 (1970).
1182. Strathmann, K.: *Electrodialysis and its application in the chemical process industry.* Separ. Purif. Methods, 14(1) 41-66 (1985).
1183. Street, G. (Ed.): *Highly selective separations in biotechnology.* Blackie Academic & Professionals: London, 1994.
1184. Strelow, F.W.E.: *Distribution coefficients and cation-exchange behaviour of 45 elements with a macroporous resin in hydrochloric acid methanol mixtures.* Anal. Chim. Acta, 160, 31-45 (1984).
1185. Strelow, F.W.E.; Rethemeyer, R.; Bothma, C.J.C.: *Ion exchange selectivity scales for cations in nitric acid and sulfuric acid media with a sulfonated polystyrene resin.* Anal.Chem., 37(1) 106-11 (1965).
1186. Struppe, H.G.: *Gas-Chromatographie.* Akademie Verlag: Berlin, 1968.
1187. Stromholm, D.; Svedberg, T.: *Untersuchungen uber die Chemie der radioaktiven Grundstoffe.* Z. Anorg. Chem., 61, 338-46 (1909).
1188. Subramanian G. (Ed.): *Chiral separation techniques: A practical approach.* 3rd Edition, Wiley-VCH: Weinheim, 2007.
1189. Sucha, L.; Kotrly, S.: *Solution equilibria in analytical chemistry.* Van Nostrand Reinhold: London, 1972.

1190. Summer, J.B.: *Isolierung und Krystallisation des Enzymes Urease.* J Biol.Chem., 69, 435 (1926).
1191. Suthanthiraraj, P.P.A., Graves, S.W.: *Fluidics.* Curr.Protoc. Cytom., Unit-1.2 (2013).
1192. Suzuki, N.; Satoh, K.; Shoji, H.; Imura, H.: *Liquid-liquid extraction behavior of arsenic(III) arsenic(V) methylarsinate and dimethylarsinate in various systems.* Anal. Chim. Acta, 185, 239 (1986).
1193. Svantesson, I.; Hagstrom, I.; Persson, G.; Liljenzin, P.O.: *Distribution ratio and empirical equations for the extraction of elements in Purex high level waste solutions.* J. Inorg. Nucl. Chem., 41, 383-9 (1979).
1194. Svarovski, L.: *Solid-liquid separation.* 4th Ed. Butterworth-Heinemann, 2000.
1195. Svedberg, T.: *Masse und Grosse von Proteinmolekulen.* Nature, 123, 871 (1929).
1196. Svedberg, T.; Eriksson, I.: *Molecular weight of erythrocruorin.* J. Am. Chem.Soc., 55, 2834-41 (1933).
1197. Svedberg, T.; Fahraeus, R.: *A new method for the determination of the molecular weight of the proteins.* J. Am.Chem.Soc., 48, 430-438 (1926).
1198. Svedberg, T.; Nichols, J.P.: *Determination of size and distribution of size of particles by centrifugal methods.* J. Am.Chem.Soc., 46, 2910-17 (1923).
1199. Svedberg, T.; Pedersen, K.O.: *The ultracentrifuge.* Univ. Press: Oxford, 1940.
1200. Svensson, H.: *Isoelectric fractionation analysis, and characterization of ampholytes in natural pH gradients. I. The differential equation of solute concentration at a steady state.* Acta Chem.Scand., 16, 325 (1961).
1201. Swamy, P.M.: *Laboratory manual on biotechnology.* Rastogi Publications: Meerut, 2008.
1202. Szilard, L.; Chalmers, T.A.: *Chemical separation of the radioactive element from its bombarded isotope in the Fermi effect.* Nature, 134, 462 (1934).

1203. Szymanowski, J.; Prochaska, K.: *Interfacial activity of hydroximes and the reaction order of copper extraction in toluene/water system.* J.Radioanal.Nucl.Chem., Articles, 129(2) 251-264 (1989).

1204. Takamine, J.: *Adrenalin - the active principle of the suprarenal glands and its mode of preparation.* Amer. Journ. Pharm., 73, 523 (1901).

1205. Takaoka, T.; Nishiki, T.; Kimura, S.; Tomioka, Y. *Batch permeation of metal ions using liquid surfactant membranes.* J. Membr. Sci., 46, 67-80 (1989).

1206. Tanaka, M.: *"Ligand buffer": a concept useful in the theoretical consideration of equilibria involving chelating agents. Precipitation and solvent extraction in ligand buffers.* Anal. Chim. Acta., 29(3) 193-201 (1963).

1207. Tanny, G.B.; Hauk, D.: *Filtration of particulates and emulsions with a pleated, thin channel, cross-flow module.* Sep.Sci.Technol., 15(3) 317 (1980).

1208. Taramasso M.: *Considerations for the design of a rotating unit for continuous production by gas chromatography and its applications.* J.Chromatog. 49, 27-35 (1970).

1209. Tavare, Narayan S.: *Industrial crystallization. Process simulation analysis and design.* Springer Science: New York, 1995.

1210. Tavlarides, L.L.; Benjamin, G.O.: *General analysis of of multicomponent mass transfer with simultaneous reversible chemical reactionts in multiphase system.* Chem. Eng. Sci., 24, 553-69 (1969).

1211. Tavlarides, L.L.; Stamatoudis, M.: *The analysis of interphase reactions and mass transfer in liquid-liquid dispersions.* Adv. Chem. Eng., 11, 199-270 (1981).

1212. Taylor, R. (Ed.): *Reprocessing and recycling of spent nuclear fuel.* Woodhead Publisher: London, 2015.

1213. Tedder, D.W.; Rudd, D.F.: *Parametric studies in industrial distillation. Part.I. Design comparisons.* Am.Inst.Chem.Eng. Journal, 24(2) 303-15 (1978).

1214. Terabe, S.: *Electrokinetic chromatography: an interface between electrophoresis and chromatography.* Trends Anal.Chem., 8(4) 129-134 (1989).
1215. Terabe, S.; Otsuka, K.; Ichikawa, K.; Tsuchiya, A.; Ando, T.: *Electrokinetic separations with micellar solutions and open-tubular capillaries.* Anal.Chem., 56(1) 111-113 (1984).
1216. Teramoto, M.: *Model of the permeation of copper through liquid surfactant membranes.* Sep.Sci.Technol., 18(8) 397 (1983).
1217. Teramoto, M.; Matsuyama, H.; Takaya, H.; Youeharo, T.; Miyake, Y.: *Separation of benzene from cyclohexane by flowing liquid membrane and removal of heavy metal ions by spiral-type flow in liquid membrane.* In: Proceedings ISEC'88, vol.Ill, p.110. Academy of Science USSR: Moscow, 1988.
1218. Teramoto, M.; Sakai, T.; Yanagava, K.; Ohsuga, M.: *Modeling of the permeation of copper through liquid surfactant membranes.* Sep.Sci.Technol., 18(8) 735-64 (1983).
1219. Teramoto, M.; Sakuramoto, T.; Koyama, T.; Matsuyama, H.; Miyake, Y.: *Extraction of lanthanides by liquid surfactant membranes.* Sep.Sci.Technol., 21(3) 229-50 (1986).
1220. Theodore, L.; Buonicore, A.J.: *Industrial air pollution control equipment.* CRC Press: Ohio, 1976.
1221. Theuerer, H.C.: *Purification of $SiCl_4$ by absorption techniques.* J. Electrochem.Soc., 107, 29-32 (1960).
1222. Thibaut Brian, P.L.: *Staged cascades in chemical processing.* Prentice Hall, 1972.
1223. Thieu, M.P.; Hatton, T.A.; Wang, D.I.C.: *Liquid emulsion membranes and their applications in biochemical separations.* ACS Symp.Ser. (Sep.Rec., Pur.Biot.) 314, 67-77 (1986).
1224. Thode, H.G.; Urey, H.C.: *The further concentration of* ^{15}N. J.Chem.Phys., 7, 34-9 (1939).
1225. Thomas, H.: *Heterogeneous ion exchange in a flowing system.* J.Am.Chem.Soc., 66, 1664 (1944).

1226. Thomson, R.W.; King, C.J.: *Systematic synthesis of separation schemes.* Am.Inst. Chem. Eng. Journal, 18, 941 (1972).
1227. Thormann, W.; Arn, D.; Schumacher, E.: *Determination of transient and steady states in electrophoresis.* Electrophoresis, 5, 109-18 (1984).
1228. Thormann, W.; Mosher, R.A.: *Electrophoretic transport equations: Electrophoretic models based on migration only and their interrelatioships.* Electrophoresis, 6, 413-18 (1985).
1229. Thormann, W.; Mosher, R.A.; Bier, M.: *Separations by capillary electrophoresis.* In: Chemical Separation. v. 1, p. 159. C.J. King, J.D. Navratil, Eds., Litarvan: Denver, 1986.
1230. Thyagarajan, K.; Ghatak, Ajoy: *Lasers. Fundamentals and applications.* 2nd ed., Springer: Heidelberg, 2010.
1231. Tien, H.S.: *Bilayer lipid membranes: Theory and practice.* Marcel Dekker: New York, 1974.
1232. Timashev, S.F.: *Fizikokhimiya membrannykh protsessov (Physical chemistry of membrane processes).* Khimiya: Moskva 1988.
1233. Timerbaev, A.R.; Petrukhin, O.M.: *Zhidkostnaya adsorptsionnaya khromatografiya khelatov (Liquid adsorption chromatography of chelates).* Nauka: Moskva, 1989.
1234. Timofeeev, D.P.: *Kinetika adsorptsii (The kinetics of adsorption).* Izdat. Akad.Nauk SSSR: Moscow 1962.
1235. Tipeon, R.S.: *Crystallization and recrystallization.* In: Technique of organic chemistry, Vol. 9., p. 963-484. A. Weissberger (Ed.), Interscience: New York, 1950.
1236. Tirmyaev, A.F.; Kulikov, R.V.; Potashnikov, A.K.; Sysoev, E.V.: *Enhancing the selectivity of the X-ray luminescence separation of diamonds by digital processing of signals.* J.Mining Sci., 43(5) 555 (2007).
1237. Tiselius, A.: *A new apparatus for electrophoretic analysis of colloidal mixtures.* Trans. Farad. Soc., 33, 524 (1937).

1238. Tiselius, A.: *Electrophoresis of serum globin.* Biochem. J., 31, 313 (1937).
1239. Tkac, A.; Cvengros, J.: *Continuous processes in wiped films. 1. Multistage molecule distillation in an arrangement with a single convex-shape evaporator body.* Ind.Eng.Chem. Process Des. Dev., 17(3) 242-245 (1978).
1240. Toda, F. (Ed.): *Enantiomer Separation: Fundamentals and Practical Methods.* Springer Science & Business Media 2007.
1241. Tolgyessy, J.; Kyrs, M.: *Radioanalytical chemistry. I.:* Ellis Horwood: Chichester, 1989.
1242. Tomar, B.S.; Steinebach, O.M.; Terpstra, B.E.; P. Bode, P.; Wolterbeek, H.Th.: *Increasing the specific activity of medically useful isotopes through Szilard Chalmers reaction.* Annual Meeting on Nuclear Society, June 8-12, 2008, Anaheim, California. Trans.Am.Nucl.Soc., 98, 880-881 (2008).
1243. Tompkins, E.R.; Khym, J.X.; Cohn, W.E.: *Ion exchange as a separation method. I. The separation of fission -produced radioisotopes indicating individual rare earths, by complexing elution from Amberlite resin.* J.Am.Chem.Soc., 69, 2769-77 (1947).
1244. Tondre, C.; Xenakis, A.: *Use of microemulsions as liquid membranes.* Faraday Discuss. Chem. Soc., 115-26 (1984).
1245. Toth, J.: *State equations of the solid-gas interface layers.* Acta Chim.Ac.Sci.Hung. 69, 311 (1971).
1246. Toth, J.: *A uniform interpretation of gas/solid adsorption.* J.Colloid Interf.Sci. 79 (1) 85-89 (1981).
1247. Treybal, R.E.: *Mass transfer operations.* 2nd ed. McGraw Hill: New York, 1968.
1248. Tromp, K.W.: *Neue Wege für die Beurteilung der Aufbereitung von Steinkohle.* Gluckauf, 73(6) 125-31 (1937).

1249. Trubert, D.; Le Naour, C.: *Fundamental aspects of single atom chemistry.* In: *Chemistry of superheavy elements.* M. Schädel (Ed.) pp.95-116. Springer: Heidelberg, 2003.
1250. Tsysin, G.L.; Kovalev, I.A.; Nesterenko, P.N.; Penner, N.A.; Filippov, O.A.: *Application of linear model of sorption dynamics to the comparison of solid phase extraction systems of phenol.* Sep.Pur.Technology 33, 11-24 (2003).
1251. Tsuji, A.; Sekiguchi K.: *Adsorption of nicotine acid hydrazine on cation exchanger of various metal forms.* Nippon Kagaku Zasshi, 81, 84 (1960).
1252. Tsvet, M.S.: *Khromatographicheskii absorptsionyi analiz (A chromatograph absorption analysis).* Izd. AN SSSR: Moskva, 1946.
1253. Tswett, M.S.: *Adsorptionanalyse und chromatographische Methode. Anwendung auf die Chemie des Chlorophylls.* Ber. Dtsch. Botan. Ges., 24, 384-93 (1906). 1037.
1254. Tswett, M.S.: *O novoi kategorii adsorptsionnykh iavlenii i o primenenii ich k biokhimicheskomu analizu.* Proc. Warsaw Soc. Nat. Sci., Biol. Sect., 14 (6) (1903).
1255. Tucker, W.D.; Greene, M.W.; Murrenhoff, A.P.: *Production of carrier-free tellurium-131, iodine-131, molybdenium-99 and technetium-99m from neutronirradiated uranium by fractional sorption on aluminium oxide.* Atompraxis, 8, 163 (1962).
1256. Turkova, J.: *Affinity chromatography.* Elsevier: Amsterdam, 1978.
1257. Ungarish, M.: *Hydrodynamics of suspensions.* Springer: Berlin 1993.
1258. Urbain, G.: *On the lutecium und neoytterbium.* C.R.d.Acad.d.Sci., 146, 406-8 (1908).
1259. Urey, H.C.; Brickwedde, F.G.; Murphy, G.M.: *A hydrogen isotope of mass 2 and its concentration.* Phys. Rev., 40, 1-15 (1932).

1260. Urey, H.C.; Fox, M.; Huffman, J.R.; Thode, H.G.: *A concentration of N^{15} by a chemical exchange reaction.* J.Am. Chem.Soc., 59, 1407-8 (1937).
1261. Vacik, J.; Kopecek, J.: *Specific resistances of hydrophilic membranes containing ionogenic groups.* J. Appl.Polym. Sci., 19(11) 3029-44 (1975).
1262. Valcarel, M.; Luque de Castro, M.D.: *Non-chromatographic continuous separation techniques.* Royal Soc.Chemistry: London, 1991.
1263. Valenzuela, D.; Myers, A.: *Adsorption equilibrium data handbook.* Prentice Hall: New York, 1989.
1264. Valeton, I.I.P.: *Wachstum und Auflösung der Krystalle.* Z. Krystall., 59, 335 (1923).
1265. Valko, K. (Ed.): *Separation methods in drug synthesis and purification.* In: *Handbook of analytical separations. Vol.1*, R.M.Smith (Ed.), Elsevier: Amsterdam, 2000.
1266. Van Deemter, J.J.; Zwiderweg, F.J.; Klinkenberg, A.: *Longitudinal diflusion and resistance to mass transfer as causes of nonideality in chromatography.* Chem. Eng. Sci., 5, 271-89 (1956).
1267. Van Krevelen, D.W.; Hoftijzer, P.J.: *Kinetics of gas-liquid reactions. I. General theory.* Ree. Trav. Chem., 67, 563-86 (1948).
1268. Van Laar, I.I.: *On the theory of vapor pressures of binary mixtures.* Z. Physik.Chem., 83, 599-608 (1913).
1269. Van Laar, I.I.: *The vapor pressure of binary mixtures.* Z.Phys.Chem., 72, 723-51 (1910).
1270. Van Leeuwen, H.P.: *Dynamic aspects of metal speciation in aquatic colloidal systems*, In: *Environmental Particles*, V.I., p.497-521. J.Buffle, H.P.van Leewen (Eds.). Lewis: Boca Raton, 1989.
1271. Van Ness, H.C.; Albolt, M.M.: *Classical thermodynamics of nonelectrolyte solutions with application to phase equilibria.* McGraw Hill: New York, 1982.

1272. Van Nieuwenhuijzen A.; Van der Graaf, J. (Eds.): *Handbook on particle separation processes*. IWA Publishing: London, 2011.
1273. Van Oss, C.J.: *Ultrafiltration membranes*. Progr.Sep.Purif., 3, 97-132 (1970).
1274. Van Oss, C.J.; Absolom, D.R.; Neumann, A.W.: *Repulsive van der Waals forces II. Mechanism of hydrophobic chromatography*. Sep.Sci.Technol., 14(4) 305-17 (1979).
1275. Van Rysselberghe, P.; Eisenberg, S.: *Activity coefficients in concentrated aqueous solutions of strong electrolyte described by a formula containing the mean ionic diameter as single parameter*. J.Am.Chem.Soc., 61, 3030-7 (1939).
1276. Van Nieuwenhuijzen, A.; Van der Graaf, J.: *Handbook of particle separation processes*. IWA Publishing: London, 2011.
1277. Van't Hoff, J.H.: *Études de dynamique chimique*. Frederik Muller: Amsterdam, 1884.
1278. Van Zeggeren, F.; Storey, S.H.: *The computation of chemical equilibria*. Cambridge University Press: Cambridge, 1970.
1279. Varteressian, K.A.; Fenske, M.R.: *The system methylcyclohexane-aniline-n heptane. Graphical design methods applied to the fractional extraction of methyl cyclohexane and n-heptane mixtures*. Ind.Eng.Chem., 29, 270-7 (1937).
1280. Vazquez-Duhalt, R.; Quintero-Ramirez, R.: *Petroleum biotechnology: Developments and perspectives*. Elsevier Amsterdam 2004.
1281. Veasey, T.J.; Wilson, R.J.; Squires, D.M.: *The physical separation and recovery of metals from wastes*. Gordon&Breach Science Publisher: Yverdon, 1993.
1282. Verway E.J.W.: *Ionenadsorption und Austausch*. Kolloid-Z. 72, 187 (1935).
1283. Vesterberg, O.: *Synthesis and isoelectric fractionation of carrier ampholytes*. Acta Chem.Scand., 23(8) 2653-66 (1969).

1284. Vetter, K.J.: *Electrochemical kinetics.* Academic Press: New York, 1967.
1285. Vigdorovich, V.N.; Marychev, V.V.: *Certain special case of zone recrystallization* Russ. J.Phys.Chem., 39(8) 1087-1089 (1965).
1286. Volmer, M.; Esterman, J.: *Ueber den Mechanismus der Molekuelabscheidung an Kristallen.* Z. Physik, 7, 13 (1921).
1287. Wadsworth, M.E.: *Reactions at surfaces.* Phys.Chem., 7, 4.13-72 (1975).
1288. Wakabayashi, T.; Oki, S.; Omori, T.; Suzuki, N.: *Some applications of the regular solution theory to the solvent extraction I. Distribution of beta-diketones.* J.Inorg.Nucl. Chem., 26, 2255-64. (1964.).
1289. Walas, S.M.: *Phase equilibria in chemical engineering.* Butterworth: Boston, 1985.
1290. Walker, W.H.; Lewis, W.K.; McAdams, W.H.: *Principles of chemical engineering. 2nd ed.* McGraw-Hill: New York, 1927.
1291. Wallis, G.B.: *One-dimensional two-phase flow.* McGraw-Hill: New York, 1969.
1292. Walters, R.R.: *Affinity chromatography.* Anal.Chem., 57(11) 1099A-1114A (1985).
1293. Walton, H.F.: *Ligand exchange chromatography.* In: Ion exchange and solvent extraction, Vol.4., p.121-159. J. Marinsky, Y.Marcus, (Eds.), Marcel Dekker: New York, 1973.
1294. Walton, H.F.: *Liquid chromatography of organic compounds on ion-exchange resins.* Sep. Purif. Method., 4, 189 (1975).
1295. Walton, H.F. (Ed.): *Ion-exchange chromatography.* Dowden: Stroudsburg, 1976.
1296. Wang, L.K.; Chen J.P.; Hung, Y-T. (Eds.): *Membrane and desalination technologies.* Spriner: New York, 2011.
1297. Wankat, P.C.: *Rate-controlled separations.* Elsevier Applied Science: London 1944.

1298. Wankat, P.C.: *Calculations for separation with three phases. I. Staged systems.* Ind.Eng.Chem. Fundam., 19(4) 358-63 (1980).
1299. Wankat, P.C.: *Improved preparative chromatography: moving port chromatography.* Ind.Eng.Chem.Fundam., 23(2) 256-60 (1984.).
1300. Wankat, P.C.; Noble, R.D.: *Calculations for separation with three phases. II. Continuous contact systems.* Ind.Eng.Chem. Fundam., 23, 137 (1984).
1301. Wankat, P.C.: *Equilibrium-staged separations.* Prentice Hall PTR, 1988.
1302. Wankat, P.C.: *Separation process engineering: Includes mass transfer analysis.* Prentice Hall: Upper Saddle River, 2012.
1303. Ward, W.J. III.: *Analytical and experimental studies of facilitated transport.* Am.Inst.Chem.Eng. Journal, 16, 405-410 (1970).
1304. Wardius, Ds.; Hatton, T.A.: *A module for liquid membrane extraction with instantaneous reaction in cascaded mixers.* Chem. Eng.Commun., 37, 159-71 (1985).
1305. Warner, B.F.; Naylor, A.; Duncan, A.; Wilson, P.D.: *A review of the suitability of solvent extraction for the reprocesing of fast reactor fuel.* In: Proceedings ISEC'7,4, Vol. 2, p. 1481-97. Soc. Chem. Industry: London, 1974.
1306. Warren, D.C.: *New frontiers in membrane technology and chromatography: Applications for biotechnology.* Anal. Chem., 56(14) 1529 (1984).
1307. Waseda, Y.; Isshiiki, M.: *Purification process and characterization of ultra high purity metals: Application of basic science to metallurgical processing.* Springer Verlag: Berlin, 2002.
1308. Wasserman, E.; Levine, L.: *Quantitative micro-complement fixation and its use in the study of antigenic atructure by specific antigen-antibody inhibition.* J.Immunol., 87, 290-295 (1961).

1309. Weber, von H.B.: *Zur Phosphoreszenz organischer Substanzen.* Z.Naturforsch., 4, 124 (1949).
1310. Webster, G.K.: *Supercritical fluid chromatography: Advances and applications in pharmaceutical analysis.* CRC Press: Boca Raton, 2014.
1311. Weinberger, R. (Ed.) *Practical capillary electrophoresis.* Academic Press: New York, 1993.
1312. Weissberger, A. (Ed.): *Technique of organic chemistry. Vol.4. Distillation.* Interscience: New York, 1951.
1313. Weller, S.; Steiner, W.A.: *Separation of gases by fractional permeation through membranes.* J. Appl. Phys., 21, 279-83 (1950).
1314. Wesselingh, J.A.; Krishna, R.: *Mass transfer.* Prentice Hall: New York, 1990.
1315. West, D.R.F.: *Ternary equilibrium diagrams.* 2^{nd}ed. Chapman&Hall: London, 1982.
1316. Wheaton, R.M.; Baumann, W.L.: *Nonionic separations with ion-exchangers.* Ann. N. Y. Acad.Sci., 57, 159-76 (1953).
1317. Whitaker, J.R.: *Paper chromatography and electrophoresis. Vol.1. Electrophoresis in stabilizing media.* Academic Press: New York, 1967.
1318. White, G.E.: *Graphical construction of transfer units in packed towers. Estimation of ratio of heights of liquid- and gas-transfer units.* Trans. Am. Inst. Chem. Engrs., 36, 359 (1940).
1319. Whitman, W.G.: *The two-film theory of gas absorption.* Chem. Met. Eng., 29(4) 146-52 (1923).
1320. Wichterle, O.; Mikes, O.: *Simple apparatus for countercurrent distribution.* Chem.Listy, 51, 1569-74 (1957).
1321. *Wicke, E.: Empirische und theoretische Untersuchungen der Sorptionsgeschwin-digkeit von Gasen an porosen Stoffen II.* Kolloid Z. 86, 295-313 (1939).
1322. Wieme, R.J.: *Agar gel electrophoresis.* Elsevier: Amsterdam, 1965.

1323. Wilke, C.R.; Chang, P.C.: *Correlation of diffusion coefficients in dilute solution.* Am.Inst. Chem. Eng. J., 1, 264-270 (1955).
1324. Williams, S.K.R.; Caldwell, K.D. (Eds.): *Field-flow fractionation in biopolymers analysis.* Spriner Verlag: Wien, 2011.
1325. Williamson, B.; Craig, L.C.: *Identification of small amounts of organic compounds by distribution studies. V. Calculation of theoretical curves for use of counter-current distribution.* J. Biol. Chem., 168, 687-97 (1947).
1326. Wilson, C.L.; Wilson, D.W. (Eds.): *Comprehensive analytical chemistry. Vol IIb. Physical separation methods.* Elsevier: Amsterdam, 1968.
1327. Wilson, G.M.: *Vapor-liquid equilibriums, correlation by means of a modified Redlich-Kwong equation of state.* Advan.Cryog.Eng., 9, 168 (1964).
1328. Wilson, I.D.; Poole, C.: *Handbook of methods and instrumentation in separation science.* Vol.I. 2nd ed. Academic Press: London, 2009.
1329. Wilson, J.N.: *A theory of chromatography.* J. Am. Chem. Soc., 62, 1583-91 (1940).
1330. Winitzer, S.: *Separation of solids at liquid-liquid interface: theory.* Sep.Sci., 8(6) 647-59 (1973).
1331. Wirth, H.E.: *Activity coefficients in sulphuric acid and sulfuric acid - sodium sulphate mixtures.* Electrochim. Acta, 16, 1345-56 (1971).
1332. Worrell, E.; Reuter, M.: *Handbook of recycling.* Elseveier: Amsterdam, 2014.
1333. Yagodin, G.A.; Kagan, S.Z.; Tarasov, V.V.: *Osnovy zhidkostnoi ekstraktsii (Principles of liquid extraction).* Khimiya: Moscow 1981.
1334. Yagodin, G.A.; Tarasov, V.V.: *Interfacial phenomena in liquid-liquid extraction.* Solvent Extr. Ion Exch., 2(2) 139-78 (1984).

1335. Yagodin, G.A.; Tarasov, V.V.; Ivakhno, S.Yu. *Condensed interfacial films in metal extraction systems.* Hydrometallurgy, 8(3) 293-305 (1982).
1336. Yalow, R.S.; Berson, S.A.: *Immunnoassay of endogeneous plasma insulin in man.* J.Clin.Invest., 39, 1157-75 (1960).
1337. Yashin, Ya., I.: *Fiziko-khimicheakie osnovy khromatographicheskogo razdeleniya (Physico-chemical principles of chromatographic separation).* Khimiya: Moscow, 1976.
1338. Yatsimirskii, K.B.; Vasilyev, V.P.: *Konstanty ustoichivosti khimicheskikh soyedinenii (Stability constant of chemical substances).* Izd. AN SSSR: Moscow, 1959.
1339. Yegorov, Yu.V.: *On some semiempirical sorption rules expressed by mass action law.* Radiokhimiya., 13(3) 370 (1971).
1340. Yon, C.M.; Thurnock, P.H.: *Multicomponent adsorption equilibria on molecular sieves.* Am.Inst.Chem.Eng.Symp. Ser 117 75-83 (1971).
1341. Zabezhinskii, Ya.L.: *Kinetics of sorption and desorption.* Zh. Fiz. Khim., 14, 139 (1940).
1342. Zahradnik, R.; Hobza, P.; Slanina, Z.: *Calculations of Henry constants and partition coefficients.* Coll. Czechosl. Chem. Commun., 40, 799-802 (1975).
1343. Zakgeim, A.Yu.: *A limit of the applicability of the thermodynamic laws.* Russ.J.Phys.Chem., 40(8) 104.2-1043 (1966).
1344. Zeldovich, Ya.B. (Seldowitsch): *Ueber die Theorie der Freundlichschen Adsorptionsisotherme.* Acta physicochimica U.R.S.S.1, 961-74 (1935).
1345. Zeldovich, Ya.B.: *Proof of singularity of the solution of mass law equation.* Zh.Fiz.Khim., 11(5) 685-688 (1938).
1346. Zelvenskii, M.Ya.; Solovkin, A.S.: *Activity coefficients of components of the system: macroconcentration* $Pu(OH)$-HNO_3-H_2O. Radiokhimiya, 22, 642-652 (1980).

1347. Zelvenskii, Ya.D.; Shalygin, V.A.; Golubkov, Yu.V.: *Purification of silicon tetrachloride from the phosphor trichloride impurity.* Khim.Prom., (5) 34.7-52 (1962).

1348. Zhou, Q.J.; Gu, Z.M.: *Studies on the extraction of Eu^{3+} by means of electrostatic pseudo liquid membranes.* Water Treatment, 3(2) 127-35 (1988).

1349. Zhukova, L.A.: *Teoriya soosazhdeniya i dinamicheskogo osazhdeniya ionov (Theory of coprecipitation and dynamic precipitation and coprecipitation of ions).* Energoizdat: Moscow, 1981.

1350. Zhuravskaya, N.A.; Kiknadze, E.V.; Antonov, Yu.A.; Tolstoguzov, V.B.: *Concentration of proteins as a result of the phase separation of waterprotein polysaccharide system. 2. Concentration of milk protein.* Nahrung, 30(6) 601-613 (1986).

1351. Zimakov, I.E.; Rozhavskii, G.S.: *Repeated radioactive dilution for the determination of small amounts of impurities.* Trudy Komissii Anal.Khim.AN SSSR, 9, 231-9 (1958).

1352. Zimens, K.E.: *Kinetics of heterogeneous exchange reactions. I. Study of solid reactions by means of isotope exchange.* Arkiv Kemi, Mineral. Geol., 120(18) 1 (1945).

1353. Zolotov, Yu.A.: *Ekstraktsia vnutrikompleksnykh soyedinenii (Extraction of inner complex compounds).* Nauka: Moskva, 1968.

1354. Zolotov, Yu.A.: *Preconcentration in inorganic trace analysis.* Pure Appl.Chem., 50(2) 129-48 (1978).

1355. Zolotov, Yu.A.; Kuzmin, N.M.: *Ekstraktsionnoye kontsentrirovaniye (Preconcentration by extraction).* Khimiya: Moskva, 1971.

1356. Zolotov, Yu.A.; Petrukhin, O.M.; Gavrilova, L.G.: *Synergetic effects in the solvent extraction of chelate compounds. Influence of the donor atoms of the reagent.* J.Inorg.Nucl. Chem., 32(5) 1679-88 (1970).

1357. Zsigmondy, R.: *Uber* einige *fundamentalbegriffe der Kolloidchemie. II. Elektrische Teilchenladung und der neue Begriff "Mizelle".* Z.Phys.Chem., 101, 292-322, (1922).
1358. Zuber, K.: *Separation of mercury isotopes by a photochemical method.* Nature, 136, 796 (1935).
1359. Zvara, I.: *Experiments on chemistry of element 104, kurchatovium. V. Adsorption of kurchatovium chloride from the gas stream on surfaces of glass and potassium chloride.* J.Inorg.Nucl.Chem., 32(6) 1885-94 (1970).
1360. Zvara, I.; Belov, V.Z.; Domanov, V.P.; Korotkin, Yu.S.; Chelnokov, L. P.; Shalaevskii, M.R.; Schegolev V.A.; Yussomnua, M.: *Chemical isolation of kurchatovium.* Radiokhimiya, 14(1) 119-22 (1972).
1361. Zvara, I.; Zvarova, Ts.; Tsaletka, R.; Chuburkov, Yu.S.; Shelaevskii, M.R.: *Rapid continuous separation of groups III and IV transition elements by using chlorides.* Radiokhimiya, 9 (2) 231 (1967).

List of Symbols

Tables contain the exponents valid for the basic dimensions of the values:

$M \equiv$ mass,
$N \equiv$ mole,
$L \equiv$ length,
$T \equiv$ time,
$K \equiv$ temperature,
$I \equiv$ electric current.

For example:
(i) The electromotive force has the dimension: $M^1L^2T^{-3}I^{-1}$
(ii) The extraction constant K_{ex} has the dimension: M^mL^l where m and l depend on the stoichiometry of extraction reaction.

Numerical values of **fundamental constants** are as follows:

Constant	Symbol	Value
Avogadro's number	N_A	6.022×10^{23} mol^{-1}
Bohr magneton	μ_B	9.274×10^{-24} J T^{-1}
Boltzman's constant	k	1.381×10^{-23} J K^{-1}
Electron charge	e	1.602×10^{-19} C
Faraday constant	F	9.649×10^4 C mol^{-1}
Gas constant	R	8.314 J K^{-1} mol^{-1}
Gravitational acceleration for sea level at zero latitude	g	9.780 m s^{-2}
Electrical permitivity of a vacuum	ε_0	8.854×10^{-12} C^2 J^{-1} m^{-1}
Planck's constant	h	6.626×10^{-34} J s

		Dimensions					
		M	N	L	T	K	I
< >	average						
[ ]	equilibrium concentration						
A	cross-area			2			
A	total area			2			
Ab	antibody						
Ag	antigen						
C	constant in similitude relations						
CMC	critical micellar concentration			-3			
D	diffusion coefficient (diffusivity)			2	-1		
D	distribution ratio						
D	radiation dose			2	-2		
D^t	thermodiffusion coefficient	1	-1	-1			
D_i	distribution ratio of the i-th species						
E	electromotive force	1		2	-3		-1
F	fraction of isotope exchange						
F	Faraday constant		-1		1		1
F	feed amount		1				
F	thermodynamic potential	1		2	-2		
F	transfer function						
F_S	Small attraction constant		-1	3			
Fo	Fourier number						
G	gravitational constant	1		3	-2		
G	weight of substance	1					
G	Gibbs energy	1		2	-2		
H	enthalpy	1		2	-2		
HETP	height equivalent to theoretical plate			1			
HLB	hydrophilic-lipophilic balance						
HTU	height of transfer unit			1			
I	ionic strength		1	-3			
I	ionization potential	1	-1	2	-2		
I	intensity of turbulence			1	-1		
J	flux		1		1		

		Dimensions					
		M	N	L	T	K	I
K	cation						
K	permeability coefficient			2			
K	transfer coefficient			1	-1		
K^D	distribution constant						
K_S	solubility product		n	m3			
K_s	amplifying constant						
K_{ex}	extraction constant		n	1			
L	Laplace operator						
L	ligand						
L	correction coefficient						
L	molar heat of evaporation	1	-1	2	-2		
L	phenomenological (kinetic) coefficient	-1	1	-1	-1		
L	permeability coefficient			1	-1		
M	molar mass	1	-1				
M	molarity			1	-3		
M	relative molecular mass						
M	metal ion						
M	Madelung constant						
N	number of entities (atoms, etc.)						
N	number of transfer units						
N_A	Avogadro's number		-1				
NHE	normal hydrogen electrode						
Ox	oxidized form of substance						
P	product amount		1				
Pe	Péclet diffusion number						
Pr	Prandtl diffusion number						
Q	amount of substance		1				
Q	selectivity coefficient of ion exchange						
Q	quadrupole moment	1		4	-2		
R	gas constant	1	-1	2	-2	-1	
R	hydrodynamic resistance			-1			
R	yield fraction						
R_F	retention factor						

		Dimensions					
		M	N	L	T	K	I
Re	Reynolds number						
Red	reduced form of substance						
S	amount of solid phase	1					
S	Svedberg unit				1		
S	entropy	1		2	-2	-1	
Sc	Schmidt number						
Sh	Sherwood number						
St	Stanton number						
T	temperature					1	
U	enzyme unit		1		-1		
U	impulse						
$\bar{V}$	molar volume		-1	3			
V	volume			3			
V	total variance						
VdW	Van der Waals						
W	thermodynamic probability						
W	waste amount		1				
X	complexity function						
X	gradient of thermodynamic potential	1		1	-2		
Y	product of activity coefficients						
Y	response function						
a	constant of Freundlich isotherm						
a	specific surface	-1		2			
a	thermodynamic activity						
b	constant of Freundlich isotherm						
c	molar concentration		1	-3			
d	thickness			1			
e	electron						
e	electron charge				1		1
erf	error function						
erfc	error function fraction						
f	activity coefficient on molar scale		-1	3			
f	molar frictional coefficient	1	-1		1		
g	gravitational acceleration			1	-2		

		Dimensions					
		M	N	L	T	K	I
r	number of dimensionless parameters						
r	phase ratio						
r	polar coordinate			1			
r	radius			1			
r	reflux ratio						
s	complex variable						
t	time				1		
u	controlling variable						
u	energy of intermolecular interaction	1		2	-2		
u_i	ionic mobility	1		2	-3		-1
u	velocity			1	-1		
u_V	volume interaction energy	1		-1	-2		
v	normalized concentration						
v	variance (degrees of freedom)						
v	volume fraction						
w	linear velocity			1	-1		
w	mass fraction						
x	Cartesian coordinate			1			
x	distance			1			
x	mole fraction						
y	activity coefficient on molality scale	1	-1				
y	Cartesian coordinate			1			
y	process variable						
z	Cartesian coordinate			1			
z	charge number						
z	separation coordinate			1			
Δ	decrement						
Γ	surface concentration		1	-2			
Λ	molar conductance	-1	-1		3		2
Π	osmotic pressure	1		-1	-2		
Π	ratio of free and protonated forms						
Φ	parameter of association						

		Dimensions					
		M	N	L	T	K	I
Φ	quantum yield						
Θ	fraction of output						
α	coefficient of side reactions						
α	degree of dissociation						
α	exponent in dimensional equations						
α	Harned coefficient		-1	3			
α	polarizability of molecule		-1	3			
α	specific filter cake resistance	-1		1			
α	temperature expansion coefficient					-1	
β	pressure expansion coefficient	-1		2			
β	stability constant		n	1			
β	thermodiffusion factor						
γ	activity coefficient						
δ	solubility parameter	$\frac{1}{2}$		$-\frac{1}{2}$	-1		
δ	thickness of diffusion layer			1			
ε	depth of potential energy	1		2	-2		
	of molecule	1					
ε	dielectric permittivity (dielectric constant)			1	-2		
ε	porosity						
ζ	dimensionless separation path						-2
ζ	electrokinetic (zeta) potential	1		2	-3		-2
η	dynamic viscosity	1		-1	-1		
η	overvoltage	1		2	-3		-2
κ	dimensionless transfer coefficient						
κ	specific conductivity	-1		-3	3		2
λ	crystallization coefficient						
λ	decay constant				-1		
λ	dimensionless thickness						
λ	obstruction factor						
μ	chemical potential	1	-1	2	-2		
μ	dipole moment	1		3	-2		
μ	electric magnetic moment	1					
ν	kinematic viscosity			2	-1		
ν	stoichiometric coefficient						

SUPERSCRIPTS

*	boundary
*	for the standard state of a pure substance
·	time derivative (rate)
·	standard state for a unit concentration
·	hypothetical pure standard state
–	molar
0	thermodynamic standard state
B	Born
BF	Bernal-Fowler
DH	Debye-Hückel
E	eddy
E	thermodynamic excess function
F	in feed
I	phase I
II	phase
M	membrane
M	molecular
S	surface
S	Stockmayer
W	weighted
W	in waste
ef	effective
elst	electrostatic
id	ideal
m	exponent for Reynolds number
m	maximal
n	exponent for Schmidt number
o	pure; at zero of some parameter
o	standard state

o	thermodynamic
rep	repulsion
st	standard
t	total (gross)
+	cation
-	anion

SUBSCRIPTS

0	for zero concentration of a component
1	1st component
2	2nd component
A	anion
A	component, substance A
B	component, substance B
D	diffusion
D	Debye
I	belonging to phase I
II	belonging to phase II
K	cation
L	ligand L
L	macromolecular anion
LJ	Lennard-Jones
M	central atom M
S	planar
S	solvent
S	surface
b	boiling point
c	concentration in molarity scale
c	concentration
diff	diffusion
dip	dipole
disp	dispersion
ex	extraction
hydro	hydrodynamic
i	i-th component, type of species
ind	induction
j	stoichiometric coefficient of anion

k	type of central atom
m	maximal
m	mass related
m	molality scale
m	melting point
m	stoichiometric coefficient
n	stoiochiometric coefficient
mix	mixing, mixture
p	stoichiometric coefficient of OH ligand
r	number of reaction
r	relative
sep	separation
visc	viscous
w	water
x	concentration in mole fraction scale

ABOUT THE AUTHORS

Fedor Macášek

born 1937, Žilina, Czechoslovakia, graduated from the Moscow State University (U.S.S.R) in 1962 in Physical Chemistry and Radiochemistry. Since his graduation he has been working at the Faculty of Natural Sciences of Comenius University in Bratislava; 1968 PhD., 1984 DSc; 1985 full professor of nuclear chemistry, 2004 Professor Emeritus.

At Comenius University for over 30 years he was lecturing Nuclear Chemistry for science students and Separation Chemistry for radiochemists and analytical chemists, and is an author and co-author of several text-books in the fields. From 1972 to 1983 he served as head of the Department of Nuclear Chemistry and from 1981 to 1985 as vice-dean and dean of the Faculty of Natural Sciences.

From 1986 to 1989 Dr. Macášek was vice-minister of the Slovak Ministry of Education and in 1994 became state secretary of the Ministry of Education and Science of the Slovak Republic.

In the years 2003-2004 he worked as a Senior Scientific Worker at the Slovak Office of Standards, Metroology and Testing within the project of Cyclotron Center of the Slovak Republic, and was a National Coordinator of the technical cooperation project of the IAEA on radiochemical facilities for production of medical radionuclides (1997-2004). Since 2005, he has been a member of the board of directors and since 2006 a Quality Manager of BIONT, a.s. (Bratislava Ion Technologies).

His main research interests concerned separation chemistry and the recovery of valuable fission products, e.g. cesium-137, technetium-99 and palladium from spent nuclear fuel (in cooperation with the Institute of Nuclear Research in

Řež-Prague), radiation chemistry of solvents, solvent extraction and emulsion membranes extraction, and recently the production and quality control of PET radiopharmaceuticals. He also worked in the field of chemical speciation of radionuclides in food and soils, radiochemical analysis of environmental samples, as well as the utilization of magnetic sorbents for contaminated soil remediation, particularly with Dr. Navratil.

He is the author or co-author of over 110 scientific papers, 70 research reports, 14 patents as well as 2 books on nuclear and separation chemistry, *Nuclear Chemistry* (Academia, Prague and E.Horwood, New York, 1992) and *Separation Chemistry* (Ellis Horwood, New York, 1992). He has delivered nearly 50 lectures at scientific meetings abroad. He is a member of the editorial boards of the *Journal of Radioanalytical and Nuclear Chemistry* (1988) and of *Solvent Extraction and Ion Exchange* (1983). He was also the organizer and chairman of 10 biennial international Conferences on "Separation of Ionic Solutes – SIS" (1985-2003).

James D. Navratil

born in Denver, Colorado, in 1941, was trained as an Analytical Chemist at the University of Colorado (PhD, 1975). He is Professor Emeritus of Environmental Engineering and Science at Clemson University; he retired in May 2006 after joining the university in January 2000. His other teaching experiences include serving as a Chemical Training Officer in the U.S. Army Reserve (1964-70), teaching general chemistry at the University of Colorado (1976-78), and teaching chemical engineering and extractive metallurgy subjects at the University of New South Wales, Australia (1987-90), where he also served as Head of the Department of Mineral Processing and Extractive Metallurgy. In addition, he was an Affiliate Professor at the Colorado School of Mines (1985-87), University of Idaho (1998-2000), and Clemson University (1998-2000).

Dr. Navratil has more than 40 years of experience in environmental management, waste and water treatment research and development, separations science and technology, and actinide chemistry and radiochemistry, acquired primarily at the U.S. Department of Energy (DOE) Rocky Flats Plant and through his assignments with the International Atomic Energy Agency (IAEA), Chemical Waste Management, DOE's Energy Technology Engineering Center, the Idaho National Engineering and Environmental Laboratory, Rust Federal Services, and currently, Hazen Research, Inc., where he is a senior technical advisor.

Accomplishments in the development of separation, recovery, and waste treatment processes have earned Dr. Navratil numerous honors including the annual award of the Colorado Section of the American Chemical Society (ACS), Rockwell International Engineer of the Year, two IR-100 Awards, and three society fellowships. He was a member of the IAEA team awarded the 2005 Nobel Peace Prize, and in 2006 received the Lifetime Achievement Award for Commitment to the Waste-management, Education and Research Consortium (WERC) and to WERC's International Environmental Design Contests.

Dr. Navratil has several patents to his credit and has given more than 300 presentations, including lectures in more than 80 countries. He has co-edited or co-authored 21 books (most recently with Jiri Hala, *Radioactivity, Ionizing Radiation, and Nuclear Energy*, in 2012), published more than 250 scientific publications, and has served on the editorial boards of over a dozen journals. He was instrumental in the founding of the journals *Solvent Extraction and Ion Exchange* (serving as co-editor for many years) and *Preparative Chromatography* (serving as editor), as well as the ACS's Subdivision of Separation Science and Technology and its award in Separation Science and Technology, and DOE's Actinide Separation Conferences and its Glenn Seaborg Award in Actinide Separations. Dr. Navratil has also organized or co-organized many conferences, symposiums, and meetings for the ACS, DOE, and IAEA.

www.ingramcontent.com/pod-product-compliance
Lightning Source LLC
Chambersburg PA
CBHW020718180526
45163CB00001B/22